Developments in Laser Techniques
and Fluid Mechanics

Springer

Berlin
Heidelberg
New York
Barcelona
Budapest
Hong Kong
London
Milan
Paris
Santa Clara
Singapore
Tokyo

R. J. Adrian · D. F. G. Durão · F. Durst · M. V. Heitor
M. Maeda · J. H. Whitelaw (Eds.)

Developments in Laser Techniques and Fluid Mechanics

Selected Papers from the 8th International Symposium,
Lisbon, Portugal, 8-11 July, 1996

With 367 Figures

Springer

Editors:

R. J. Adrian
Dept. of Theoretical and
Applied Mechanics
University of Illinois
at Urbana - Champaign
Urbana, Illinois 61801, USA

D. F. G. Durão
Dept. of Mechanical Engineering
Instituto Superior Técnico
Av. Rovisco Pais
1096 Lisboa Codex, PORTUGAL

F. Durst
Lehrstuhl für Strömungsmechanik
University of Erlangen - Nuremberg
Cauerstraße 4
91058 Erlangen, GERMANY

M. V. Heitor
Dept. of Mechanical Engineering
Instituto Superior Técnico
Av. Rovisco Pais
1096 Lisboa Codex, PORTUGAL

M. Maeda
Dept. of Mechanical Engineering
Keio University
1-14-1 Hiyoshi, Kohuku
Yokohama 223, JAPAN

J. H. Whitelaw
Imperial College of Science,
Technology and Medicine
Dept. of Mechanical Engineering
Exhibition Road
London SW7 2BX, ENGLAND

ISBN 3-540-63572-6 Springer-Verlag Berlin Heidelberg New York

Cip data applied for
Die Deutsche Bibliothek - CIP-Einheitsaufnahme
Developments in laser techniques and fluid mechanics : selected papers from the 8th inter-
national symposium, Lisbon, Portugal, 8-11 July, 1996 / R. J. Adrian ... (ed.).-
Berlin ; Heidelberg ; New York ; Barcelona ; Budapest ; Hong Kong ; London ; Milan ; Paris ;
Santa Clara ; Singapore ; Tokyo : Springer, 1997
    ISBN 3-540-63572-6

© Springer-Verlag Berlin Heidelberg 1997
Printed in Germany

Typesetting: Camera-ready by authors
Printing: Mercedesdruck, Berlin; Binding: Lüderitz & Bauer, Berlin
SPIN:10537261   61/3020-5 4 3 2 1 0 - Printed on acid-free paper

Preface

This volume includes revised versions of a series of selected papers presented at the **Eighth International Symposium on Applications of Laser Techniques to Fluid Mechanics** held at *The Calouste Gulbenkian Foundation in Lisbon*, during the period of July 8 to 11, 1996.

The papers describe *Novel Ideas for Instrumentation, Instrumentation Developments*, results of measurements of *Wall-Bounded Flows, Free Flows and Flames* and of the flow and combustion in *Engines*. The papers demonstrate the continuing and healthy interest in the development of understanding of new methodologies and implementation in terms of new instrumentation.

The prime objective of this Eighth Symposium was to provide a forum for the presentation of the most advanced research on laser techniques for flow measurements, and communicate significant results to fluid mechanics. The applications of laser techniques to scientific and engineering fluid flow research was emphasized, but contributions to the theory and practice of laser methods were also considered where they facilitate new improved fluid mechanic research. Attention was placed on laser-Doppler anemometry, particle sizing and other methods for the measurement of velocity and scalars such as particle image velocimetry and laser induced fluorescence.

We would like to take this opportunity to express our thanks to those who contributed to the success of the conference. The assistance provided by the Advisory Committee, by assessing abstracts was highly appreciated. The companies who partook in the exhibition of the manufacturers' technical presentation are also acknowledged. In addition, thanks go the participants who contributed actively in discussions, learned from the presentations and were essential to the success of this symposium. And last, but not least, we are highly indebted for the financial support provided by the Sponsoring Organizations that made this Symposium possible.

SPONSORING ORGANIZATIONS
DHL, ICEP, European Research Office: United States Army, Navy and Air Force Departments, FLAD, Fundação Calouste Gulbenkian, Instituto Superior Técnico (IST), Instituto Tecnológico Para A Europa Comunitária (ITEC), Junta Nacional de Investigação Científica e Tecnológica (JNICT), INETI.

The Editors

Table of Contents

Chapter I

Novel Ideas for Instrumentation

Chapter II

Instrumentation Developments

Chapter III
Wall-Bounded Flows

Chapter IV
Free Flows and Flames

Chapter V

Engines

Chapter I

Novel Ideas for Instrumentation

I.1 Power and Sensitivity Improvement of LDA-Systems by Fiber Amplifiers

H. Többen, H. Müller, D. Dopheide

Physikalisch-Technische Bundesanstalt (PTB)
Laboratory for Fluid Flow Measuring Techniques
D-38116 Braunschweig, Bundesallee 100, Germany

Abstract. A new concept is presented where fiber amplifiers are used in fiber optical LDA-systems as booster amplifier to magnify the laser power in the measuring volume as well as preamplifier to increase the sensitivity of detector units.

Keywords. Laser Doppler anemometry, SNR improvement, fiber amplifiers

1 Introduction

Scattered light of low intensities is produced by particles with small dimensions, particles in high speed flows and in LDA-systems with a great working distance. In these cases powerful lasers like Argon-ion and Nd:YAG-laser are preferred in LDA-systems. Because the emission wavelengths of the Nd:YAG and of the Argon laser are aside of the maxima of responsivity from Si-, InGaAs- and Ge-detectors, the output power of these lasers can not be completely utilized for the LDA-system. In example, Arndt et. al. (1996) have demonstrated that twice of the laser power of a 830 nm diode laser LDA must be applied at 1064 nm Nd:YAG LDA to obtain the same sensitivity for the whole system for particles sizes smaller than 1 μm.

As shown here, a drastical increase of the sensitivity can be obtained by optical preamplification with fiber amplifiers placed in front of photodetectors. This offers a very interesting possibility of improving the performance of LDA-systems, specially of LDA-systems, like Nd:YAG LDA-systems, with working wavelengths at low responsivities of commonly used conventional photodiodes.

A further interesting application of fiber amplifiers is the optical post amplification of laser radiation produced by laser sources with small emission linewidths but low output powers like longitudinal singlemode laser diodes as DFB- (distributed feedback), DBR- (distributed Bragg reflector) or well tunable TTG (tunable-twin-guide)-lasers. These lasers are used in newly developed shift LDA-systems as presented by Müller and Dopheide (1993), where the output beams of two DFB-laser diodes are focused directly into the measuring volume.

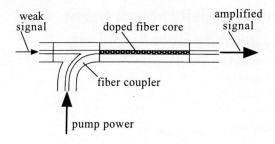

Fig. 1   Basic function principle of a fiber amplifier

The most advantage of post or booster amplification using fiber amplifiers is the increase of the signal source output power without changing the spectral characteristics by pumping with a pump source with low requirements on its spectral characteristics.

Fiber amplifiers can also be used in already existing fiber optical LDA-systems, i.e. multicomponent LDAs where fibers are used to couple the LDA sources with the measuring head. In this case it is easy to amplify the laser radiation by simply inserting fiber amplifiers.

2   Fundamentals of fiber amplifiers

The basic function principle of a fiber amplifier which is placed between standard fibers is shown in figure 1. A fiber amplifier consists essentially of a pump laser diode, a fiber coupler and a short fiber doped with rare earth ions. The pump power of the pump laser diode supplies the amplifier with energy. The doped fiber core creates the active medium, its dopant is excited to higher energy levels by the pump power.

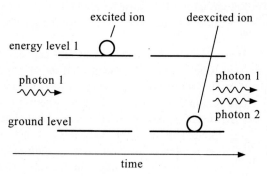

Fig. 2   Principle of stimulated emission

Amplification occurs if the excited energy level of an ion is depleted by an incident photon which induces, or stimulates, the emission of a second photon, as shown in figure 2. Because the emitted photon has the same frequency, phase and polarization as the stimulating photon it is an exact replica of the photon already present. The radiation remains coherent and the spectral emission characteristic doesn't change. A fiber coupler combines the pump and the signal radiation in the doped fiber. The coupler can be placed in front of or behind the active fiber for backward or forward pumping.

The greatest efforts in developing optical amplification with fiber amplifiers have been done in the optical telecommunication technique for realizing long distance transmission systems with transmission rates of Gbit/s at a wavelength of 1,55 μm. The mostly developed fiber amplifier is the EDFA (Erbium doped fiber amplifier). The EDFA, which can be pumped by 980 nm or 1480 nm broadband emitting multimode laser diodes, amplifies laser radiation in the wavelength range between 1520 nm and 1560 nm. A record amplification value of 54 dB (factor: 250.000) was achieved with an EDFA for small signal amplification reported by Laming et. al. (1992). Further well developed fiber amplifiers have been realized for the wavelength range around 1,3 μm with Neodymium or Praseodymium doped fibers. The experiences gathered from these fiber amplifiers, specially from the EDFA, can be transferred to other fiber amplifier set-ups to realize amplification at preferred wavelength.

3   Characteristics of fiber amplifiers

Fiber amplifiers placed behind a laser source with low output power (mW or sub-mW range) can be used as booster (post) amplifier. They work in the saturated regime and convert pump power delivered from broadband sources into signal power of high precision lasers. With an EDFA as booster amplifier a pump-to-signal conversion efficiency of 59 % was realized by Laming et. al. (1991).

Behind fiber coupled lasers, booster amplifiers give the chance to use spectral high quality sources like DFB- or DBR-laser diodes with low output power and low price. Specially to realize fiber optical multicomponent LDA-systems utilizing the optical frequency difference of separate fiber coupled laser sources for each LDA beam, booster amplifiers simply inserted as „black boxes" can be used to magnify the laser power in the measuring volume.

As preamplifiers in front of photo detectors fiber amplifiers can be used to increase the signal-to-noise ratio (SNR). Specially, aside of the maxima of responsivity at 900 nm in Si-detectors and at 1,55 μm in InGaAs- or Ge-detectors, for example at 1064 nm or 532 nm which are emission wavelengths of powerful Nd:YAG solid state lasers, the minimum detectable signal level can be drastically reduced by broadband optical preamplification.

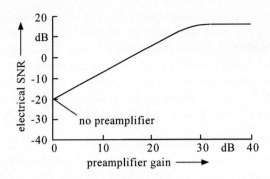

Fig. 3   SNR as function of preamplifier gain

SNR calculations made by Desurvire (1994) have shown that the SNR of the electrical current delivered from a PIN photodiode can be drastically increased by optical preamplification with an EDFA. The calculations have demonstrated that:
-   The electrical SNR of signal detection with a PIN photodiode can be increased of about 36 dB (factor: 4000) in combination with optical preamplification in an EDFA for typical values used for the amplification of 1,55 μm signals. The result is shown in figure 3.
-   If a SNR of 20 dB (factor: 100) should be required for the detection unit, an enhancement of minimum detectable power of 25 dB (factor: 316) can be achieved by using optical preamplification.
-   In comparison with an APD photodiode the PIN photodiode with optical preamplification delivers the best results.
Thus, the conclusion is that optical preamplification technique is a very valuable tool in laser Doppler anemometry.

4   First Experiments

Figure 4 shows the experimental set-up used for our first examinations of optical amplification with a fiber amplifier as booster amplifier to be applied in a fiber optical frequency shift LDA-system. An Erbium-doped silica fiber of 15 m length was used as active part in the system. The Erbium-doped fiber was connected to the standard singlemode fiber components with mechanical splices (SP).

A low power tunable TTG-laser diode with an emission wavelength near 1,55 μm and an emission linewidth of a few MHz was used as the signal source. A 980 nm laser diode was used as pump. The output radiation of the pump and the signal laser source was focused into the core of standard fiber components. The WDM-fiber coupler (WDM: wavelength division multiplexer) selects or combines the pump and signal radiation. Due to backscattered light and amplified spontaneous emission radition generated in the doped fiber, the optical isolator

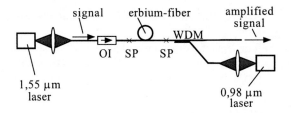

Fig. 4 Set-up of a booster fiber amplifier for 1,55 µm laser radiation in an Erbium-doped
fiber

(OI) with fiber pigtails in front of the 1,55 µm-laser diode ensures a stable
operation of the signal source.

Figure 5 shows one of the first results of booster amplification with the set-
up shown in figure 4. The launched 1,55 µm-signal power was 340 µW. It can be
observed that the amplified signal power at 1,55 µm increases approximately
linear with increasing pump power at 0,98 µm. At a launched pump power of
93 mW the input signal power of 0,34 mW was amplified up to 44 mW. This
corresponds to an amplification factor of 129 (21 dB) and a pump-to-signal
conversion efficiency of 34 %. By optimization of the fiber parameters (numerical
aperture, fiber core diameter, doping concentration, fiber length) and the pump
configuration improvements of these first results are possible.

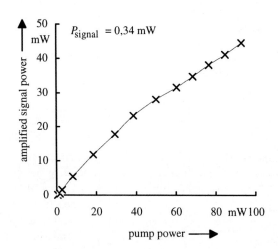

Fig. 5 Amplification of 340 µW input signal power at 1,55 µm with a 980 nm pump as
shown in fig. 4

5 Concept of a fiber optical shift-LDA with fiber amplifiers

Figure 6 shows the concept of a fiber optical shift-LDA-system which combines the post and pre amplification of optical radiation with fiber amplifiers. Two laser diodes (LD1 and LD2) with an emission frequency difference and emission linewidths in the MHz range produce the LDA radiation (P_{LD1} and P_{LD2}). Booster fiber amplifiers behind each source increase the laser power levels ($P_{LD1,ampl.}$ and $P_{LD2,ampl.}$). Therefore, low power laser diodes i.e. conventional pigtailed well tunable DFB laser diodes can be used as LDA sources.

A fiber coupler combines small parts of the laser radiation (P'_{LD1} and P'_{LD2}) to produce a beat signal in the photodiode 1 (PD1) as reference signal for the signal evaluation unit. Miniaturized focusing optics with fiber pigtails focus the amplified laser radiation into the measuring volume.

The scattered light of tracer particles in the flow is focused into the fiber core of a third fiber amplifier in front of the photodiode 2 (PD2) which is used as preamplifier. For enhancing the SNR of the measuring signal the preamplifier amplifies the launched scattered light ($P_{scatter}$) to higher power levels ($P_{scatter,ampl.}$).

A fiber coupler splits up the output power (P_{pump}) of the pump laser (PLD) into three parts, which are coupled into the doped fibers by WDM fiber couplers. A powerful multimode laser diode can be used as pump source. To keep away disturbing radiation, e.g. pump radiation, from laser diodes and photodiodes optical filters or optical isolators can be placed in front of these devices.

By mixing the measuring signal, delivered from PD2, with the simultaneously detected reference signal, delivered from PD1, and employing

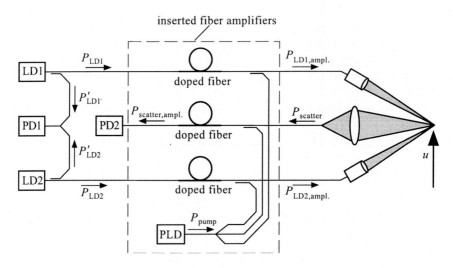

Fig. 6 Schematical set-up of a fiber optical shift-LDA system with fiber amplifiers for post and pre amplification

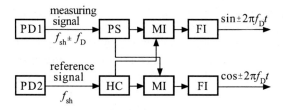

PS: power splitter   MI: mixer
HC: hybrid coupler   FI: filter

Fig. 7   Quadrature detection unit for signal processing of fiber optical shift LDA-system
according to fig. 6

quadrature signal processing techniques, presented earlier by Müller et. al. (1994), the shift frequency f_{sh} with all its frequency fluctuations and bandwidth influences is eliminated, the directional information is retained and the Doppler frequency f_D is obtained in the base band. Figure 7 shows the schematical diagram of the quadrature demodulation unit.

6   Conclusion

A new concept was presented for increasing the power in the measuring volume and improving the sensitivity of LDA-systems with fiber amplifiers.

As booster amplifiers they offer the possibility to increase the output power of DFB- or DBR-laser sources with low cost broadband emitting pump lasers. In a first experiment an amplification factor over 100 (20 dB) was realized.

As preamplifier in front of a PIN-photodiode fiber amplifiers enhance the SNR of the detected signals and allow to detect extremely weak burst signals. An enhancement over 36 dB (factor: 4000) is predicted using Erbium-doped fiber amplifiers. Such a fiber amplifier can improve the sensitivity of LDA-receivers dramatically and allows to set novel impetus to diode pumped solid state lasers and photodiode receivers.

7   References

Arndt, V., Müller, H. & Dopheide, (1996), Comparison measurements for selection of suitable photodetectors for use in Nd:YAG LDA systems, Experiments in Fluids, vol. 20, pp. 460-465

Desurvire, E., 1994, Erbium-doped fiber amplifiers, pp. 167-174, John Wiley & Sons, New-York.

10

Laming, R.I., Townsend, J.E., Payne, D.N., Meli, F, Grasso, G. & Tarbox, E.J., 1991, High-power Erbium-doped-fiber amplifiers operating in the saturated regime, Photonics Tech. Letters, vol. 3, pp. 253-255.

Laming, R. I., Zervas, M. N. & Payne, D. N., 1992, Erbium-doped fiber amplifier with 54 dB gain and 3,1 dB noise figure, Photonics Tech. Letters, vol. 4, pp. 1345-1347.

Müller, H. & Dopheide, D., 1993, Direction sensitive laser Doppler velocimeter using the optical frequency shift of two stabilize laser diodes. Proc. SPIE Vol. 2052 Laser Anemometry Advances and Applications, pp. 323-330

Müller, H., Czarske, J., Kramer, R., Többen, H., Arndt, V., Wang, H. & Dopheide, D, 1994, Heterodyning and quadrature signal generation: advantageous technique for applying new frequency shift mechanisms in the laser Doppler velocimetry. Proc. 7th Int. Symp. on Appl. of Laser Techniques to Fluid Mechanics, 11-14 July 1994, Lisbon, pp. 23.3.1-23.3.8

I.2 Fibre Optical Multicomponent LDA-System Using the Optical Frequency Difference of Powerful DBR-Laser Diodes

H. Müller, H. Wang, D. Dopheide

Physikalisch-Technische Bundesanstalt (PTB)
Laboratory for Fluid Flow Measuring Techniques
Bundesallee 100, D-38116 Braunschweig, Germany

Abstract. A fibre optical directional two component LDA utilizing the optical frequency differences of three fibre coupled 100 mW DBR laser diodes is presented. The resulting frequency shifts are used for the channel separation as well as for the directional discrimination. Additionally the use of one laser source for each LDA beam allows to multiply the laser power in the measuring volume.

First experimental results of two dimensional directional velocity measurements with the realized adjustment insensitive powerful LDA set-up are presented.

Keywords. Directional multicomponent LDA, fiber optics, frequency shift, DBR lasers

1 Introduction

The generation of a frequency shift by utilizing the optical frequency difference of two stabilized monomode laser diodes allows in principle the realization of simple and compact directional LDA set-ups (Müller and Dopheide (1993), Müller et al. (1994)), when the resulting beat signal of the laser diodes is used as reference signal which is correlated with the detected carrier frequency LDA measuring signal. Thus the resulting frequency shift is exclusively used as an auxiliary carrier frequency with drastically reduced requirements on stability and bandwidth (Müller et al. (1994)).

In any way laser diodes with almost equal emission frequencies are required. Up to now it was necessary to select appropriate monomode laser diodes by measuring the spectral characteristics of numerous laser diodes depending on the current and the temperature. Recently available powerful DBR laser diodes have output powers in the 100 mW range and additionally a wide wavelength tuning range so that the selection problem of appropriate laser diodes vanishes.

Using the optical frequency differences of three DBR laser diodes to generate two shift frequencies which are exclusively used as auxiliary carrier frequencies for the channel separation as well as for generation of two quadrature signal pairs in the baseband containing the directional velocity information one gets a powerful two component LDA set-up.

The directional two component LDA-system has been realized by employing fibre optics. By the use of three fibre coupled DBR lasers, fibre coupled photodiodes, fibre couplers for the generation of the reference signals by optical heterodyning and a back scatter optic for the detection of the LDA signal, it was possible to realize a compact and adjustment insensitive LDA-system.

2   Experimental set-up

The concept is based upon the application of three appropriate monomode laser diodes with almost equal emission frequencies and overlapping tuning ranges without mode hopping. Therefore DBR (distributed Bragg reflector) laser diodes have been used, one for each LDA beam. In contrast to conventional laser diodes having a Fabry Perot resonator structure (see figure 1a), these DBR laser diodes employ a phase grating structure (Bragg grating) in the laser resonator (see figure 1b) which guarantees a dynamic single mode operation (Amann (1991)) and allows to adjust the emission wavelength without mode hopping by varying the current or temperature of the laser diode.

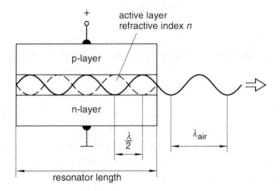

Fig. 1a. Conventional monomode laser diode having a Fabry Perot resonator structure.

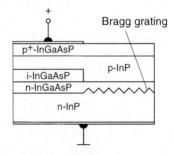

Fig. 1b. Resonator structure of a DBR laser diode.

A typical wavelength tuning characteristic depending on the temperature for different diode currents is shown in figure 2 for both laser diode resonator structures shown in figure 1.

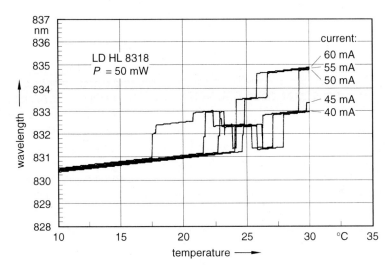

Fig. 2a. Wavelength characteristic of a conventional monomode laser diode with a Fabry Perot resonator structure showing a critical wavelength tuning behaviour caused by mode hopping effects.

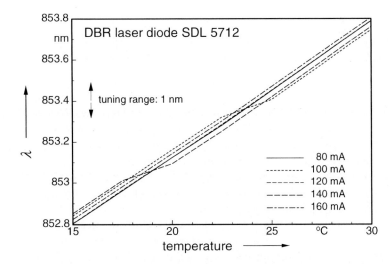

Fig. 2b. Wavelength characteristic of a DBR laser diode for several diode currents depending on the temperature. The wavelength characteristic shows wide tuning ranges without any mode hopping.

14

Figure 2b shows a typical wavelength dependence upon the working temperature for commercially available DBR laser diodes with well determined wavelengths near 582 nm. For any laser diode current the wavelength can be tuned by means of the working temperature over a range of about 1 nm. Thus one easily can get overlapping wavelength ranges of different DBR laser diodes.

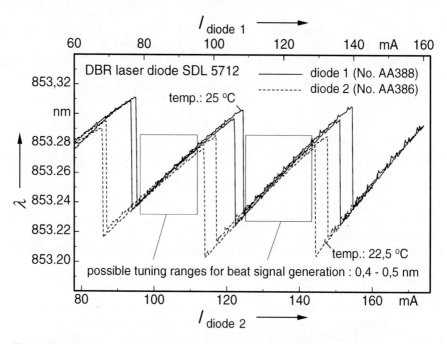

Fig. 3. Wavelength characteristic of two arbitrarily chosen DBR laser diodes, almost equal emission frequencies are marked as possible tuning ranges for the generation of beat signals.

Figure 3 shows useable tuning ranges for the generation of a beat signal with an arbitrarily chosen DBR laser diode pair of the SDL 5712 type (wavelength 852 nm, optical output power 100 mW). Without having to perform any laser diode selection one easily can get a beat signal in the hundred MHz range which can be used as an auxiliary carrier frequency for the quadrature signal generation.

To realize a two component directional LDA system using the optical frequency differences of three LDA beams as carrier frequencies for the channel seperation and the quadrature signal generation, fibre coupled optical devices can advanteageously be applied (Wang et al. (1994)). To avoid external resonator effects by reflections at fibre ends or connectors, optical isolators should be inserted between the DBR laser diode and the fibre coupling device. Figure 4 shows the frequency spectra of the beat signal of two DBR laser diodes whose output beams were superimposed by a fibre coupler a) without and b) with the use of an optical isolator. Thus, optical isolators should be integrated to ensure a stable operation of the system.

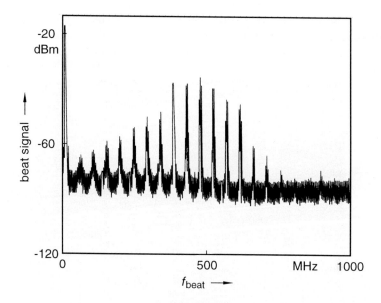

Fig. 4a. Beat signal generation of two DBR laser diodes using a fibre coupler without optical isolator between laser diode and fibre coupling device. Reflections at the fibre facets cause external resonator effects resulting in multiple longitudinal modes given by the optical path lengths in the fibre.

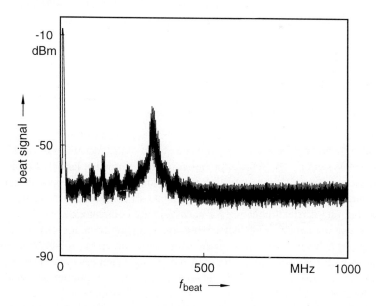

Fig. 4b. Beat signal generation of two DBR laser diodes using a fibre coupler with optical isolator between laser diode and fibre coupling device. By eliminating reflections into the DBR laser diode one gets a stable beat frequency operation.

Figure 5 shows the employed fibre coupled LDA measuring head containing the transmitting and receiving optics.

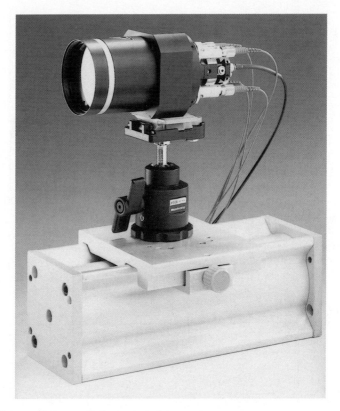

Fig. 5. Fibre coupled transmitting and receiving optics.

Figures 6a and 6b show the experimental set-up of the two component frequency shift LDA system based on the application of three fibre coupled DBR laser diodes and the utilization of their optical frequency differences. The optical frequency differences between the LDA beams result in a carrier frequency measuring signal containing one carrier frequency for each velocity component to be measured. In order to use the carrier frequencies exclusively for the separation of the different components as well as for the quadrature signal generation in the baseband containing the directional information, the LDA measuring signal has to be correlated with one carrier frequency reference signal for each velocity component (see reference 1 and reference 2 in figure 6b).

The reference signals were easily obtained by optical heterodyning of the LDA beams employing fibre couplers and pigtailed PIN photo diodes (see figure 6a) and the correlation was performed by the mixer/correlator unit shown in figure 6b.

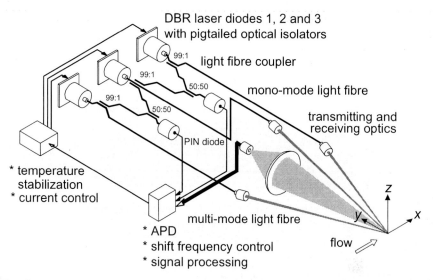

Fig. 6a. Block diagram of a directional two component LDA system based on fibre optics utilizing the optical frequency difference of DBR monomode laser sources as frequency shift.

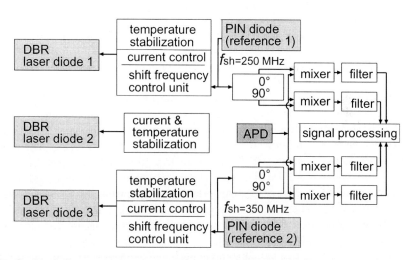

Fig. 6b. Block diagramm of the laser diode driver and the correlator unit.

As the resulting carrier frequencies are completely eliminated by the correlator unit when separating the components and generating quadrature signal pairs in the base band containing the directional information in the phase relationship of the quadrature signal pair, arbitrary frequency shifts higher than the occuring Doppler shifts can be used for a channel separation without having any influences on the evaluation of the measuring information.

3   Experimental results

Figure 7 represents typical burst signal pairs from the measurement of two orthogonal velocity components in a free jet.

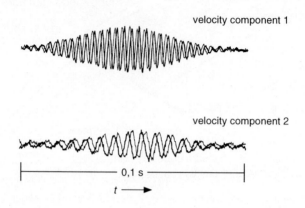

Fig. 7. Typical quadrature signal bursts of the two orthogonal velocity components at the output of the mixer unit in figure 6b measured simultaneously in a free jet.

Figure 8 shows the quadrature signals of each velocity component resulting from particles having nearly the same velocity but different directions when passing through the measuring volume. The velocity direction was changed about 180 degrees by a revolving nozzle in four successive steps (see figure 8a, b, c, d) as indicated by the velocity vectors in the scheme of the measuring volume. The measuring volume is schematically represented by two orthogonal fringe systems one for the u-component in x-direction indicated by the fringe spacing Δx and one for the w-component in z-direction indicated by the fringe spacing Δz. The phase relationship between the signals within the quadrature signal pairs (+90° or -90°) directly corresponds to the sign of the velocity component to be measured. In the first two steps (see figure 8a, b) the phase relationship within the measured quadrature signal pairs for the u- and the w-component remains, then it changes for the u-component (see figure 8c) and at least for both velocity components corresponding to a reversal of the flow (see figure 8d).

Especially if at least two velocity components have to be determined simultaneously, the evaluation of the phase angle time function additionally allows to determine velocities for low numbers or fractions of LDA signal periods resulting from particles passing nearly parallel through the fringe systems in the measuring volume (Müller et al. (1995)).

It has to be pointed out that having quadrature signals in the baseband also quadrature demodulation techniques based upon the evaluation of the phase angle time function can be applied advantageously (Czarske et al. (1993), Czarske et al. (1994)).

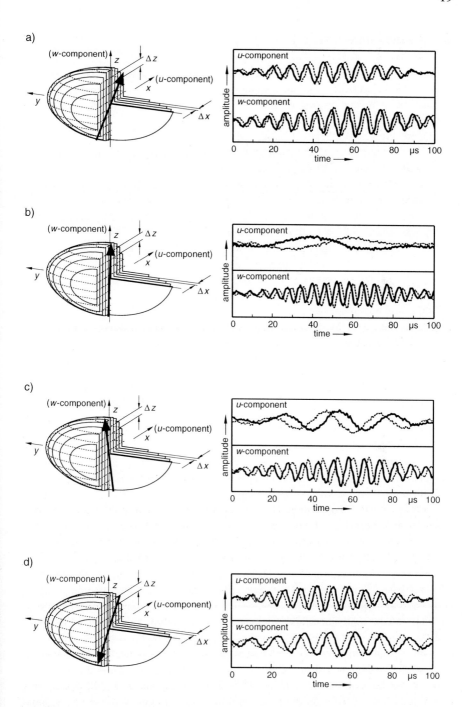

Fig. 8a, b, c, d. Quadrature signal pairs of the two measured velocity components for different flow directions

4　Conclusions

The advantages of a novel LDA concept using separate monomode laser diodes for each LDA beam and quadrature demodulation techniques for the burst signal evaluation have been presented. Some typical experimental results show the application of the LDA system in a two component flow.

The advantages of described LDA system consist in
- the increase of the laser power in the measuring volume by the use of one laser source for each LDA beam,
- the utilization of the beat signal of the laser diodes which is exclusively used as auxiliary carrier frequency for the quadrature signal generation,
- the possible application of quadrature demodulation techniques in the base band concerning: 　* 　directional discrimination
　　　　　　　　　* 　the measurement of very small velocities
　　　　　　　　　* 　the drastically reduced requirements on number of signal periods which has to be evaluated for a precise velocity determination.

Additional advantages are:
- the enlarged velocity measuring range starting　from quasistatic up to high speed flows,
- the simple extension to multicomponent systems by applying fibre optics separating the velocity components by carrier frequency techniques,
- the utilization of fluctuating frequency shifts without deteriorating the measuring accuracy.

5　References

Amann, M. Ch. 1991, Wavelength tuneable single mode laser diodes, Advances in Solid State Physics, Vol 31, pp. 201–218

Czarske, J., Hock, F. and Müller, H. 1993, Quadrature demodulation – A new LDV - Burstsignal Frequency Estimator, Proc. 5th Int. Conf. Laser Anemometry – Advances and Applications, Koningshof, Veldhoven, SPIE VOL. 2052, pp. 79–86

Czarske, J., Hock, F. and Müller, H. 1994, Minimierung der Meßunsicherheit von Laser-Doppler-Anemometern, tm – Technisches Messen, Vol. 61, no. 4, pp. 168–182

Müller, H. and Dopheide, D. 1993, Direction sensitive Doppler velocimeter using the optical frequency shift of two stabilized laser diodes, Proc. 5th Int. Conf. Laser Anemometry – Advances and Applications, Koningshof, Veldhoven, SPIE VOL. 2052, pp. 323–330

Müller, H.; Többen, H.; Arndt, V.; Strunck, V.; Wang, H.; Kramer, R.; Dopheide, D. 1994, New frequency shift techniques in laser anemometry using tunable semiconductor lasers and solid state lasers. In: Modern Techniques and Measurements in Fluid Flows: The Proceedings of The Second International Conference on Fluid Dynamic Measurement and Its Applications (edited by Shen Xiong; Sun Xijiu, International Academic Publishers Beijing, 1994), pp. 3–19

Müller, H.; Czarske, J.; Kramer, R.; Többen, H.; Arndt, V.; Wang, H.; Dopheide, D. 1994, Heterodyning and quadrature signal generation: advantageous techniques for applying new frequency shift mechanisms in the laser doppler velocimetry, Proc. 7th Int. Symposium on Application of Laser Techniques to Fluid Mechanics, Lisbon, 1994, Vol. I, pp. 23.3.1–23.3.8

Müller, H., Többen, H., Strunck, V., Arndt, V., Dopheide, D. 1995, Quadrature demodulation in novel frequency shift LDV systems as an alternative to fringe biasing. In: Laser Anemometry – 1995 – presented at the 1995 ASME/JSME fluid engineering and laser anemometry conference and exhibition, Hilton Head, South Carolina, FED-Vol. 229, pp. 455–458

Wang, H.; Müller, H.; Arndt, V.; Strunck, V.; Dopheide, D. 1994, Experimental realization of a direction sensitive two-dimensional LDA using frequency-shifted diode lasers. In: Modern Techniques and Measurements in Fluid Flows: The Proceedings of The Second International Conference on Fluid Dynamic Measurement and Its Applications (edited by Shen Xiong; Sun Xijiu, International Academic Publishers Beijing, 1994), pp. 126–129

I.3 Two-Component Directional Laser Doppler Anemometer Based on a Frequency Modulated Nd:YAG Ring Laser and Fiber Delay Lines

J.W. Czarske[1] and H. Müller[2]

1: Laser Zentrum Hannover e.V., Department of Development, Hollerithallee 8, D-30419 Hannover, Germany
2: Physikalisch-Technische Bundesanstalt, Laboratory for Fluid Flow Measuring Techniques, Bundesallee 100, D-38116 Braunschweig, Germany

Abstract. A novel method for the directional multicomponent laser Doppler anemometry, LDA, based on the generation of different carrier frequencies, one for each velocity component, is presented. The carrier frequencies are generated by a chirp laser frequency modulation in conjunction with fiber delay lines of different lengths. Since the carrier frequency generation is realized without involving additional frequency shift elements like Bragg cells the LDA arrangement can be significantly simplified. Accurate velocity measurements without influence of carrier frequency fluctuations are accomplished by correlating the generated measuring Doppler signal with reference signals, given by the same carrier frequencies. The employed quadrature demodulation signal processing technique enables the measurement of the momentary Doppler frequency in the baseband. This novel LDA system is demonstrated by a directional measurement of two orthogonal velocity components of fluid flows

Keywords. Fiber-optic laser Doppler anemometry, heterodyne technique, frequency multiplexing, quadrature demodulation technique

1 Introduction

The investigation of complex fluid flows by laser Doppler anemometry, LDA, requires the measurement of at least two velocity components with directional discrimination. Such LDA systems conventionally involve the heterodyne technique by using Bragg frequency shift cells in order to achieve the directional discrimination. However, Bragg cells have a significant optical loss, require a considerable alignment effort and can not be sufficiently miniaturized. The measurement of multiple velocity components is usually achieved by applying a wavelength multiplex technique. This technique is conventionally realized by using a multiple-line argon ion laser, emitting at 488 nm and 514.5 nm wavelength, respectively and by employing components for the wavelength separation of the laser beam as well as of the scattered light. Hence, these multicomponent directional LDA systems exhibit a high alignment effort, a low portability and are in consequence not suited for special measuring applications. The introduction of fiber-optics has improved this situation by the realization of flexible LDA measuring heads [Stiegelmeier 1992, Jentink 1994], but the problem of a high alignment effort, especially by applying Bragg cells in combination with monomode fibers remained.

To overcome these drawbacks, we propose a multicomponent directional LDA system, based on an all-fiber heterodyne technique without having to use Bragg cells or a wavelength multiplex technique, so that the LDA system can be significantly simplified. The applied heterodyne technique is based on a method first proposed by [Jones 1982], using a frequency modulated laser diode and an optical arrangement providing a path length difference for the generation of one carrier frequency. In this contribution we demonstrate a straight-forward development of this method for the generation of different carrier frequencies by employing several delay lines of different optical path length differences. The different carrier frequencies enable a separation of the different velocity components within one detection signal [Adian 1975, Drain 1980], so that the wavelength multiplexing components can be saved. Furthermore, the following novel components are used for the implementation of this method: (i) A diode-pumped Nd:YAG ring laser as frequency modulated light source, having a single-mode power of up to several watts [Freitag 1995] and an efficient compact design, so that the realized LDA system can have a high portability, (ii) cheap telecommunication monomode fibers for an alignment-insensitive delaying and transferring of the laser waves into the LDA measuring volume and (iii) a quadrature demodulation signal processing technique for a precise velocity determination without any influence of carrier frequency fluctuations.

The realized powerful portable fiber-optic LDA system, having an alignment insensitive optomechanical robust design is demonstrated by a directional measurement of two velocity components of a fluid flow.

2 The Principle

The carrier frequencies of the heterodyne LDA system can be generated by four different methods: (1) The application of external frequency shift elements, usually acousto-optic modulators, i.e. Bragg cells, (2) the application of a dual-frequency laser, which emits two waves of slightly different frequencies and orthogonal polarizations [Czarske 1994], (3) the application of several lasers, stabilized on constant difference frequencies, i.e. carrier frequencies [Müller 1996], and (4) the application of a frequency modulated laser in combination with realized optical path length differences between the LDA waves [Czarske 1995b, Czarske 1996b].

The standard arrangement of heterodyne LDA systems is based on method (1), realized usually by employing Bragg cells. The other heterodyne methods (2) - (4) have the advantage to save this additional optoelectronic frequency shift element, so that the LDA system can be simplified. The method (2), well-known from the laser interferometry [Williams 1993], was introduced into the LDA by the employment of a diode-pumped birefringent Nd:YAG microchip laser [Czarske 1994]. However, such laser types emits only a few 10 mW power. Furthermore, only one carrier frequency can be generated, so that the method (2) is not suitable for the implementation of the frequency multiplexing scheme in order to achieve a multicomponent velocity measurement. These limitations can be overcomed by method (3), already realized by powerful diode-pumped Nd:YAG ring lasers [Kramer 1994] and by compact diode-lasers [Müller 1996]. However, the realizing of a two-dimensional heterodyne LDA system by method (3) requires three highly stabilized lasers. A further reduction of the technical effort can be achieved by method (4), since only one laser source is necessary for the realization of the multicomponent heterodyne LDA system. In this contribution, this novel heterodyne method is used for the realization of a directional two-dimensional LDA system.

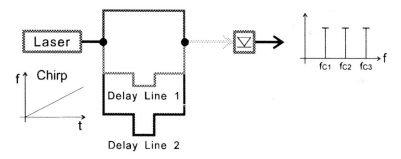

Fig. 1. Principally scheme of the directional two-dimensional LDA system. The chirp frequency modulated laser wave is splitted up into three LDA waves, having different delay times. In the measuring volume the three chirp waves are combined, so that three different carrier frequencies f_{C1}, f_{C2}, f_{C3} are generated.

In Fig. 1 the schematic of this LDA system is shown. The generated three carrier frequencies allow the directional measurement of three Doppler frequencies, see Fig.2. In Fig.3 the generation of the carrier frequencies by means of a chirp frequency laser modulation is explained. The chirp frequency modulated laser wave is splitted up into three chirp waves, having relative delay times of T_i, i=1,2. Assuming a chirp laser frequency modulation, given by $f_L = f_0 + at$ with a as

26

slope, the optical frequencies of the three chirp waves result to: $f_1=f_0+at$, $f_2=f_0+a(t-T_1)$, $f_3=f_0+a(t-T_2)$. The frequency differences between the LDA beam pairs results to

$$\Delta f_1=f_3-f_2=-a\Delta t_1, \Delta f_2=f_2-f_1=-a\Delta t_2, \Delta f_3=f_3-f_1=-a\Delta t_3, \tag{1}$$

where the relative delay times are given by $\Delta t_1=T_2-T_1, \Delta t_2=T_1, \Delta t_3=T_2$.

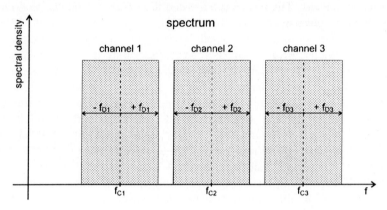

Fig. 2. Scheme of the frequency multiplex technique. Three generated different carrier frequencies enable the separation of the Doppler frequencies $\pm f_{D1}, \pm f_{D2}, \pm f_{D3}$ of three measuring components.

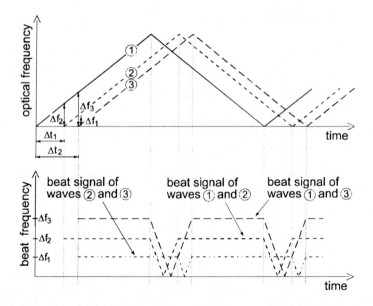

Fig. 3. Principle of the carrier frequency generation. The optical frequency difference between pairs of LDA waves is equal the beat frequency, which is the carrier frequency of the different measuring components.

The employed Nd:YAG ring laser is frequency modulated by a piezo-electrical element, mounted on the monolithic Nd:YAG crystal. The chirp frequency modulation is accomplished by applying a triangular modulation voltage. The modulation slope in Eq.(1) results to $a=2Uf_M df_L/du$ with U as the modulation peak to peak voltage, f_M as the modulation frequency and df_L/du as the laser modulation coefficient. The delay times in Eq.(1) are given by $\Delta t_i = n_i l_i/c$, $i=1,2,3$ where l_i are the relative geometrical delay lengths, n_i is the refraction index and c is the light velocity. The resulting frequency differences, i.e. carrier frequencies f_{Ci}, $i=1,2,3$ are given by

$$f_{Ci}=\Delta f_i = \Delta t_i \frac{df_L}{dt} = \frac{2n_i l_i f_M U}{c} \frac{df_L}{du}, \, i=1,2,3 \qquad (2)$$

It should be noted, that the different carrier frequencies can be generated simply by different delay line lengths. As further advantage, the amount of the carrier frequencies can be adjusted, e.g. close to the occurring Doppler frequencies, by the choice of the laser modulation parameters, i.e. modulation frequency f_M and modulation voltage U.

3 Experimental

3.1 Optical Arrangement

In Fig. 4 the arrangement of the heterodyne LDA system is shown. The used diode-pumped monolithic Nd:YAG miniature ring laser emits 500 mW at 1319 nm wavelength and is piezo-electrically frequency modulated [Freitag 1995]. The laser wave is launched into a monomode fiber with a efficiency of about 80%. The fiber guided laser wave is splitted up by fused fiber couplers and delayed by fiber coils of different lengths. These delay times result in constant frequency differences between the generated three laser waves, see Fig. 3. The laser waves are focused into the measuring volume by a fiber-optic LDA measuring head, manufactured by Schäfters & Kirchhoff, Hamburg, Germany. The Mie back-scattered light waves from scattering particles passing through the measuring volume, is imaged by the measuring head into a multimode fiber of 125 μm core diameter. The guided scattering wave is transferred onto an avalanche InGaAs photo diode, so that the measuring signal is generated. In Fig. 5 the spectrum of the measuring signal for a non-moving scattering particle is shown. The measuring frequencies f_{Mi} of the occurring three measuring components are given by $f_{Mi} = f_{Ci} \pm f_{Di}$, $i=1,2,3$, where f_{Ci} are the carrier frequencies and f_{Di} are the Doppler frequencies of the different velocity components. The Doppler frequencies are given by $f_{Di}=v_i/d_i$, $i=1,2,3$, with d_i as fringe spacings of the fringe systems. The three carrier frequencies shown in Fig. 5 are generated by a triangular laser frequency modulation with a modulation frequency of f_M= 2 kHz, a modulation voltage of U= 500 V and a laser coefficient of about df_L/du= 2.2 MHz/V. According to Eq.(2), the carrier frequency amounts can be calculated, recognizing that the fiber delay lengths are l_1= 118 m, l_2= 221 m, l_3=339 m, see Fig.4, and assuming that the refraction index is n= 1.451 for the germanium doped fiber and n=1.447 for the fluoride doped fiber, respectively. Compared to the first realization of the described heterodyne technique [Czarske 1995b, Czarske 1996b], the used fiber lengths are more than one magnitude shorter. Besides the reduced technical effort, further advantages of the shorter fiber lengths are the smaller disturbances at the switching times of slopes of the triangular modulation signal and the suppression of stimulated Brillouin scattering, occurring in long fiber lines [Cotter 1982].

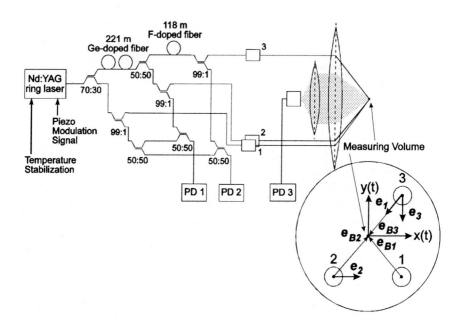

Fig. 4. Optical arrangement of the realized two-dimensional directional heterodyne laser Doppler anemometer. The heterodyne technique is realized by a piezo-electrical laser frequency modulation in combination with fiber-optical delay lines (PD1 and PD2: reference signal pin detectors, PD3: measuring signal avalanche detector).

The measuring directions e_1, e_2, e_3 of the three velocity components are given by $e_1 = e_{B3} - e_{B2}$, $e_2 = e_{B2} - e_{B1}$, $e_3 = e_{B3} - e_{B1}$, where e_{Bi}, i=1,2,3 are the vectors of the three LDV laser beams. Only two of the three measuring directions are linear independent, since the three measuring vectors lie inside a plane. Hence, a two-dimensional velocity measurement can be realized. The evaluated two components 2,3 are corresponded to the carrier frequencies 2,3, which have a higher amount than the carrier frequency 1, so that advantages in the signal processing occur. The carrier frequency amounts has to fulfill the demand, that the difference between the carrier frequencies of different measuring components is greater than the twice Doppler shift, see Fig. 2. Hence, the carrier frequency amounts should be maximized. According to Eq.(2), this can be achieved by the enhancement of the modulation frequency, the modulation amplitude and the fiber length. The enhancement of the modulation frequency results in smaller modulation period durations, so that the LDA burst measurement time is reduced. The fiber length is limited by e.g. the occurring attenuation in the fiber lines. In contrast to these parameters, the laser modulation amplitude principally can be significant enhanced. This will be investigated in the next time.

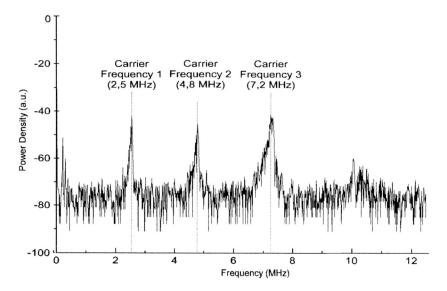

Fig. 5.   Frequency spectrum of the measuring signal for a non-moving scattering particle. The laser modulation parameters are chosen as follows: Modulation frequency f_M= 2 kHz, modulation voltage U= 500 V.

As shown in Fig. 5, the carrier frequencies have fluctuations. To eliminate the influence of the carrier frequency fluctuations on the accuracy of the Doppler frequency measurement, the measuring signal is correlated with reference signals, given by the carrier frequencies for both measuring components 2 and 3, see Fig. 6. The reference signals are generated by out-coupling of a small part of the LDA waves onto pin InGaAs photo diodes, see Fig. 4.

In principle, different modulation signal forms can be used for the described heterodyne technique. However, the employed chirp modulation, based on a triangular modulation signal enables the generation of nearly constant carrier frequencies. The sinusoidal modulation results in a high carrier frequency

30

fluctuation from zero to maximum, which can not be tolerated in the signal processing. The sawtooth modulation form has different slopes, resulting in different carrier frequencies, which can disturb the signal separation, then several velocity components have to be evaluated. Additionally a sawtooth modulation allows only the half carrier frequency amount compared to triangular modulation, assuming the same laser frequency modulation amplitude and period.

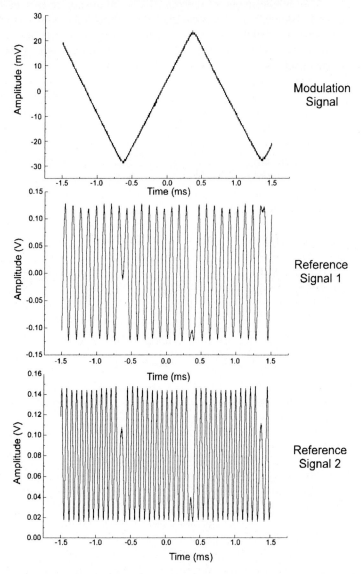

Fig. 6. Time functions of the triangular laser modulation signal and the two reference beat signals. The laser modulation frequency is f_M=500 Hz and the signal voltage is about U= 50 mV.

3.2 Signal Processing Technique

The determination of the Doppler frequency can be achieved by different signal processing techniques:

(1) The generated burst signals can be evaluated in different frequency bands, (i) the carrier frequency band, i.e. the original frequency range of the applied heterodyne technique. (ii) The intermediate frequency band, where the intermediate frequency is lower than the carrier frequency and can be adjusted by an electronic oscillators and mixer units. (iii) The baseband, i.e. the frequency range without carrier frequency. However, due to the occurring carrier frequency fluctuations, see Fig. 5, the measuring signal has to be correlated with generated reference signals. This correlation process can be accomplished by reference signals, having (a) the same carrier frequency or (b) a different carrier frequency compared to the measuring signal. The methods (a) and (b) result in baseband signals (iii) and intermediate frequency signals (ii), respectively. However, the generation of baseband signals has several advantages, e.g. Doppler frequencies with a low amount, e.g. quasi static particle movement, can be measured with a high resolution. Hence, in this paper the generation and evaluation of baseband signals will be described, see below.

(2) For the determination of the burst Doppler frequency many LDA signal processing techniques are currently available, see e.g. [Dopheide 1995, Tropea 1995]. The FFT processor technique and the counter technique are often used in LDA. However, the time-resolution of the Doppler frequency measurement is low, since they are based on an averaging process. An alternative signal processing technique is the quadrature demodulation technique [Czarske 1995a, Czarske 1996a, Czarske 1996b, Müller 1996]. This novel technique is based on the time-resolved measurement of the burst signal phase and their momentary frequency. In the following, the quadrature demodulation technique will be applied for the determination of the Doppler frequency of the generated baseband quadrature signal pairs.

In Fig. 7 the block diagram of the used heterodyne technique is shown. The measuring signal is correlated by two-stage mixer and flow-pass filter units with the reference signals 1 and 2, given by the carrier frequencies of the measuring components 2 and 3, respectively. In consequence, the occurring carrier frequency fluctuations are eliminated. This will be demonstrated for a one-dimensional LDA system by the following expressions [Müller 1994]: The measuring signal $a_M(t)=A_M sin[2\pi f_M(t)t]$, with $f_M(t)=f_C(t) \pm f_D(t)$ is mixed with the reference signal $a_R(t)=A_R sin[2\pi f_R(t)t]$, where $f_R(t)=f_C(t)$, i.e. the reference signal frequency is equal the carrier frequency $f_C(t)$. The mixing product is given by $A_M A_R cos[2\pi f_D(t)t]-cos[2\pi(2f_C(t)\pm f_D(t))t]/2$. After a suitable low-pass filtering one gets the baseband signal $A_M A_R cos[2\pi f_D(t)t]/2$. Obviously, the carrier frequency $f_C(t)$ with all its fluctuations is eliminated and the Doppler frequency $f_D(t)$ can be precisely determined. One assumption of this result is a symmetrical setup for the generation of reference and measuring signal, respectively, in order to ensure that both signals contain the same carrier frequency. In consequence, two demands have to be fulfilled: (i) The relative difference between the optical delay lengths, see Eq.(2), of the measuring path $n_M l_M$ and the reference path $n_R l_R$ should be negligible compared to their common fiber coil delay line length, see Fig. 4. The measuring path length difference $n_M l_M$ is given by the path from the 99:1 fiber couplers, acting as wave dividers, to the measuring volume, where the three LDA beams are combined, see Fig.5. The reference path length difference $n_R l_R$ is given by the path from the same 99:1 fiber couplers to the 50:50 fiber couplers, where the optical reference waves are combined, before they interfered at the photo detectors PD1 and PD2. (ii) The absolute difference between the measuring and reference path lengths, respectively, should be negligible compared to the correlation length of the carrier frequency fluctuations, which are mainly caused by the non-linearity of the laser modulation coefficient.

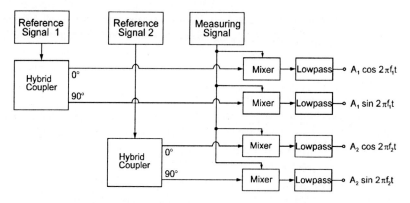

Fig. 7. Scheme for the generation of two quadrature signal pairs in the baseband.

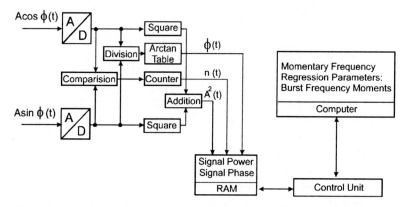

Fig. 8. Quadrature demodulation technique, shown for one measuring component. After digitizing the quadrature signal pair the wrapped phase $\phi(t)$, the incrementally counted periods $n(t)$ and the signal power $A^2(t)$ are determined. Based on these parameters, the momentary burst frequency as well as the burst frequency moments, like the averaged burst frequency can be determined.

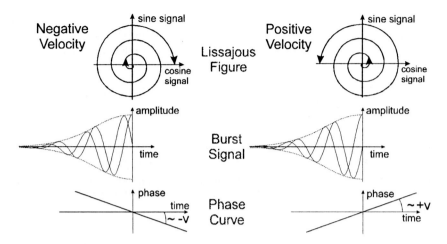

Fig. 9. First half of a quadrature signal pair for opposite moving directions of the scattering particle, passing the measuring volume. The demodulated phase of the quadrature signal pair determines the velocity amount and the moving direction of the scattering particle.

By recognizing these demands, sufficient LDA baseband signals are generated. To retain the directional discrimination when eliminating the carrier frequencies, quadrature signal pairs are generated [Müller 1994]. A commercial available hybrid coupler is used to achieve a broadband 90^0 phase shift, see Fig. 7. The bandwidth for ensuring this quadrature phase shift defines the tolerable fluctuations of the carrier frequencies.

Since the signal processing is accomplished in the baseband it should be noted that the carrier frequency signals are auxiliary signals for the realization of (i) the frequency multiplexing of the different measuring components and (ii) ensuring a constant 90^0 phase shift, necessary for the quadrature signal pair generation.

The quadrature signal pair can be evaluated by different signal processing techniques, e.g. by a complex FFT processor technique or a directional counter technique. However, as already mentioned, in this contribution the quadrature demodulation technique is applied, see Fig. 8: This technique is based on the measurement of the relative signal phase by a division of the quadrature signal pair and an arctan calculation. Together with an incremental directional counting of the signal periods absolute signal phases are achieved. Additionally the signal power is determined, so that a weighting of the signal phases can be accomplished. By a differentiation procedure the momentary Doppler frequency can be determined. Furthermore, the measured phase time curve can be evaluated by the regression method: The fitting of e.g. a regression polynom curve provides the determination of the average burst frequency and higher frequency moments, like the frequency change within the burst signal. In consequence, the averaged accelerations in a burst signal can be measured.

In Fig. 9 an overview on the different representations of quadrature signal pairs is shown. In the top and the middle the Gaussian burst quadrature signal pairs are described in parameter-form and as time functions, respectively. The resulting phase time curve contains the information on the movement of the scattering particle, since the phase is proportional to the particle position. It is important to note, that the neighboring half periods of triangular chirp frequency laser modulation have slopes of different signs, see Fig. 3, so that the generated frequency difference of the LDA beams changes their sign. Hence, the different slopes have to be taken into account in the signal processing. This was accomplished by a multiplication of the determined regression slopes by ±1, where the sign is dependent on the modulation slope. In consequence, the finally resulting sign of the regression slope determines the fluid flow direction correctly.

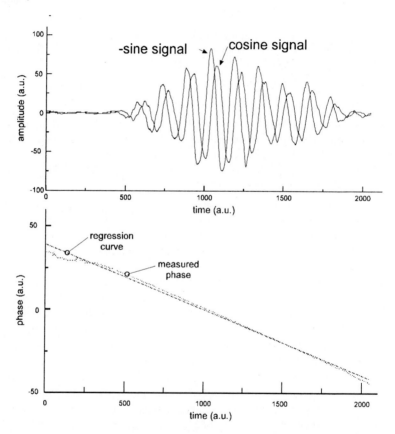

Fig. 10. Measured quadrature signal pair. The Gaussian burst signal form results from the one-particle scattering and Gaussian intensity distribution in the measuring volume. At the ends of the burst signal the measured phase has a high phase noise, caused by the poor signal-to-noise ratio. In consequence, a significant deviation of the straight regression curve and the phase curve occurs.

In Fig. 10 a typical burst signal pair of the measuring component 2 is shown. This signal pair was generated by an air flow with injected scattering water particles. Since a low particle concentration was chosen, it results the observed one-particle scattering. At higher particle concentrations, a multiple-particle scattering signal can be observed, see Fig. 11. As shown, the straight regression line has very small deviations compared to the measured phase curve. Hence, it can be concluded that also multiple-scattering signals can be evaluated by the quadrature demodulation technique. The opposite sign of the regression slope of Fig. 11 compared to Fig. 10 is caused by the opposite fluid flow direction. The higher amount of the regression slope of Fig. 11 compared to Fig. 10 has its reason in the higher flow velocity.

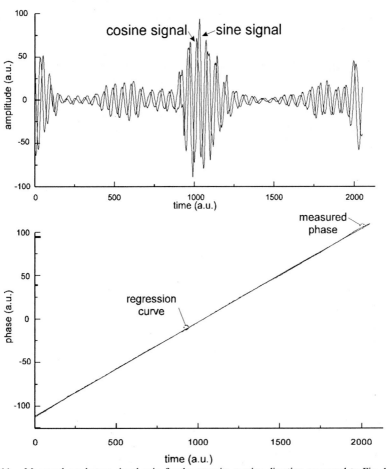

Fig. 11. Measured quadrature signal pair, for the opposite moving direction compared to Fig. 11. In consequence, the phase difference between the signal pair is changed and the slope of the phase curve has a different sign. Due to the occurring multiple-particle scattering, it results the observed intermitting signal form.

4 Conclusions

A novel fiber-based heterodyne laser Doppler anemometer, LDA, for the directional two-dimensional velocity measurement of fluid flows has been realized. The presented LDA system has the following advantages:

(i) The employed diode-pumped Nd:YAG ring laser type delivers a high single-frequency power of multiple Watt [Freitag 1995], which is significantly higher than the optical power of single-mode diode lasers [Durst 1990, Müller 1996]. Hence, a high signal-to-noise ratio and in consequence a high measurement accuracy can be achieved. Compared to diode-pumped fiber lasers [Czarske 1997], the Nd:YAG ring laser enables a defined frequency modulation of the single-mode laser emission, one demand of the proposed chirp heterodyne technique.

(ii) The compact design and the high efficiency of the Nd:YAG ring laser results in a high portability of the LDA system, especially compared to conventional LDA systems, employing argon ion lasers.

(iii) The employed fiber-optics enables the alignment insensitive realization of transferring, defined delaying, splitting and combining of the LDA waves, necessary for the employed heterodyne technique. Since the carrier frequency generation is achieved without additional optoelectronic elements, the heterodyne LDA system can be significantly simplified. The generation of stimulated Brillouin scattering [Cotter 1982] in the fiber lines, as one of the most important mechanisms causing optical losses, is eliminated by the use of small fiber lengths of only a few hundred meters.

(iv) The applied heterodyne technique enables a precise velocity measurement without influence of carrier frequency fluctuations. Due to the quadrature demodulation processing in the baseband, quasistatic scattering particle movements can be measured with high velocity resolution. Additionally, velocity fluctuations can be directionally evaluated with a high time resolution, since the phase angles of the generated quadrature signal pair and hence the momentary Doppler frequency can be determined.

Acknowledgments

The authors thank Dipl-Ing. O. Dölle, Dr. I. Freitag, Dr. A. Tünnermann, Prof. H. Welling and Prof. D. Dopheide for support.

References

Adrian, R.J., 1975, A biploar, two component laser-Doppler velocimeter, J. o. Physics E: Scientific Instruments, **8**, pp.723-726.

Cotter, D., 1982, Observation of stimulated Brillouin scattering in low-loss silica fibre at 1.3μm, Electron. Lett., **18**, pp.495-496.

Czarske, J. & Müller, H., 1994, Heterodyne interferometer using a novel two-frequency Nd:YAG laser, Electron.Lett., **30**, pp.970-971.

Czarske, J. & Müller, H., 1995a, Birefringent Nd:YAG microchip laser used in heterodyne vibrometry, Optics Commun., **114**, pp.223-229

Czarske, J. & Müller, H., 1995b, Multicomponent heterodyne laser Doppler anemometer using chirp-frequency modulated Nd:YAG ring laser and fiber delay lines, Electronics Lett., **31**, pp.970-971.

Czarske, J., 1996a, Verfahren zur Messung und Auswertung der Interferenzphase in der Laser-Doppler-Velocimetrie, VDI-Fortschritt-Berichte **8**, 530, VDI, Düsseldorf, Germany.

Czarske, J., 1996b, Method for the Analysis of the Fundamental Measuring Uncertainty of Laser Doppler Velocimeters, Opt. Lett., **21**, pp.522-524.

Czarske, J., Zellmer, H., Tünnermann, A., Welling, H. & Müller, H., 1997, Novel high-power laser Doppler anemometer using a diode-pumped fiber laser Applied Physics B (Rapid Communication), **B64**, pp.119-123.

Dopheide, D., 1995, Neue Halbleitermeßverfahren für komplexe Strömungen, Habilitationsschrift, Fachbereich Maschinentechnik, Universität-Gesamthochschule Siegen.

Drain, L.E., 1980, The laser Doppler technique, Wiley, Chichester, pp.204.

Durst, F., Müller, R. & Naqwi, A., 1990, Measurement accuracy of semiconductor LDA systems, Exp. in Fluids, **10**, pp.125-137.

Freitag I., Tünnermann, A. & Welling H., 1995, Power scaling of monolithic miniature Nd:YAG ring lasers to optical powers of several watts Optics Commun., **115**, pp.511-515.

Jentink, H.W., Stieglmeier, M. & Tropea, C., 1994, In-flight measurements using laser Doppler anemometry, J. Aircraft, **31**, pp. 444-446.

38

Jones, J.D.C., Corke, M., Kersey, A. & Jackson, D.A., 1982, Miniature solid-state directional laser Doppler velocimeter, Electron. Lett., **18,** pp. 968-969.

Kramer, R., Müller, H. & Dopheide, D., 1994, The realization of a continuously tunable optical frequency shift LDA-system at green wavelength for highly turbulent flows using diode-pumped Nd:YAG lasers and monolithic ring frequency doublers, Proc. 7th international symposium on applications of laser techniques to fluid mechanics, Lisbon, 11.-14. July 1994 (Portugal: Instituto Superior Tecnico), paper 14.5.

Müller, H., Czarske, J., Kramer, R., Többen, H., Arndt, V., Wang, H. & Dopheide, D., 1994, Heterodyning and quadrature signal generation: advantageous techniques for applying new frequency shift mechanisms in the laser Doppler velocimetry, Proc. 7th international symposium on applications of laser techniques to fluid mechanics, Lisbon, 11.-14. July 1994 (Portugal: Instituto Superior Tecnico), paper 23.3.

Müller, H., Wang, H. & Dopheide, D., 1996, Fiber optical multicomponent LDA-system using the optical frequency difference of powerful DBR laser diodes, Proc. 8th international symposium on applications of laser techniques to fluid mechanics Lisbon, 8-11 July 1996 (Portugal: Instituto Superior Tecnico), paper 32.3, also in this proceeding.

Stieglmeier, M. & Tropea, C., 1992, A miniaturized, mobile laser-Doppler anemometer, Appl. Opt. **31,** pp.4096-4105.

Tropea, C., 1995, Laser Doppler anemometry: recent developments and future challenges, Meas. Sci. Technol., **6,** pp. 605-619.

Williams, D.C. (Ed.), 1993, Optical methods in engineering metrology, Chapman & Hall, London, chapter 5 and 6.

I.4 Integrated Laser Doppler Velocimeter made by Ion-Exchange in Glass Substrate

P.L.-Auger[1], A. Cartellier[2], P. Benech[1] and I. Schanen Duport[1]

[1] LEMO, UMR 5530 INPG-UJF-CNRS, 23 Avenue des Martyrs, BP 257, 38016 GRENOBLE CEDEX, France
[2] LEGI, UJF-CNRS-INPG, BP 53 X, 38041 GRENOBLE CEDEX, France

Abstract. A fully integrated laser Doppler velocimeter made by ion-exchange in a glass substrate is presented. Thanks to the integration of both the emission and reception systems, no optical adjustment are required. Beside, when flush mounted on a wall, this LDV does not collect the unwanted reflected light caused by the proximity of that wall. It is thus well suited for velocity measurements in boundary layers. In this paper, the fabrication technique is thoroughly described and some qualifications are commented. In the prospect of wall shear stress measurements, it is also shown that, by optimizing the waveguiding structure, highly localized velocity measurements are accessible.

Keywords. Integrated Velocimeter, LDV, LDA, Integrated Optics, Ion-Exchange

1 Introduction

Measurements of velocity gradients in the viscous sublayer is not an easy task. Among the non intrusive possible techniques, laser Doppler anemometry is probably the best one.[1-3] Such apparatus made in classical volume optic are found in research laboratories and industries despite their elevated cost. For such systems, both the emission and the reception system are located outside the fluid flow and requires the presence of transparent walls. Their spatial resolution depends mainly on the size of the measuring volume formed by the intersection of the two lasers beams. The smallest ones are comprise between 30 x 20 x 20 μm^3 and 200 x 200 x 300 μm^3 [4-7] and requires rather complex lenses system which is relatively big. The shortest distance between the measuring volume and the channel wall doesn't go under 100 to 200 μm.[8-10] One of the reason for this is that diffused light is reflected from the wall and degrades the optical signals received. Also, if the measuring volume touches the wall, then unwanted reflection of the volume itself will also cause a degradation of the signals.[11-12]

Partially integrated velocimeter were also developed.[13] In that case, the emission system was introduced through the channel wall but the reception system was still exterior to the fluid flow. Direct velocity gradient were measured with such a partially integrated captor.[13] The fringe spacing varied linearly with the distance. However, rather long optical adjustments were required and the uncertainty was quite large (± 50 %).

In the present paper, we propose a first prototype of a fully integrated velocimeter, i.e. including the emission and the reception, able to be introduced inside a channel wall. Reception is done on the same side as the emission. For this velocimeter, no adjustments of any kind by an operator are required. The final aim of the project is to produce a velocimeter with a very high spatial and temporal resolution. The measuring volume will be located inside the viscous sublayer ($y^+ \leq 6$). So indirect measurements of the velocity gradient and the skin friction by the same way could be done.

The fabrication of the velocimeter is first presented. Afterwards, a description of it is given. Measurements obtained with an optical fiber, as the reception element of the optical signals, are then presented and discussed. Finally, possibilities offered by a new configuration of interferometer are discussed.

2 Fabrication

The main part of the velocimeter is the system producing the fringes. It is realized by integrated optic techniques on a glass substrate. The choice of that substrate was governed by the fact that glass is well suited to realize passive components.[14] It is indeed a cheap materiel with good optical properties. Also, the technique of ion-exchange, the most widely used technique to produce waveguides on glass, is well controlled.[14-16]

To realize the present velocimeter prototype, Mentzel soda-lime glass is utilized. It is first clean in a tensio-actif industrial soap solution (Decon soap) that takes out organic as well as inorganic contaminants. Then, a layer of approximately 150 nm of aluminum is evaporated on it. Afterwards, standard photolithography process employed in the fabrication of integrated circuits is used. Positive photoresist is spinned over the aluminum (0,5 µm thick). The guiding structure is reproduced from the photolithographic mask to the resist by exposition to U. V. light, as it is shown on figure 1. Exposed resist will be dissolved in a developer solution, contrarily to the unexposed one. Aluminum can then be locally chemically attacked with an acid solution. After that, all of the remaining photoresist is removed with acetone. As the final result, the guiding structure is present on the aluminum layer.

The glass plate with his aluminum coating is immersed in a pure molten salt of potassium nitrate (KNO^3). An exchange between the potassium ions present in the salt and the sodium ions contained in the glass takes place over specific regions where there is no aluminum present, see figure 1, the latter blocking the diffusion of ions. A specific waveguiding pattern is thus created in the glass by this way. Indeed, due to the difference between the size of the sodium and the potassium ions, local constraints are created in the glass that give rise to an increase of the refractive index.

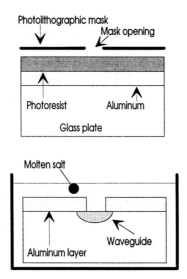

Fig. 1. The photolitographic process. Top) U. V. exposition. Bottom) Ion-exchange step

For the sample fabricated, the exchange took place for 4 hours at a temperature of 350°C. In those conditions, the maximum index change is 0.11.

Finally, the aluminum layer is removed and both edges of the glass plate , corresponding to the entrance and exit sides of the light trajectory, are mechanically polished at a precision of 0.1 μm.

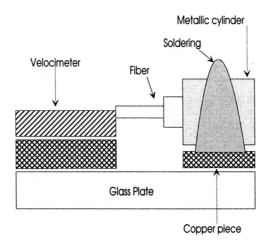

Fig. 2. Schematic representation of the connection between the optical fiber and the glass sample

A singlemode optical fiber connected to a pigtail laser diode is glued inside a small metal cylinder which is attached to the sample by soldering it to a small copper piece glued on a support plate, like can be seen in figure 2. The procedure is relatively easy to do but rather long. Nevertheless, the attachment is mechanically very solid, it resists to water and to humidity and it is quite stable in time. No coupling power change could be observed for a 12 months old sample.

3 Description of the Integrated Velocimeter

The integrated guiding structure used for the emission of the fringes is shown on figure 3. The first part of it is a singlemode straight waveguide which is coupled to the optical fiber of the pigtail laser diode which wavelength is 832 nm. The guided light is then divided into two equal parts in the Y-junction. The latter plays the role of a beam splitter. Each arm of the interferometer that follows has the same length and is terminated by a taper structure.[17] the purpose of the latter is to excite only the fundamental mode of a multimode waveguide of 40 µm width (this width is the one of the photolitographic mask opening). The distribution of light intensity is almost Gaussian. The size of a single waveguide, 2 µm, is small and the beam produced by such an optical waveguide would strongly diverge in free space while the beam issued from the fundamental mode of a 40 µm waveguide will act like a quasi-collimated beam. The divergence of the latter is on order of 1.5°.

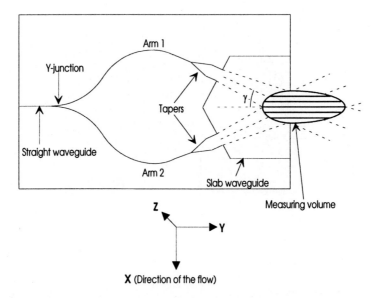

Fig. 3. Schematic representation of the integrated emission system

The output beams of the two tapers will then travel in a slab waveguide, where light is confined in one dimension only, and will produce the interference pattern.[18] Half of the latter exist inside the glass plate while the other half is outside the exterior medium. That does not have to be the case : the interference pattern could completely be outside the glass plate. Still, the present waveguiding geometry imposes that fact. In both mediums, it is easy to demonstrate that the fringe spacing (Δx) is given by :

$$\Delta x = \frac{\lambda_0}{2 N_{eff} \sin \gamma},$$ (1)

where λ_0 is the wavelength in free space, N_{eff} is the effective index of the slab waveguide and γ is the half angle between the two beams (see fig. 3). It is worth noting that the fringe spacing is independent of the refractive index of the exterior fluid. The position of the measuring volume (in the case where it would entirely be outside the glass plate) and its dimensions do depend however upon the fluid index. Ulrich and Martin[19] have indeed demonstrated that light, when guided in a slab waveguide, follows Snell law of refraction with the effective index instead of the refractive index of the bulk materiel.

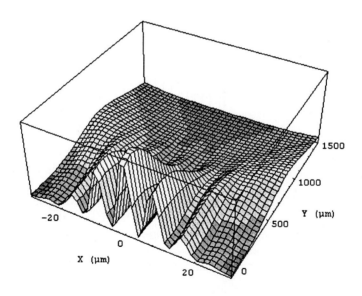

Fig. 4. Computed interference fringe of the measuring volume at z = 0 in water

For the present prototype, the angle between the two beams inside the glass is 1.8°, the fringe spacing 8.77 μm. The width of the output beam at $1/e^2$ of the intensity produced by the taper is 36 μm. In those conditions, the size of the measuring volume in water (half of the whole interference region) is approximately 60 μm x 1000 μm x 50 μm, as shown in the calculation of figure 4. The present configuration is therefore not adapted for measurements in the viscous sublayer. Nevertheless, it demonstrates the feasibility of such a velocimeter.

An image of the fringe pattern observed with a digital camera on the edge of the glass is given in figure 5 as well as the corresponding intensity profiler. The contrast ratio being only 0.9 is explained by the fact that the middle of the fringe pattern is not exactly located on the edge of the glass. The geometry of the tapers causes the envelope of the fringes (fig. 5) not to be a Gaussian type. Other taper produces Gaussian type envelope, like shown on fig. 6, but due to the small angle available at the present time between the beams, the number of fringes is not sufficient enough for a LDV. This will be corrected in future configurations.

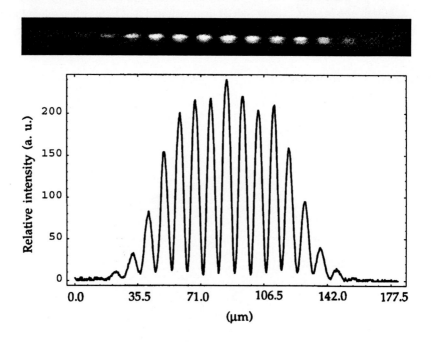

Fig. 5. Top) Image of the fringe pattern on the edge of the glass. Bottom) Corresponding intensity

The total optical output power of the sample for an input power of 30 mW is 0,95 mW, corresponding to 15 dB optical loss in the sample. The propagation losses are estimated to 1 dB/cm, so the major part of those losses (11 dB) comes from the connection between the optical fiber and the sample. Great care must then be taken when connecting the fiber to the sample. Total losses on the order of 8 dB can reasonably be expected.

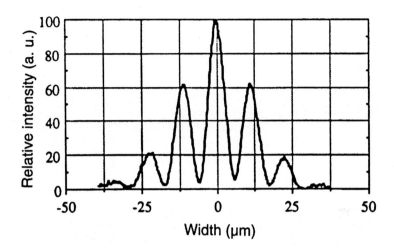

Fig. 6. Fringe pattern produced with a different taper and having a Gaussian envelope

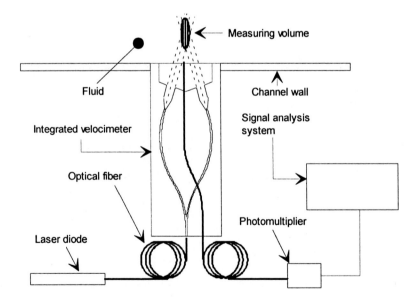

Fig. 7. Scheme of the complete integrated velocimeter

Finally, an optical fiber, that can easily be glued on the glass plate, collects the light scattered by particles passing in the measuring volume and transmits it to a photomultiplier. The scheme of the complete integrated velocimeter is given in fig. 7. Because the laser diode and the photomultiplier can both be several meters apart from the head of the velocimeter, measurements can be done in places of difficult access.

4 First Experimental Validations

To test the velocimeter, a metallic cylinder of 500 μm diameter is used as the diffusing particle. It is placed on the side of a rotating disk of 4.3 ± 0.2 cm diameter, making two complete revolutions per minute. The speed of the diffusing element is thus 0,90 ± 0.04 cm/s. The extremity of the emitting system was placed at a distance of approximately 0.5 mm away from the cylinder.

First, a plastic fiber was used to collect the retrodiffused light. An example of the signals obtained on an oscilloscope is given in fig. 8. A low-pass filter was used to lower the noise produced by the amplification circuit following the photomultiplier. The measured Doppler frequencies ranging from 1.02 to 1.08 kHz correspond to velocities comprise between 0.894 and 0.947 cm/s, which are comparable to the one calculated above

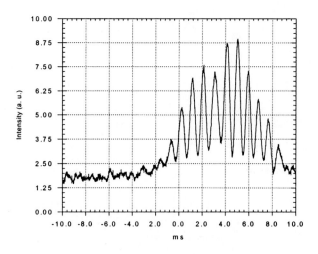

Fig. 8. Example of Doppler burst obtained with the plastic fiber as the reception element. Longitudinal distance is 24 mm. A low-pass filter (2.5 kHz) was used

The reception fiber was allowed to move freely in regard to the glass plate. Study of the Doppler signal as a function of the position of the fiber was done that way. The longitudinal distance between the fiber extremity and the glass edge was varied. For each position, the signal visibility (η), defined as :

$$\eta = \frac{I_{max} - I_{min}}{I_{max} + I_{min}}, \tag{2}$$

where I_{max} and I_{min} are respectively the maximum and minimum intensities of the signal measured, is calculated. Figure 9 shows the results of that study. It can be seen from the graph that a distance of 24 mm gives the maximum ratio, 0.73. For greater distances, it becomes difficult to discern the signal from the noise. Also, one has to keep in mind that the center of the measuring volume is located very near the side of the glass plate, so a ratio of 1 is impossible to get in those conditions. The conclusion of that first study is that plastic fiber is inadequate for measures in the subviscous layer because its extremity cannot be placed next to the edge of the glass. This is essential if the velocimeter is to be placed inside the wall of a channel. Nevertheless, it should be considered in the case where someone would like the measuring volume to be at a distance from the wall in the order of 1 or 2 cm.

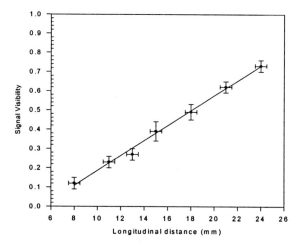

Fig. 9. Visibility as a function of the longitudinal distance between the fiber and the glass

The same study was done but with a multimode silica fiber (100/140 μm) this time. In that case, Doppler frequencies could be observed for a distance varying from 0 to at least 4 mm between the fiber extremity and the glass side. Example of filtered and non filtered signals are given in figure 10. The lateral position of the fiber is however critical for a null longitudinal distance. It has to be placed within ±100 μm from the symmetry axis of the integrated guiding structure. Signal merging from a measuring volume located between 0 and 500 µm of the glass side can be collected with that multimode fiber glued on the glass plate.

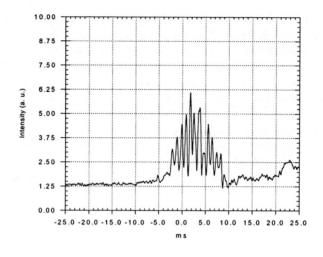

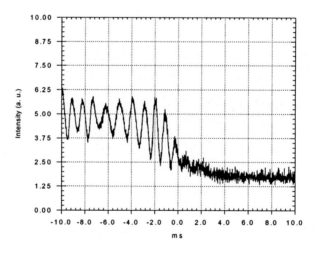

Fig. 10. Example of signals obtained with the silica multimode fiber. Top) low-pass filter (2.5 kHz) used. Bottom) No electric filter used

5 New Generation of Prototype

The measuring volume of our first prototype is however too large for accurate velocity measurements in the viscous sublayer. The actual configuration but with a different geometry could produce measuring volumes of 50 x 65 x 10 μm³ in water (75 x 50 x 15 μm³ in air) with a fringe spacing of 0.5 μm. The distance from the glass side is 100 μm. This seems to be the limit of the taper configuration.

To further improve the spatial resolution, a new design of the interferometer has been imagined using a convergent structure. Optical simulations show that it should provide a probe volume of 15 x 20 x 40 μm³ in water, located at a distance ranging between 0 up to 500 μm from the glass side. The fringe spacing is 0.5 μm. The probe volume extend along the direction normal to the wall is 20 μm. This design is well adapted to wall shear stress measurements. Some of these new prototypes have been manufactured and are currently under test in terms of optical performances. In the near future, their performance with respect to wall shear stress measurements will be checked in various well controlled flow conditions, both in air and water flows.

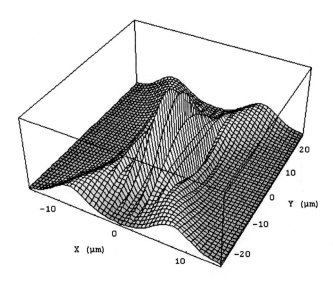

Fig. 11. Example of computed fringe pattern of the measuring volume of the new prototype at z = 0 in water

6 Conclusion

A first prototype of an integrated laser Doppler velocimeter has been presented. Its emission system has been made by integrated optics techniques on a glass substrate. Using a reception consisting of either a multimode silica fiber or a plastic fiber, Doppler bursts from retrodiffused light and with good visibility have been recorded. The feasibility of this technique has thus been demonstrated, indicating that the combination of an integrated optical interferometer and a multimode optical fiber is an easy and efficient way to produce cheap velocimeters free of optical adjustments. Future words will be devoted to the qualification of wall shear stress measurements using a new generation of prototypes allowing a much better spatial resolution.

7 Acknowledgments

The first author would like to thank the FCAR organization of the Quebec government for a doctoral research grant. Authors are grateful to PIR Ultimatech for financial support under grant 24 N-II.4.1.

8 References

[1] Hanratty TH.J., Campbell J.A., 1983, "Chapter 11 : Measurement of wall shear stress", in Fluid Mechanics Measurements, Golstein R.J. Ed., Hemisphere Publishing Corp., distributed by Springer-Verlag, pp. 559-611.

[2] Saetran L.R., 1987, "Comparison of five methods for determination of the wall shear stress, AIAA J., 25, 11, pp. 1524-1527.

[3] Gasser D., Thomann H., Dengel P., 1993, "Comparison of four methods to measure wall shear stress in a turbulent boundary layer with separation", Experiments in Fluids, 15, pp. 27-32.

[4] Martin S.R., Bates C.J., 1992, "Small-probe-volume laser Doppler anemometry measurement of turbulent flow near the wall of a rib-roughened channel" Flow Meas. Instrum. Vol. 3, No. 2, pp. 81-88.

[5] Leprince F., Riethmuller M.L., 1986, "LDV Measurements in a viscous sublayer : determination of skin friction", Laser Techniques and Applications in Fluids Mechanics - Proceedings of the 3rd International Symposium, Lisbon, Portugal, paper 1-6.

[6] Aizu Y., Ushizaka T., Asakura T., 1985, "Measurement of the Velocity Gradient Using a Laser Doppler Phenomenon", Appl. Phys. B 36, pp. 155-161.

[7] Keveloh C., Staude W., 1983, "Determination of velocity gradients with scattered light cross-correlation measurements" Applied Optics, Vol. 22, No 2, pp. 333-338.

[8] Wernet M.P., Edwards R.V., 1986, "Implementation of a new time of flight laser anemometer", Applied Optics, Vol. 25, No 5, pp. 644-648.

[9] Dybbs A., Edwards R.V., 1987, "Refractive index matching for difficult situations", 2nd Int. Conf. Laser Anemometry - Advances and Applications, Glascow, UK, BHRA, Cranfield, paper 1-22.

[10] Durst F., Keck T., Kleine R., 1985, "Turbulence quantities and Reynolds stress in pipe flow of polymer solutions", 1st Int. Conf. Laser Anemometry - Advances and Applications, Manchester, UK, BHRA, Cranfield, paper 35-52.

[11] Mishina H., Vlachos N.S., Whitelaw J.H., 1979, "Effect of wall scattering on SNR in laser Doppler anemometry", Applied Optics, Vol. 18, No 14, pp. 2480-2485.

[12] Cline C., Deutsh S., 1993, "On elevated RMS levels in wall-bounded turbulent flows when measured by laser Doppler velocimetry", Experiments in Fluids, Vol. 15, pp. 130-132.

[13] Petrik S., Naqwi A.A., Durst F., 1992, "Fiber optic dual-cylindrical wave sensor for measurement of flow velocity gradients", 2nd French-German workshop on Optical Measurement Techniques - Fibers Optics and Instrumentation, 13-14 October, St. Etienne, France.

[14] Johansson J., Djanta G., and Coutaz J.-L., 1992, "Optical waveguide fabricated by ion exchange in high-index commercial glasses", Applied Optics, Vol. 31, No 15, pp. 2796-2799.

[15] Ramaswamy R.V. and Srivastava R., 1988, "Ion-exchanged glass waveguides : a review", Journal of Lightwave technology, Vol. 6, No 6, pp. 984-1001.

[16] Findakly T., 1985, "Glass waveguides by ion exchange : a review", Optical engineering, Vol. 24, No 2, pp. 244-250.

[17] Duport I., Benech P., Kahlil D., Rimet R., 1994, Study of linear tapered waveguides made by ion exchange on glass", J. of Physics, D : Applied Physics, 25, pp. 913-918.

[18] Schanen Duport I., Benech P., Rimet R., 1984, "New integrated-optics interferometer in planar technology", Applied Optics, Vol. 33, No 35, pp. 5954-5958.

[19] Ulrich R. and Martin R.J., 1971, "Geometrical optics in thin film light guides", Applied Optics, Vol. 10, pp. 2077-2085.

I.5   Sparse Array Image Correlation

D. P. Hart

Massachusetts Institute of Technology
Department of Mechanical Engineering, Room 3-246
Cambridge, MA 02139-4307, USA

Abstract. With the development of Holographic PIV (HPIV) and PIV Cinematography (PIVC), the need for a computationally efficient algorithm capable of processing images at video rates has emerged. This paper presents one such algorithm, sparse array image correlation. This algorithm is based on the sparse format of image data - a format well suited to the storage of highly segmented images. It utilizes an image compression scheme that retains pixel values in high intensity gradient areas eliminating low information background regions. The remaining pixels are stored in sparse format along with their relative locations encoded into 32 bit words. The result is a highly reduced image data set that retains the original correlation information of the image. Compression ratios of 30:1 using this method are typical. As a result, far fewer memory calls and data entry comparisons are required to accurately determine tracer particle movement. In addition, by utilizing an error correlation function, pixel comparisons are made through single integer calculations eliminating time consuming multiplication and floating point arithmetic. Thus, this algorithm typically results in much higher correlation speeds and lower memory requirements than spectral and image shifting correlation algorithms.

This paper describes the methodology of sparse array correlation as well as the speed, accuracy, and limitations of this unique algorithm. While the study presented here focuses on the process of correlating images stored in sparse format, the details of an image compression algorithm based on intensity gradient thresholding is presented and its effect on image correlation is discussed to elucidate the limitations and applicability of compression based PIV processing.

Keywords. PIV, DPIV, correlation, image compression.

NOMENCLATURE

Φ	Correlation function
β	Characteristic image pixel size $[m]$
Δ	Correlation search length $[pixels]$

Δs	Imaged particle displacement [m]
Δt	Time between image exposures [$sec.$]
$\Delta i, \Delta j$	Difference in pixel image [$pixels$]
∇	Gradient operator
γ	Image compression ratio
$\bar{v}$	Flow velocity [m/s]
D	Particle image diameter [m]
G_∇	Relative flow divergence
I	Pixel intensity
i,j	Image coordinates [$pixels$]
m,n	Data array indices
l	Variable-length encoded data entry length [$pixels$]
M	Image magnification
M, N	Interrogation image diameter [$pixels$]
u,v	Pixel displacement in x and y directions
x,y	Pixel image coordinates

1. INTRODUCTION

Until recently, Particle Image Velocimetry, PIV, has been limited to applications in which two-dimensional, instantaneous velocity measurements are of interest. Most flows, however, are unsteady and three-dimensional in nature and thus, there has been a growing effort to develop three-dimensional velocity measurement techniques and techniques to quantitatively resolve unsteady flows. This effort has resulted in the development of Holographic PIV (HPIV) and PIV Cinematography (PIVC). Both these techniques are highly computationally intensive often requiring the determination of millions even tens of millions of vectors. With present software processing speeds, a single experimental run using HPIV or PIVC can take several hours of computer time to obtain results. Because of this, dedicated coprocessors are often utilized in these applications. These costly coprocessors, although significantly faster than present PC software processing, are still slower than desired. Ideally, one would like to process HPIV and PIVC images at a rate faster than they can be acquired. This negates the need to store the images requiring only that the results be stored. It also allows an investigator to observe PIVC results in near real-time and potentially use the information for system feedback control in much the same way LDV systems are now being used in industry.

At present, electronic imaging systems operate with pixel transfer rates on the order of 10 million pixels per second. At 8 bits per pixel, this is roughly twice the speed at which most PC's can stream uncompressed data to a hard disk. Even compressed by a factor of ten, more than one megabyte of storage is needed for each second of video signal. A typical statistical correlation with 64x64 pixel windows and 50% overlap requires more than 75 million multiplications and 225 million memory calls per second to process data at this rate - far faster than the capabilities of present PC technology. Fourier correlation techniques require significantly fewer operations but due to multiple memory calls and floating-point calculations, their processing requirements are still well beyond present PC capabilities for real time PIVC or video rate processing of HPIV images. Thus, if video rate PIV processing is to be achieved without the need for a dedicated coprocessor, an algorithm must be developed that significantly reduces the number of memory calls and arithmetic operations. This paper introduces one such algorithm, sparse array image correlation.

2. METHODOLOGY

Sparse array image correlation is based on storing and correlating a compressed data set that retains the particle displacement information from the original PIV image. By reducing coding and interpixel redundancy, far fewer memory calls and calculations must be made to correlate the image. PIV images typically contain significant data redundancy. Compression ratios of 30:1 or greater are normal. Thus, since the time required to correlate an image is proportional to the square of the number of data entries, significant gains in processing speed are possible.

Background

The simplest form of data reduction that can be made to a PIV image is to eliminate the low intensity pixels from the image file. Since the low intensity pixels contribute little to no information about particle displacement, their elimination has very little effect on the accuracy of the image correlation. Several high-speed algorithms have been developed based on this type of data reduction. The most recent of these algorithms is the one by Hatem and Aroussi (1995) in which a probability histogram of possible particle displacements is used to determine the velocity vector. Unlike Hatem and Aroussi's algorithm, sparse array image correlation relies on a true correlation of the image - it is not a particle tracking type algorithm and it does not rely on the binary (0,1) representation of particles. The relative intensity difference between pixels is maintained despite the utilization of an image compression scheme. A more pertinent algorithm to the present algorithm is the one by Landreth and Adrian (1987) in which each section of an image is orthoganally compressed and the low intensity pixel combinations are eliminated from the data before it is correlated. Like the present algorithm, Landreth and Adrian's algorithm processes the data in a sparse format. This is the basic scheme by which the present algorithm correlates images. Unlike Landreth and Adrian's algorithm, however, both coding redundancy and interpixel redundancy are reduced during image preprocessing without decoupling the x and y correlations. The two-dimensional spatial relationship and the relative intensity variation between pixels are maintained. Significant speed is gained by encoding the remaining data specifically for 32-bit processing and utilizing an error correlation function to eliminate multiplication and division operations.

As with all correlation schemes that require preprocessing of images, a tradeoff is made between the time required to reduce the data set and the time required to correlate the reduced data set. The original intent of the sparse array image correlation algorithm was to process PIV images at video rates. Therefore, the algorithm presented here uses a relatively simple data compression scheme to facilitate the processing of a video signal as it is being downloaded from a CCD camera. This allows a data set from a previous frame to be analyzed at the same time the video data from a camera is being compressed. It is desired to perform both preprocessing and correlation of the images at roughly the same rate, 1/30 of a second. The result is an image compression algorithm that is not necessarily optimized for data reduction but allows pipelining of the original image data set to reduce image preprocessing time and data transfer latency.

Image Compression

The first step in sparse array image correlation is to generate a data array that contains enough information to determine the displacement of particles in a PIV image or between two images in the case of cross-correlation. In order to facilitate processing, it is desired to retain the minimum amount of data to obtain a specified resolution in the final results. Unfortunately, it is difficult to determine *a priori* the exact information that is needed to achieve this. It can be shown, however, from the statistical correlation function,

$$\Phi_{\Delta i, \Delta j} = \frac{\sum_{m=1}^{M} \sum_{n=1}^{N} \left[I_{m+\Delta i, n+\Delta j} \cdot I_{m,n} \right]}{\sqrt{\sum_{m=1}^{M} \sum_{n=1}^{N} I_{m,n}^2} \cdot \sqrt{\sum_{m=1}^{M} \sum_{n=1}^{N} I_{m+\Delta i, n+\Delta j}^2}}$$

that pixels with high intensity contribute more to the overall value of the correlation coefficient than pixels of low intensity. This characteristic of the statistical correlation function adversely affects the ability to determine the subpixel displacement of tracer particles in a PIV image by unduly weighting the significance of high-intensity pixels. Much of the information contained in a PIV image that allows subpixel resolution of tracer particle movement resides in the intensity of pixels representing the edges of the particle images. It is not the level of pixel intensity in a PIV that allows the displacements to be determined through correlation. It is the relative *change* in intensity between the background and the tracer particle images that makes this possible. In much the same way two blank pieces of paper are aligned on a desk, image correlation relies on the change in intensity around the edges of the objects being aligned and not the featureless, low intensity gradient, regions. Thus, in principle, all pixels in low intensity gradient regions can be eliminated from a PIV image with only a slight loss in correlation information as long as the relative positions and intensities of the remaining pixels are maintained. Except for a small number of pixels representing tracer particles, PIV images are predominantly blank. Therefore, the data size necessary to determine tracer particle movement within PIV images can be significantly reduced with little or no loss in accuracy. This is the basis by which sparse array image correlation works. Eliminating pixels that have little effect on the determination of tracer particle movement reduces the data set representing a PIV image. The remaining pixel intensities are recorded in sparse format along with their relative positions. This sparse data set is then used to determine movements of the tracer particles in the fluid.

Segmentation

PIV images are strongly bimodal, composed of light particle images on a dark background, Figure 1. It is, therefore, relatively easy to eliminate low intensity, background, pixels from the data. The simplest technique to accomplish this is to set a threshold level and retain only those pixels with intensities above the threshold. A relatively robust and accurate technique for setting the appropriate threshold level is to perform a histogram concavity analysis [Rosenfield and De La Torre, 1982]. A simpler and somewhat faster technique is to generate an intensity distribution curve that indicates the number of pixels with intensities above a specified level. Since the curve is an accumulation of pixel numbers, it is piecewise smooth, at least to the resolution of the CCD camera and thus, it is a simple matter to select a threshold level that corresponds to a specific slope on the curve. This technique is not as robust or accurate as the histogram concavity analysis but, because the pixel intensities in PIV images are so strongly bimodal, the precise threshold level is often not critical.

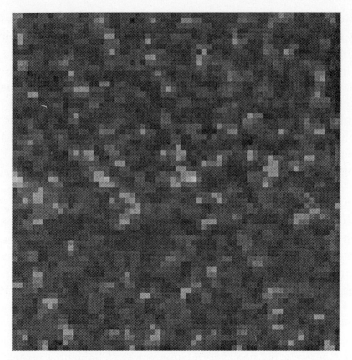

Figure 1. *- Typical 64x64 pixel region of a PIV image.*

Several PIV and Particle Tracking Velocimetry (PTV) algorithms have been developed based on the intensity thresholding of images [Hart 1996, Hatem and Aroussi, 1995.]. While at first this appears to be a simple and robust way of reducing a PIV image, there are a number of difficulties with this method that makes it inappropriate for poor quality images. Consider a double exposed one-dimensional intensity plot of two tracer particles in a flow, Figure 2(a). The intensity profile of the particle images appear Gaussian with a spot diameter that depends on the, particle diameter, image magnification, imaged wave length, pixel size, focal length, and aperture of the camera recording the image. As illustrated by Δs_1 and Δs_2 in this figure, any gradient in the flow, $\nabla \vec{v}$, over the observed region results in unequal displacements between the first and second exposures of the tracer particles. If $(\Delta s_1 - \Delta s_2)$ is small relative to the particle image spot diameter, D, then the peak correlation of the sub-window is an average of the displacements represented by the double exposure of the two particles, $(\Delta s_1 - \Delta s_2)/2$, Figure 2($b$). If, however $(\Delta s_1 - \Delta s_2)$ is large relative to D, then there exists no clear peak correlation, Figure

2(c). Although algorithms exist that are highly robust to large local velocity gradients in the flow such as the spring model algorithm by Okamoto, Hassan, and Schmidl (1995), in general, large velocity gradients result in an increase in spurious vectors. Thresholding an image has the effect of reducing the spot diameter of the particle images as illustrated by the dotted line in Figure 2(a). Thus, thresholding can result in a loss in the information necessary to obtain average particle displacement information. Furthermore, most PIV images suffer from an inconsistency in the relative intensity between particle images. This is particularly true of images that are under exposed. In these cases, the information lost by thresholding to obtain significant data reduction can result in the loss of particle displacement information even for relatively small flow divergence.

A more robust, although slightly more computationally intensive, method of segmenting an image is to rely on the magnitude of the gradient in intensity of the pixels in the image. To reduce computational intensity, the magnitude of the intensity gradient is often

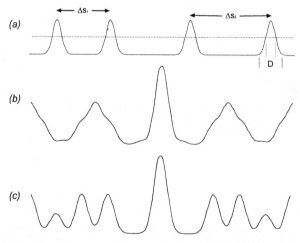

Figure 2. - *Intensity plot of two tracer particle images, 2(a). The intensity profile of each of the tracer particles is Gaussian with a spot diameter that depends on the pixel dimensions, focal length, and aperture of the camera. A gradient in the flow results in unequal spacing between exposures, $\Delta s_1/\Delta s_2 \neq 1$. If the gradient is small, the correlation of the image yields the average of the particle displacements, 2(b). If the gradient is large, the average particle displacement can not be determined, 2(c). Threshold intensity compression has the effect of reducing the image spot diameter as shown by the dotted line in 2(a).*

approximated as the absolute value of the gradients in the x and y directions, $|\nabla I| \cong \left|\frac{\partial I}{\partial x}\right| + \left|\frac{\partial I}{\partial y}\right|$. To first order, this can be calculated as, $|\nabla I| \cong \left|I_{(i+1,j)} - I_{(i,j)}\right| + \left|I_{(i,j+1)} - I_{(i,j)}\right|$. An appropriate magnitude for the cutoff in the intensity gradient can then be selected in the same manner as it is done for intensity thresholding. Pixel intensities in regions where the gradient is sufficiently high are retained and the rest are discarded (assumed to have a value of zero). The result is the compression of an image where only the pixels around the edges of tracer particles are retained. The center of the particle images which have a low intensity gradient are discarded, Figure 3. Because of this, intensity gradient segmentation of PIV images usually results in a smaller data set then images segmented by intensity thresholding. The gradient method of segmentation is the method of choice for most bimodal images [Gonzalez, Woods, 1993]. It is, however, particularly well suited to the compression of images for correlation since it is the change in pixel intensities that allows subpixel particle displacements to be determined by correlation and not the average intensity of the particle images.

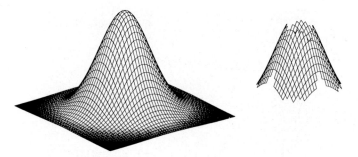

Figure 3. - *Particle image intensity plot illustrating the effect of gradient pixel segmentation.*

Data Encryption

Once an image is compressed it is stored with each pixel indices and intensity combined into a single 32-bit word. This reduces the number of memory calls that must be made when correlating. For example, $i=2$, $j=2$, $I=254$ is stored as 00000000001000000000001011111110 *binary* = 2,097,918. By masking the bits, the values of $i, j,$ and I can be extracted from this single entry in a few clock cycles of most processors.

Along with the sparse image array, an indices table is generated which contains the location in the sparse image array of the first entry representing a pixel combination in the next line of a PIV image. This line index array is used to jump to the next value of j in the sparse image array when a specified pixel separation is exceeded in the ith direction. When correlating large images, this index array significantly speeds processing.

Window Sorting

The reduction in the number of data entries in the PIV image data set by the elimination of pixels in regions with a low intensity gradient and the encoding of the remaining data greatly improves the speed at which correlation windows can be sorted from the data set. In addition, the line index array reduces the number of multiple entries into the sparse image array that must be made to extract the pixels located in a given correlation subwindow. Despite this, window sorting is a slow memory intensive task that requires considerable processing time. The present algorithm requires almost as much time to sort the correlation subwindows from the image data as it does to correlate the subwindows once they have been sorted.

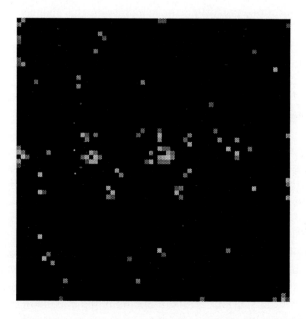

Figure 4. - Reconstructed image of Figure 1 after gradient level compression. This image has been compressed to 1/30 of its original size. Sufficient correlation information remains in this image to accurately determine particle movement even after gradient level compressing this image by a factor of 200:1. Most PIV images can be compressed 30:1 with very little loss in correlation information.

Correlation window sorting in sparse array format is considerably more difficult than it is in an uncompressed format since the spacing of the data entries is image dependent. A simple block transfer as is commonly done in an uncompressed format cannot be done in the sparse array format. A solution to this is to generate the sparse array at the same time that the correlation windows are being extracted from the image. This technique works well, as long as there is no significant overlap of the correlation windows. If there is significant overlap, the number of redundant memory calls greatly slows processing. The most computationally efficient technique is to presort all of the correlation windows as the sparse array is generated. This technique requires a significant increase in memory storage depending on the overlap in the correlation windows. A 50% overlap results in a four times increase in memory storage. The 32-bit sparse array data encryption scheme, itself, requires four times the number of bits per pixel. Therefore, there is an increase in memory storage requirement by a factor of sixteen. Image compression, however, sufficiently reduces the number of data entries such that there is a net reduction in data storage by roughly a factor of four compared with storing the entire image in memory at one time. In addition, presorting the windows in this manner moves the processing time for window sorting from the basic correlation algorithm into the image-preprocessing algorithm. This allows more time for image correlation within the 1/30 of a second video framing speed. Presorting the correlation subwindows at the same time the image is compressed is, therefore, the optimum solution in the majority of applications.

Search Length Selection

Processing speed can be further increased while, at the same time, reducing the odds of obtaining spurious correlation values by limiting the search for a maximum correlation. This is done by allowing the user to specify a maximum change in Δi and Δj based on knowledge of the image being correlated. An adaptive scheme can be used to narrow the correlation search - a scheme that predicts the range of correlation values to calculate based on previous calculations from subwindows of the same image. This procedure, however, is not particularly robust and can result in spurious errors in obtaining the maximum correlation. Because the sparse array correlation process is inherently very fast, adaptive schemes generally do not gain enough processing speed to warrant their use. It is sufficient to set a single value for the correlation range for an entire image.

Subwindow Correlation

By using the error correlation function rather than a statistical correlation function, image correlation can be carried out using integer addition and subtraction only. These are very fast operations for most microprocessors requiring only a few clock cycles. It is far faster to perform these calculations than to use a 'look-up table' scheme to avoid 8-bit or 4-bit pixel multiplication. The use of the error correlation function, therefore, significantly improves processing speed over the more commonly used statistical correlation function. A detailed analysis of the error correlation function in comparison to the statistical correlation function is presented in a paper by Roth, Hart, and Katz (1995). It was shown that the error correlation function produces essentially the same results as the more computationally intensive statistical correlation function.

The error correlation function can be expressed as,

$$\Phi_{\Delta i,\Delta j} = \frac{\sum_{m=1}^{M}\sum_{n=1}^{N}\left[I_{m,n} + I_{m+\Delta i,n+\Delta j} - \left|I_{m,n} - I_{m+\Delta i,n+\Delta j}\right|\right]}{\sum_{m=1}^{M}\sum_{n=1}^{N}\left[I_{m,n} + I_{m+\Delta i,n+\Delta j}\right]}$$

such that,

$$\Phi_{\Delta i,\Delta j} = 1 - \frac{\sum_{m=1}^{M}\sum_{n=1}^{N}\left[\left|I_{m,n} - I_{m+\Delta i,n+\Delta j}\right|\right]}{\sum_{m=1}^{M}\sum_{n=1}^{N}\left[I_{m,n} + I_{m+\Delta i,n+\Delta j}\right]}$$

The value of this correlation function ranges from 1 when the images are perfectly correlated to 0 when there is no correlation between the images. Because it relies on the difference in pixel intensities, it does not unduly weight the significance of high-intensity pixels as does the statistical correlation function. Aside from being faster to calculate than the statistical correlation function, it has the added benefit of being easier to implement in hardware without the need for a microprocessor. The error correlation function, therefore, has potential for use in hardware based PIV systems.

Unlike the more common statistical correlation function, the error correlation function used in sparse array image correlation is not computed one entry at a time. The entire correlation table is constructed by summing entries as they are found while iterating through the sparse image array. When auto-correlating subwindows, each entry in the sparse image array is compared with the entries below it and a correlation approximation between the entries is added into the correct location in the correlation table based on the difference in i and j between the array entries. If the location is out of range of the specified search length in the ith direction, the entry is ignored and processing continues with the next entry specified in the line index array. If the location is out of range in the jth direction, the entry is ignored and a new series of iterations are made starting with the next sparse image array entry. Because the sparse array is correlated from the top down, only the half of the correlation table representing the positive j direction is calculated. The auto-correlation of an image is symmetrical and thus, calculation of both halves of the correlation table is unnecessary.

Cross-correlation is accomplished by generating two sparse image arrays representing the two images being correlated. The entries of one array are then compared to *all* of the entries of the other array that are within the search length. Because the difference in array indices can be both positive and negative in the i and j directions, the entire non-symmetrical correlation table is calculated.

Once the correlation table is complete, the table is searched for the maximum correlation value. A simple bilinear interpolation scheme is then used to determine the correlation maximum within subpixel resolution. Bilinear interpolation is ideal in this application since reducing the data set by image preprocessing and using the error correlation function results in a very steep, nearly linear, correlation peak.

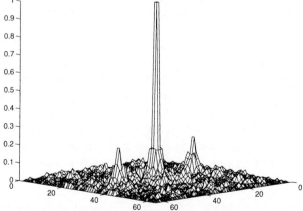

Figure 5. - *Correlation table resulting from sparse array auto-correlation of the image in Figure 4. In practice, the sparse array image data set is correlated from the top down so that only half of the correlation table is calculated. Both halves of the symmetrical correlation table are shown here for clarity. Note the steep correlation peaks. These sharp peaks aid subpixel interpolation.*

3. PROCESSING SPEED

The computational intensity of sparse array image correlation is comparable to the better known statistical correlation technique except that the image data set is compressed in preprocessing. If the data set is reduced to a fraction, γ, of the original image data set, than the number of data comparisons that must be made is, $\frac{1}{2}\gamma\Delta^2(\gamma N^2 - 1) + \gamma N^2$ for sparse array auto-correlation and $\gamma^2\Delta^2 N^2$ for cross-correlation. For PIV images where the particle seeding densities are high such that, $\gamma N^2 \gg 1$ and $\gamma\Delta^2 \gg 1$ then $\frac{1}{2}\gamma\Delta^2(\gamma N^2 - 1) + \gamma N^2$ is approximately equal to, $\frac{1}{2}\gamma^2\Delta^2 N^2$. A typical PIV data set can be reduced by a factor of 30 such that $\gamma=0.3$. Thus, a typical 64x64-pixel correlation subwindow requires a little less than one thousand data comparisons to complete an auto-correlation with a search window of 20x20 pixels. During each comparison, three memory calls are made, one to retrieve a data entry to be compared with the data entry already in the processors register, one to retrieve the value of the correlation table entry, and one to place the comparison result in memory. Memory calls require a great deal more processing time than integer addition and subtraction so that the time for each data entry comparison is essentially the time it takes to make these memory calls [Hennessy & Patterson 1990]. PCI based systems can transfer over 60*Mbytes* of data per second or about two million 32-bit data entries per second over the bus. By ordering data entries sequentially when extracting the correlation subwindows from the image data set, bus transfer rates of this speed can be achieved by block memory transfers. Thus, correlation speeds of 2,000 *vec./sec.* are theoretically possible for typical PIV images under these conditions.

4. ACCURACY AND ROBUSTNESS

The process of correlating images in sparse format using the algorithm presented here is independent of the method by which the image data set is generated. With no compression of the image, sparse array image correlation is *identical* to the more common statistical image shifting correlation method except for the use of the error correlation function. A comprehensive study of the error correlation function by Roth *et. al.* showed that there exists no significant variation between the results from the error correlation function and that of the statistical correlation function [Roth, Hart, Katz, 1995.]. There is, thus, no reason to believe that the error correlation function is any more or less accurate than statistical or spectral correlation techniques. Any inaccuracy or lack of robustness in the present algorithm can be attributed entirely to the loss of data from image compression. The speed of sparse array image

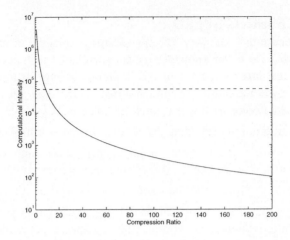

Figure 6. - *Sparse array cross-correlation processing computational intensity (solid line) relative to spectral correlation (dotted line) as a function of image compression ratio for a 64x64 pixel correlation subwindow. Note that this is a semi-log plot. The computational intensity of sparse array cross-correlation for densely seeded images is $\gamma^2\Delta^2N^2$ where FFT spectral correlation is $3N^2\log N + N^2$.*

correlation, however, is strongly dependent on the reduction in the image data set through compression. Little is gained by using this algorithm if the data set remains unchanged, Figure 6. It is therefore necessary to address the problems associated with image compression to assess the limitations of the sparse array image correlation algorithm.

Velocity Gradient Affects

A method of determining the probability that a particular correlation is valid is to perform a non-parametric correlation and observe the peak correlation value relative to the mean. This can be accomplished by ranking the pixels in an image before correlation. Pixels with the same intensity are assigned an average of the rank they would receive if they had different values. In this manner, non-parametric correlation provides a means by which the effects of image compression can be assessed. Images with a poor rank correlation value relative to the mean are more likely to produce spurious vectors and to lack information needed to obtain accurate subpixel resolution. Thus, non-parametric correlation provides, in essence, a measure of the correlation signal to noise ratio.

As discussed in Section 2., the particle image diameters are important for resolving the average particle displacement in flows where there exist large gradients in the flow velocity. This is the factor which image compression affects the most. Other parameters such as seeding density, average flow velocity and relative correlation window size that influence all PIV correlation processing are unaffected. Using rank correlation, the effects of intensity threshold compression are plotted in Figure 7 for several values of the ratio of the flow divergence to the particle image diameter based on the correlation window diameter, N,

$$G_{\nabla} = \frac{M \left|\nabla \cdot \bar{v}\right| N \Delta t}{D}$$. For *ideal* PIV images where there exists no

divergence in the flow, $G_{\nabla} = 0$, and no variation in peak image intensity, the compression ratio has little effect on the relative correlation of the image until a significant portion of the data of the tracer particle images are eliminated. These images can be compressed to a small fraction of their original size and retain enough correlation information to determine the subpixel displacement of the tracer particles. As shown in Figure 7, images where G_{∇} does not equal zero are affected to an increasingly greater extent by compression as G_{∇} increases.

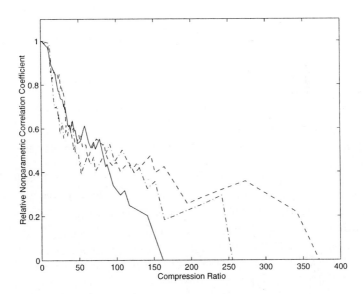

Figure 7. - *Relative non-parametric correlation coefficient plotted as a function of intensity threshold image compression for G_{∇} =0, 1, and 2. The correlation coefficient is based on non-parametric rank correlation and is normalized relative to the uncompressed image. As the imaged flow divergence increases, the compression ratio at which the correlation is lost, indicated by a zero relative non-parametric correlation coefficient, decreases.*

Intensity Variation Effects

Variations in intensity resulting from poor illumination and/or variations in particle characteristics severely affect the ability to extract particle displacement information from a PIV image using correlation. This is particularly true of auto-correlation processing, as there often exists a systematic intensity variation between the first and second exposures of the particle images. This ·type of intensity variation adversely affects both spectral correlation and compressed image correlation processing. It, however, limits the level to which an image can be compressed and thus has a much more pronounced effect on the speed and accuracy of compressed image correlation. Non-systematic intensity variations, which result from differences in tracer particle characteristics and non-uniform illumination, affect both spectral correlation and compressed image correlation to roughly the same degree.

Consider the Gaussian intensity profile of a tracer particle image. This profile can be approximated by $I_r \cong I_o e^{-(4.3r/D)^2}$ where I_o is the peak intensity of a particle centered at (x_o, y_o) and r is the distance from the center. The magnitude of the gradient in intensity is then equal to $\left| \nabla I_r \right| \cong \dfrac{37 r I_o \beta}{D^2} e^{-(4.3r/D)^2}$ where β is the characteristic size of a single pixel in the image. The maximum magnitude of the intensity gradient of a particle image occurs at a distance $r=D/4.3$ and has a value $\left| \nabla I_{r=D/4.3} \right| \cong 8.6 \dfrac{I_o \beta}{D}$. If an intensity gradient threshold level is set above this value, all correlation information for this particle will be lost. Note, however, that the minimum particle diameter is roughly $D/2$ as long as the intensity gradient threshold level is set below $8.6 \dfrac{I_o \beta}{D}$. This is not true of image segmentation based on intensity level thresholding where the particle image diameter approaches zero as the threshold level is increased. If there are significant variations in particle peak intensities within an image, however, then both methods of image segmentation adversely affect correlation although intensity thresholding to a somewhat less extent for the same compression ratio. This is illustrated in Figure 8 by plotting the non-parametric correlation peak value obtained from Figure 1 as a function of compression ratio for both gradient level compression and threshold compression. As illustrated in this figure, threshold intensity compression results in less information loss at low compression ratios. At higher compression ratios, however, gradient intensity compression results in less information loss.

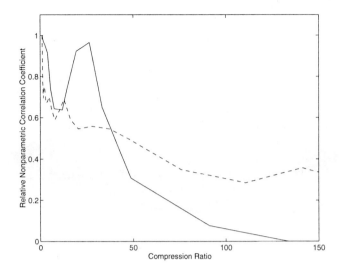

Figure 8. *- Relative non-parametric correlation coefficient plotted as a function of intensity gradient image compression (dotted line) and intensity threshold compression (solid line) for the image in Figure 1. Intensity threshold image compression results in less information at compression ratios below 40:1. At higher compression ratios, gradient image compression results in less information loss allowing the image to be compressed to less than 1/200 of its original size before particle displacement information is lost. Threshold compression, however, losses the particle displacement information at a compression ratio of less than 150:1.*

In practice, exposure levels are difficult to control and image-recording devices have limited intensity resolution. PIV images that have been over exposed have *tophat* particle image intensity profiles, Figure 9. While some information is lost because of this, these images often correlate accurately for the same reason images compressed using intensity gradient thresholding correlate accurately – it is the change in intensity at the edges of the particle images that hold the information necessary for accurate correlation and not the low gradient regions near the center of the particle images. PIV images that are slightly over exposed are often better suited to intensity gradient compression rather than intensity threshold compression. This is because the saturated intensity regions of the particle images that contain little correlation information and have a low intensity gradient are eliminated from the data. Because intensity gradient image compression is more robust to variations in image exposure, it is generally a better choice for PIV image compression even though it results in slightly more information loss at moderate compression ratios.

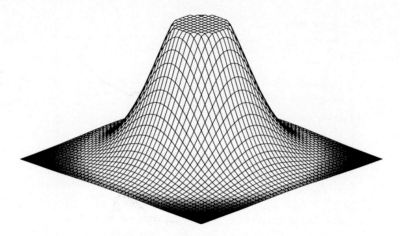

Figure 9. - *Tophat profile of a particle image which has reached the saturation level of the image recording media. PIV images that contain particle images of this type can be significantly compressed without losing correlation.*

5. Experimental Demonstration

Variations in particle image intensity and size, correlated background noise, poor contrast, insufficient illumination and optical aberrations are only a few of the many factors which effect the quality of experimental images. For this reason, a comparison was made between spectral correlation and sparse array correlation based on the processing of experimentally obtain PIVC images taken of a highly unsteady vortical flow.

Images were used from the experimental measurement made of the flow inside a Cardio Assist Device, CAD. This device is used to aid the flow of blood in patients with weakened hearts. The flow in this device is highly unsteady and exhibits strong vortical flow formations [Huang, Hart, Kamm 1997]. A frequency doubled Nd:YAG laser was used to illuminate a $10cm \ x \ 10cm$ area. The flow was seeded with $50\mu m$ florescent particles. A Pulnex, TM-9701 512x486 pixel CCD camera, recorded PIVC images at $30Hz$. Typical images are shown in Figure 10. Because of the curvature of the wall, all of the images exhibit significant variations in light intensity. These images were specifically chosen for comparison with the spectral correlation method because they exhibit features that are poorly suited to processing in sparse format. These features include significant local variation in illumination, large gradients in the flow velocity, heavy seeding densities, and very small tracer particle movement between images (less than 1 pixel on average) requiring accurate subpixel interpolation to resolve flow structures. Because of these features, the test images provide a means of illustrating the limitations of sparse array image correlation.

The experimental images were processed by cross-correlation using 64x64 pixel subwindows that overlapped by 50% in both the x and y directions. When compressed 30:1 with a maximum correlation search length of 32 pixels, the sparse array algorithm processed these images at roughly 300 *vec./sec.* on a Pentium 166*MHz* computer with 16*Mbytes* of memory. This was about sixty times faster than spectral correlation, which generated 5 *vec./sec.* on the same machine. An example of the output of both sparse array correlation and of spectral correlation of the images in Figure 10 are shown in Figure 11. Sparse array correlation yielded results that were typically within 0.05 pixels of spectral correlation. No significant variation between the two correlation algorithms was observed with image compression ratios below 50:1. At higher compression ratios, 100:1, differences in the velocity profile in low velocity areas were observed, Figure 12. These variations, on the order of 0.05 pixels, are the result of information loss due to compression. At much higher compression ratios, 200:1 and higher, sparse array image correlation generated significant spurious vectors near the wall of the test section and in other regions where high velocity gradients exist, Figure 13. This behavior is consistent with the analysis discussed in Section 4. At these extremely high compression ratios, each vector represents the correlation of less than thirty pixels.

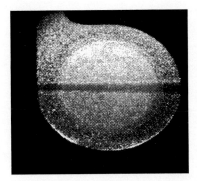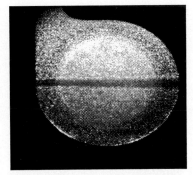

Figure 10. - Typical pair of PIV images recorded from a Cardio Assist Device, CAD. The light intensity in these images varies significantly due to the distortion caused by the three-dimensional shape of the device. Such images that exhibit significant changes in velocity and have high seeding densities are difficult to correlate in compressed format. Because of this, these images were selected to illustrate the limitations of sparse array image correlation processing.

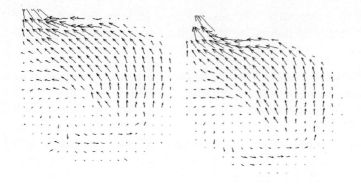

Figure 11. - Vector map obtained from sparse array cross-correlation of the images in Figure 10 with a 30:1 compression ratio (left) and by spectral correlation (right). The longest vector in these plots represent a displacement of less than 2 pixels. The majority of vectors shown represent displacements of tracer particle images of less than 1 pixel. Thus, subpixel interpolation is critical to visually resolving flow structures from these plots. Variations of less than 0.05 pixels are easily observed.

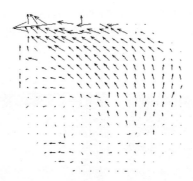

Figure 12. - *Vector map obtained from sparse array cross-correlation of the images in Figure 10 with a 100:1 compression ratio. Compare this figure to the plots in Figure 11. Note the difference in areas where there exist low x and/or y velocities. This is a result of information loss from image compression limiting subpixel resolution.*

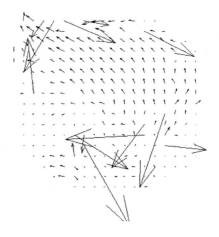

Figure 13. - *Vector map obtained from sparse array cross-correlation of the images in Figure 10 using a 200:1 compression ratio. Note the spurious vectors near the wall and in high velocity gradient regions. This is consistent with the processing limitations discussed in Section 4. At this high compression ratio, each vector represents a correlation of less than 30 pixels.*

SUMMARY AND CONCLUSIONS

Sparse array image correlation is a technique by which PIV images can be accurately processed at high-speeds. It is based on the compression of images in which the number of data set entries containing tracer particle displacement information is reduced. Very high correlation speeds are obtained by encrypting the reduced data set into 32-bit integers and correlating the data entries using an error correlation function to eliminate multiplication, division and floating point arithmetic.

The maximum correlation value associated with sparse array image correlation is characterized by a steep peak that improves subpixel interpolation. The performance of this method of image correlation, however, is largely dependent on the level to which an image can be compressed without losing significant correlation information. Thus, its performance relative to the better known spectral correlation method is image dependent.

Through an analysis of the affects of flow divergence, tracer particle image diameter, and intensity variations, it was shown that typical PIV images can be highly compressed with no significant lose in correlation information. Characteristic limitations of sparse array image correlation were illustrated by comparing results from the spectral correlation of experimental images with the results from the sparse array correlation of the same images at varying levels of compression from 30:1 to 200:1. For applications requiring extremely high correlation speeds such as holographic particle image velocimetry (HPIV) and video rate particle image velocimetry cinematography (PIVC), sparse array image correlation appears to be a viable processing technique.

REFERENCES

Adrian, R. J. (1986) "Multi-point Optical Measurement of Simultaneous Vectors in Unsteady Flow—a Review." *Int. J. Heat and Fluid Flow*, pp. 127-145.

Adrian, R. J. (1991) "Particle Imaging Techniques For Experimental Fluid Mechanics." *Annual Review of Fluid Mechanics*", Vol. 23, pp. 261-304.

Gonzalez, R. C., Woods, R. E. (1993), Digital Image Processing, Addison-Wesley Pub. Co., Reading, MA., pp. 414-456.

Hatem, A.B., Aroussi, A., (1995), "Processing of PIV Images." SAME/JSME and Laser Anemometry Conference and Exhibition, August 13-18, Hilton Head, South Carolina, FED-Vol. 229, pp.101-108.

Hennessy, J.L., Patterson, D.A., (1990), "Computer Architecture - A Quantitative Approach." Morgan Kaufmann Pub., San Mateo, CA.

Huang, H., Hart, D. P., Kamm, R. (1997), "Quantified Flow Characteristics in a Model Cardiac Assist Device", ASME Summer Annual Meeting, Vancouver, B.C., Experimental Fluids Forum, pp. .

Landreth, CC. and Adrian, R.J. (1987) "Image Compression Technique for evaluating Pulsed Laser Velocimetry Photgraphs having High particle Image Densities", FED-Vol. 49.

Okamoto, K., Hassan, Y. A., Schmidl, W. D. (1995), "New Tracking Algorithm for Particle Image Velocimetry." *Experiments in Fluids*, pp. 342-347.

Rosenfeld, A., De La Torre, P., (1983), "Histogram concavity analysis as an aid in threshold selection," *IEEE Transactions on Systems, Man, and Cybernetics,* Vol. 13(3), pp. 231-235.

Roth, Hart, and Katz, (1995) "Feasibility of Using the L64720 Video Motion Estimation Processor (MEP) to increase Efficiency of Velocity Map Generation for Particle Image Velocimetry (PIV)." ASME/JSME Fluids Engineering and Laser Anemometry Conference, Hilton Head, South Carolina, pp. 387-393.

I.6 Simultaneous Multiple Pixel Processing Algorithms for PTV and PIV

Tomomasa Uemura[1], Motonobu Yoshimoto[1],
Masataka Tatumil[1] and Akikazu Kaga[2]

[1] Department of Environmental Engineering, Osaka University,
2-1 Yamadaoka, Suita, Osaka 565, Japan

[2] Department of Industrial Engineering, Kansai University,
3-3 Yanmate, Suita, Osaka 564, Japan

Abstract.   A fast algorithm for pattern maching is proposed and some examples are shown. The new algorithm consists of two different calculation methods. One of the methods is a technique of simultaneous multiple pixel calculation, and the other method is a fast pattern matching algorithm, which can efficiently extract the most similar pattern from hindreds of candidates by making used of the statistical F-inspection technique. The calculation technique of simultaneous multiple-pixel processing can be widely applicable to both PTV and PIV analysis's.

As an application of the new algorithm, velocity distribution of a flow visualized by smoke and flow visualized by a particulate tracer are measured. The new algorithm can process PIV analysis obtaining 1000 velocity vectors in within 10 seconds using a typical PC.

1 Introduction

According to the advancement of the microprocessor technology, most image analysis of PIV/PTv measurements are processed by personal computers, and the processing time has been shorten without paying any improvements of calculation techniques. The performance of micro-processors has been improved by higher clock frequencies, the increased register with and improved data flow using multiple pipelines. From the image processing view point, the increased register which is not utilized, since pixels of most digital pictures are expressed by eight bits. By noticing the above fact, authors have developed a new calculation method which simultaneously processes multiple pixel data utilizing a wide register width. For example, four eight-bit pixel data are loaded into one 32-bit register, and can be binarized simultaneously by using the present algorithm. The method can accelerate calculations such as binarization, absolute difference of pixels, pixel counting, which are frequently appears in both PIV and PTV analysis's.

In the PTV analysis, the algorithm can be integrated in the binary image correlation method(BIC), and effectively works in some analysis procedures such as binarization, measurement of particle size, and matching of binarized picture patterns.

In the PIV analysis, the algorithm is combined with the successive abandonment method(SAM) , which is developed by one of the authors(Kaga et al., 1993). In the SAM analysis, a lot of absolute differences between a template and candidates have to be calculated. The present method is very effective for the calculation.

2   Simultaneous Multiple Pixel Processing

In the most image processing, no matter how many pixels are contained in a picture, pixels are processed pixel by pixel. In other words, the existing algorithms for the image processing do not fully utilize the capabilities of   32-bit micro-processors as far as the calculation unit is one pixel, which is usually eight-bit data. A new calculation method can improve such situation by processing multiple pixels simultaneously. In the following part, calculations are explained assuming simultaneous 4-byte processing, since most micro-processors presently have 32-bit registers.

2.1   Simultaneous Calculation Algorithm

In the calculation of integer data, negative values are expressed as compliments of two and a over-flow in a calculation brings a carry bit to the left digit or into a flag register. From a view point of fully utilization of computer power, it is preferable if four byte data packed in one 32-bit register are calculated separately. Such a calculation is realized at only a limited condition, for example, when one binary data occupies one byte(eight bits), it is a condition of a binary picture. In such case, four pixel data can be summed up at least 255 times without worrying about interference between neighboring data.

In the case of gray level pixel data, both additions and subtractions can not be done so easy. In order to perform simultaneous calculations, carries and borrows should be processed not only adequately but also within a shorter time than that of four independent calculations. The proposed method containing processes of detection of interference, carries or borrows, and the correction, can perform some operations to gray level pixels in a shorter time as byte-by-byte calculations consume.

$$L_S = \{ (\overline{X \wedge Y}) \& Z \} \mid (\overline{X} \& Y) \ \& \ 0x80808080 \qquad (1)$$

$$\overline{} : NOT, \ \wedge : XOR, \ \& : AND$$

Where, X:minuend, Y:subtracter, and Z:remainder are 32-bit integer

In the case of a subtraction, borrow bits can be detected by checking MSBs(Most Significant Bit) in a minuend, a subtracter and a remainder. Borrows occur in a subtraction between two four-byte-data can be detected by checking MSBs in corresponding bytes in Ls (equation (1)). If a MSB in Ls is '1', it means a remainder is negative, and '0' means positive remainder.

2.2 Application of the Algorithm

Binarization. The above logical equation can be utilized in 4-pixel simultaneous binarization, by substituting X and Y by threshold values and pixel values, respectively. In other words, binarization can be interpreted as a following a pixels value conversion. If a result of the subtraction (threshold) - (pixel value) is positive, the pixel value is set to "0", and for a negative remainder, pixel is set to "1".

Then MSBs of corresponding pixels in Ls exhibit the results of binarization. The four-byte simultaneous binarization can be obtained by extracting MSB-bits using a logical 'AND' operation between Ls and &h80808080. The numerical examples are shown in the upper half of figure 1.

In order to calculate each remainder, interference caused by borrow bits have to be corrected. The correction can be done by incrementing an upper byte of a negative byte. It is done by the following calculation, where "<<" means bit-shift to the left by the right side number.

$$Z = Z + L_s << 1 \qquad (2)$$

Calculation of absolute differences. As a results of four byte simultaneous subtraction, results often contain negative values. Although the equation (1) can detect occurrence of negative remainder, it is not simple to convert the negative values to positive(absolute value). In the case of an independent integer, it is well known, converting a negative integer to a corresponding absolute value can be accomplished by conjugating all the bits then adding 1. In the present case, four 8-bit integers are packed in a 32-bit register without space. The proposed algorithm offers an efficient method to convert them to positive irrespective of their locations. Utilizing the value of equation (1), a mask bits, which enable to conjugate only negative bytes, can be generated by an operation shown in equation (3).

$$L_W = L_S << 1 - L_S >> 7 \qquad (3)$$

By calculating a logical "XOR" between the packed remainder and the above mask Lw, only the negative bytes, which are marked with non-zero MSBs in L_S, are automatically conjugated. The figure 1 explains the method using numerical example of packed two bytes data.

Simultaneous Subtractions

x1-y1=100-64=32,   x2-y2=100-128=-28

	Packed 2 Bytes		Decimal		MSBs in the 2 Bytes	
X	01100100	01100100	100	100	0xxxxxxx	0xxxxxx
Y -)	01000000	10000000	64	128	0xxxxxxx	1xxxxxx
Z	00100011	11100100	35	228(-28)	0xxxxxxx	1xxxxxx

Binarization and Conversion to Absolute Value

(1) From Eq.(3) L_S=00000000 10000000

(2) Make L_1= L_S<<1 L_1=00000001 00000000

      L_7= L_S>>7 L_7=00000000 00000001

(3) L_W=L_1-L_7 L_W=00000000 11111111

(4) Zabs = Z^L_W + L_1^L_7

    = 00100011 00011100 = 35   28(Decimal)

Figure 1. Calculation of simultaneous binarization and
absolute difference between two pixels.

2.3   Applications of the Algorithm

The present method works effectively in pixel-by-pixel operations such as counting pixels, binarization, and pattern matching. In this paragraph, some applications of the simultaneous multiple pixel processing are shown.

Binarization. In order to binarize a picture, every pixel is compared to a threshold value then replaced either '1' or '0'. By introducing the present method, four pixels can be simultaneously binarized and the processing time is shorter than that of ordinary method.

A binarization can be interpreted as a conversion of a pixel value either "1" or "0" according to a sign of the remainder of subtraction a pixel value from a threshold value. A four pixel simultaneous binarization can be done by the following procedure. As the first step, make two 32-bit integers X and Y from four thresholding values and four 8-bit pixel data respectively. As the second step, substitute a remainder of subtraction Y from X into Z. Using X, Y and Z, calculate *Ls* from equation (1). When the *Ls* is separated into four 8-bit integers, a value of each integer is set either 0 or 128 whether the corresponding pixel value is smaller than the threshold value or not. This is a result of a MSB of each 8-bit data being either "0" or "1". Shifting the *Ls* by seven bits to the right, L_7 in figure 1, each 8-bit value becomes either "0" or "1".

In order to examine the present method, a picture of small particles is binarized and the results of operation time are listed in table 1. As the operating

Table 1. Binarization Time for one 512x480 pixels Picture (ms)

	Personal Computer [1]	Work Station[2]
4-Byte Simul.	190.5	44.6
Byte by Byte	412.0	82.5

[1]: Intel 80486DX2-66MHz, 16bit OS,   [2]: R4400SC-67MHz

system of the personal computer was 16-bit version, the algorithm do not show it full capability.

Counting pixels. The simplest and most effective application of the present method is counting pixels in a given area of a binary picture, usually it measures area in a binarized digital picture. Since pixel values in a binarized picture are usually set either "1" or "0", only one bit is used for the data in 8-bit capacity. Supposing four such pixel data make into one 32 bit variable, it is clear that sums in each pixel values are not influenced by the neighboring pixel data. In other words, each binary pixel data in a 32-bit variable can be independently and simultaneously summed up at least 255 times.

One of authors had developed a fast PTV analysis method(Binary Image Correlation method; BIC) which make use of a correlation coefficient simplified for binary data (T.Uemura et al, 1990). As the pixel counting is a fundamental operation in the method, under the most preferable situation, the above method works almost four times faster than the pixel-by-pixel addition. As a result, in the PTV analysis using BIC method, integration of the above method can reduce the processing time by 30%, though, the net calculation time becomes less than 50%.

Pattern matching by pursuing minimum difference. A calculation of cross correlation coefficient is a common method in evaluating a resemblance of two patterns, although it takes a lot of computation. D.I. Barnea et al.(1972) and A.Kaga et al.(1993) proposed fast algorithms for evaluation of similarities of gray level patterns by accumulating absolute differences of pixel values. As those algorithms are based on the calculation of absolute differences of two pixel values, calculations can be performed faster than calculating correlation coefficients, which compute a lot of multiplication and addition. Adding to such efficient calculations, the methods can distinguish less similar patterns before calculating all the pixels, since such patterns have relatively large accumulated values in a early stage of calculation. Before explaining the new algorithm in the following chapter, it is better to comment that the most time in the analysis is consumed in the calculation of absolute differences of pixels. A numerical example of simultaneous calculation of two absolute differences is shown in figure 1.

The effect of the new method has been found as about 25% reduction of the processing time using a 32-bit workstation. In the case of using a personal computer with a 32-bit operating system, similarly the 4-byte method can reduce the computation time by about 30 to 20%.

3   Algorithm of Successive Abandonment

D.I.Barnea and H.F.Slverman(1972) proposed a SSDA(Sequential Similarity Detection Algorithm) for fast template pattern matching of gray level pictures. Kaga et al. developed a similar method independently, then later the method is further improved in both reliability and calculation time by integrating the F-inspection technique. it is named the successive abandonment algorithm. By integrating the simultaneous multiple pixel calculation of absolute differences into the algorithm, the analysis can be accelerate further(A.Kaga et al.(1993).

The summation is executed within a interrogation area, and the sum is examined by F-inspection at every step of a partial summation. and further calculations are terminated for unpromising candidates. The ratio of the number of vectors which are obtained as the unique solution to the total number of vectors, is defined as the reliability. The algorithm seems to be better than any other PIV algorithms with respect to computation time and the capability in estimating reliability of results.

3.1   Absolute Differences and F-inspection

In the PIV analysis, a template pattern and candidate patterns are compared, and the most similar candidate is regarded as the template pattern moved to the other. In the present method, the total accumulation of gray level difference between corresponding pixels over examined pattern T_d is taken as a index of resemblance of two patterns.

$$T_d = \sum_{i=1}^{N} |f_i - g_i| \qquad (4)$$

Where, f_i, g_i represent the gray level pixel values in two comparing patterns, N is the total number of pixels.

In the calculation of the above value, when it exceed a certain critical value, the two patterns can be considered as different ones. In many cases, the calculations can be terminated rather early stage of the accumulations, if the selection of the critical value is adequate.

Suppose T_j is a partial accumulation of the absolute differences, among many T_j's calculated from all candidates, patterns with smaller values are more promising. So as a criterion to abandon less similar candidate, a following relationship can be used.

$$T_j > \alpha_j \cdot T_{j,min} \qquad (5)$$

The question is set as a best method to give the coefficient α_j. In the proposed method, the α_j can be determined by the F-inspection method. The F-inspection offer a technique to discriminate a population of a sampled data sets from a ratio of the variances. Supposing pixel value differences between pixels sampled from two independent pictures follows a normal distribution with an axis at the origin. The distribution of the absolute differences can follow a positive side of a normal distribution. As the two pictures becomes more similar, the variance of absolute differences becomes smaller. The F-inspection gives a threshold value of a variance ratio as a function of number of sampled data and a confidence level.

3.2 Distribution of Absolute Differences

A statistical model to determine an adequate value of parameter α_j is introduced as follows.

Supposing that gray levels of pixels follow a normal distribution, and assume ensemble means of pixels in the picture fragments f and g are m_f and m_g, and standard deviations being s_f and s_g respectively. The difference between two pixel values also follows a normal distribution with an ensemble mean $m = m_f - m_g$, and a standard deviation $s^2 = s_f^2 + s_g^2$. Accordingly, thinking about a distribution of $|f_i - g_i|$, they follow an distribution with which negative part of a distribution of $f_i - g_i$ is folded and superimposed on to the positive part. Assuming that an image is statistically homogeneous everywhere, the distribution of $|f_i - g_i|$ is expected to be one side of a normal distribution as shown in figure 2. In this case, a square mean value of $f_i - g_i$ $(=\sigma^2)$ can be approximately related to an average of $|f_i - g_i|$ by the following expression (6).

$$\overline{\left(f_i - g_i\right)^2} = \sigma^2 \propto \overline{\left|f_i - g_i\right|}^2 \qquad (6)$$

The above expression implies that a square of an ensemble mean of the absolute differences is proportional to the variance.

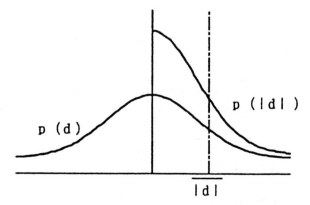

Figure 2 Positive half of a normal distribution of absolute differences.

3.3   Estimation of Parameter α_j using F-inspection

Supposing $T_{j,k}$ is calculated up j-th pixels for k-th candidate pattern, it is proportional to a variance of j-samples of absolute difference between a template and a k-th candidate pattern. Since those absolute differences can be estimated as being sampled from a population following a positive half of a normal distribution, variances calculated from the samples can be identified by F-inspection. The F-inspection distinguishes whether any two set of data are sampled from a same source by examining a ratio of the two variances. If the ratio is close to one, the two sets of data can be estimated as sampled from a same population. The F-inspection gives a criterion of the ratio as a function of a significance level and a number of samples.

The statistical technique is applied to the extraction of the most similar pattern from many candidates. Since most candidates are quite different from the template, variances obtained from such pictures may follow the positive side of a normal distribution with a relatively large variance. A candidate, which has the smallest variance of absolute differences, could be the most similar pattern. So, if a variance ratio, which is defined as a ratio of a larger variance to the minimum one, exceeds a certain value, the candidate is discarded, since it can be regarded as a quite different pattern from the one with the minimum variance. In the actual pattern identification, the level of significance is taken as 1/4000, which is empirically determined. The coefficient a in the equation (4) can be found from either a F-inspection table or definition of a function F as a function of the level of significance and a number of sampled data.

4   Analysis of Experimental Figures

As an examination of the proposed method, a flow passing through a vertical plate is measured from a pair of pictures (figure 3), in which the flow visualized with smoke. By analyzing the sequential pictures, the performance of the analysis method such as reliability, analysis time, and adequacy of parameters are examined. An example of measured velocity distribution is shown in figure 4, in which velocities are measured at 16x23 grid points, and 350 vectors are obtained in 3.05 seconds using a ordinary personal computer. The analysis conditions and results are summarized in the table 2.

Table 2 Analysis conditions and results of figure 4.

Picture size	Interrogation size	Search area size
371x399 pixels	32x32 pixels	37x37 pixels
Level of significance		1/4000
Error correction		Median filter
Number of Vectors		350
Analysis time	3.05 sec/sheet	8.7 ms/vector
CPU	Intel Pentium	133 MHz

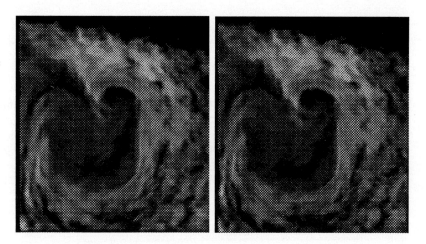

Figure 3 Flow against a vertical plate visualized by smoke.

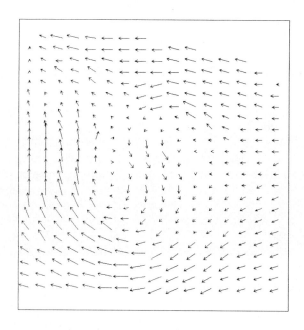

Figure 4 Velocity distribution of figure 3 measured by PIV.

A wake behind a elastic cylinder visualized by small particles are shown in figure 5, and figure 6 is an example of an instantaneous velocity distribution measured by the PIV. Instantaneous velocity distributions along vertical lines are easily recognized. The analysis conditions and results are summarized in the table 3.

When a sequence of the pictures are processed by PTV analysis, particle trajectories can be measured in a moment. Figure 7 is an example of a diagram of particle trajectories measured from six sequential pictures. The analysis finishes within 4 seconds including all processes from loading picture data to the final vector drawing. The net processing time is 1.85 seconds.

Table 3 Analysis conditions and results of figure 6.

Picture size	Interrogation size	Search area size
512x480 pixels	32x32 pixels	33x33 pixels
Level of significance		1/4000
Error correction		Median filter
Number of Vectors		484 vectors
Analysis time	4.28 sec/sheet	8.8 ms/vector
CPU	Intel Pentium	133 MHz

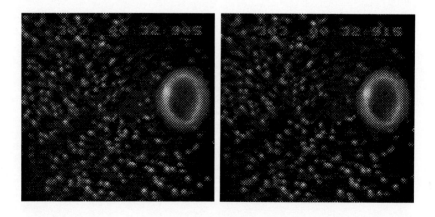

Figure 5 Wake behind an elastic cylinder visualized by particle tracer.

5 Conclusion

A new analysis method is proposed. The method enables simultaneous operations to multiple pixels. The algorithm can be utilized in some basic image processing and realize accelerated processing. Some examples of PTV and PIV analysis are shown.

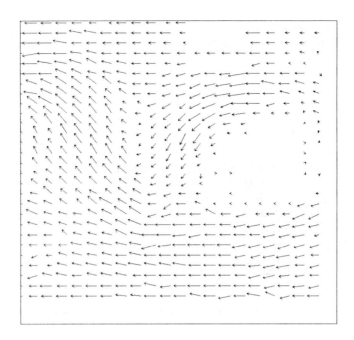

Figure 6 Velocity distribution of figure 5 measured by PIV.

Figure 7 Particle trajectories measured by PTV from consecutive six pictures.

Acknowledgment
This work is partly supported by the Science Research Promotion Fund from the Japan Private School Promotion Foundation.

Authors wish to acknowledge the help of professor Ohba at Kansai University, who provide pictures of visualized flow in Figure 5.

References

Barnea I., H.F.Silverman, "A Class of Algorithms for Fast Digital Image Registration, IEEE Trans. Computers, C-20-2,(1972), pp.179-186.

Kaga, A., Inoue, Y., Yamaguchi, K., Pattern Tracking Algorithms using Successive Abandonment, J. Flow Visualization and Image Processing, 1-2(1993), pp.283-296.

Uemura, T., Kawahara, G., Yamamoto, F., High Speed Image Analysis for Measuring Velocity Distribution from Pictures of Particles, Flow Visualization V(Proc. 5th Intl. Symp. Flow Visualization), (1989), pp.228-233, Hemisphere.

Uemura, T., Iguchi, M., Yoshimoto M., and Tatumi, M., An Efficient Calculation Technique for PTV and PIV, Seventh International Symposium for Flow Visualization, (1995), pp.641-646.

I.7 Optimization of the Shape of Receiving Aperture in a Phase Doppler System

A. Naqwi

TSI Incorporated
P.O. Box 64394, St. Paul, MN 55164-0394, USA

Abstract. The *phase centroid* of an aperture used in a phase/Doppler velocimeter (PDV) is introduced. It is defined as the point detector whose phase response is identical to the finite-area aperture of PDV. The phase centroid is fixed only for certain aperture shapes and certain particle size ranges. Guidelines are provided to obtain a spatially invariant phase centroid.

It is also shown that displacement of the phase centroid with the particle size can be used to advantage, so as to extend the size range of a system without sacrificing its sensitivity to sizing small particles. Using special aperture shapes, the phase-diameter relationship may be made non-linear (even if the corresponding point detectors have a linear response), so as to have a high sensitivity for small particles and reduce it gradually to cover a large size range within 360° range of phase shift. This arrangement leads to a rather uniform percent-of-the-reading sensitivity of the system. Furthermore, the ambiguity associated with the particle trajectory can be suppressed using shaped apertures.

Keywords. Phase Doppler, Particulate Two-Phase Flow, Optical Diagnostics

1 Introduction

The phase Doppler technique is well-known for simultaneous measurement of particle size and velocity in a fluid flow. The present work is primarily concerned with a two-receiver arrangement, such as the system shown in Fig. 1. For the sake of completeness, operation of this system is described here briefly.

As shown in Fig. 1, the output beam of an Argon-Ion laser is transmitted to the beam conditioning optics, where the laser beam is split into two beams. The laser beams exiting the beam conditioning assembly are coupled into single-mode polarization preserving fibers, connected to the transmitting probe, which focuses a pair of monochromatic beams to intersect and produce interference fringes in a small volume.

The particles under investigation scatter a fringe-like moving pattern onto the receiving apertures, as they cross the fringes in the measuring volume, see Fig. 2. The scattered light is integrated over the receiving apertures of the phase/Doppler receivers and coupled to multi-mode fibers attached to them. These optical signals consist of oscillations, such that the frequency of oscillations depends upon the particle velocity and the phase shift between the signals is a measure of particle size (see e.g. Naqwi and Durst, 1991). The optical signals are transmitted to the transducer module, where they are converted into electrical signals using photomultipliers.

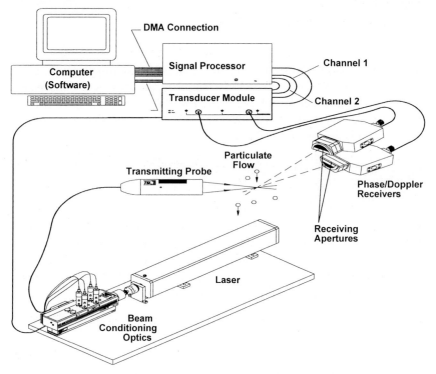

Figure 1: A standard phase Doppler setup

The electrical signals are transferred to the signal processor for determining the frequency and the phase shift. These results are delivered to a computer, where a software package reduces large ensembles of data and displays the statistical parameters, such as histograms of particle size and velocity.

The present work is concerned with optimization of the shape of the receiving apertures. In a conventional arrangement, a receiving aperture is formed by masking a portion of the circular receiving lens, such that the effective receiving area is nearly rectangular. In order to examine the effect of variations in the receiving aperture shape, the phase Doppler response is formulated for an arbitrary shaped aperture in the next section. This formulation is applied to rectangular and trapezoidal apertures in Sec. 3 and 4.

The inverse problem, i.e. determination of the aperture shape for a desirable phase response, is solved in Sec. 5, which indicates that an aperture defined by log-normal curves has optimal response, i.e. extends the measurable size range and provides a rather uniform percent-of-the-reading sensitivity. This aperture shape is examined in detail in Sec. 6 and test results are reported in Sec. 7. The key outcomes of the study are summarized in the final section.

2 The Basic Formulation

A receiving aperture with an arbitrarily varying width $w(x_A)$ is considered in Fig. 2, where x_A-axis is perpendicular to the scattered fringes and has its origin in the plane of symmetry of the phase Doppler system (i.e. a plane that bisects the laser beams and lies perpendicular to the plane of the beams). The scattered light signal collected by a segment of thickness dx_A may be expressed as

$$dP_s = C_P w(x_A)\left[1 + \cos(\omega_D t + \Delta\Phi)\right]dx_A, \tag{1}$$

assuming that the *scattered fringes* have 100% visibility. This assumption is reasonable for standard phase Doppler applications, such as spray diagnostics with receivers near Brewster angle. The constant C_P represents the average intensity of the scattered fringes. Assuming that the scattered fringe spacing s_f is uniform for a given particle diameter, the signal phase $\Delta\Phi$ is the phase shift for a point detector at x_A and is given as

$$\Delta\Phi = 2\pi x_A / s_f. \tag{2}$$

The scattered fringe spacing s_f is inversely proportional to the particle diameter and is replaced later with a non-dimensional diameter.

The total scattered light signal is obtained by integrating Eq. (1) over the entire aperture, so that

$$P_s = P + F\cos\left(\omega_D t + \overline{\Delta\Phi}\right), \tag{3}$$

where P and F are the *pedestal* and the *amplitude of fluctuation* respectively. These are given as

$$P = C_P \int_{x_{A\min}}^{x_{A\max}} w(x_A)dx_A \tag{4}$$

and

$$F = C_P \sqrt{\left(\int_{x_{A\min}}^{x_{A\max}} w(x_A)\cos(2\pi x_A/s_f)dx_A\right)^2 + \left(\int_{x_{A\min}}^{x_{A\max}} w(x_A)\sin(2\pi x_A/s_f)dx_A\right)^2} \tag{5}$$

In most of the PDV processors, the signals are *high-pass* filtered prior to measurement of phase shift and frequency, so that the signals can be represented by Eq. (3) without the first term on the right-hand side. Hence, the amplitude of fluctuation F represents the signal strength as seen by the processor.

90

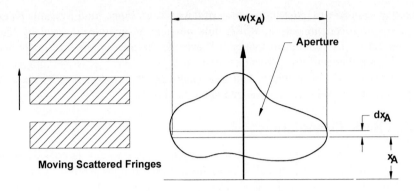

Figure 2: Arbitrary shaped aperture

The total phase shift $\overline{\Delta\Phi}$ is represented by

$$\sin\overline{\Delta\Phi} = \frac{C_p}{F}\int_{x_{A\min}}^{x_{A\max}} w(x_A)\sin(2\pi x_A/s_f)dx_A \tag{6}$$

and

$$\cos\overline{\Delta\Phi} = \frac{C_p}{F}\int_{x_{A\min}}^{x_{A\max}} w(x_A)\cos(2\pi x_A/s_f)dx_A . \tag{7}$$

The signal *visibility*, defined as the ratio of the fluctuation amplitude to the signal pedestal, may be expressed as

$$= \frac{\sqrt{\left(\int_{x_{A\min}}^{x_{A\max}} w(x_A)\cos(2\pi x_A/s_f)dx_A\right)^2 + \left(\int_{x_{A\min}}^{x_{A\max}} w(x_A)\sin(2\pi x_A/s_f)dx_A\right)^2}}{\int_{x_{A\min}}^{x_{A\max}} w(x_A)dx_A} \tag{8}$$

The phase difference between two symmetrically located receivers (see Fig. 1) is given as

$$\Delta\Phi_{12} = 2\overline{\Delta\Phi} . \tag{9}$$

For small values of phase angles, i.e. for large s_f or small particles,

$$\overline{\Delta\Phi} \approx 2\pi\overline{x}_A/s_f , \tag{10}$$

where $\overline{x}_A$ represents the geometrical centroid of the aperture, given by

$$\overline{x}_A = \frac{\int_{x_{A\min}}^{x_{A\max}} w(x_A)x_A dx_A}{\int_{x_{A\min}}^{x_{A\max}} w(x_A)dx_A} . \tag{11}$$

The phase-diameter relationship represented by Eqs. (10) & (11) is obtained by representing the tangent and sine of small angles by the angles themselves and setting the cosine to 1. Similarly, using first two terms in the power expansion of sine and cosine functions the signal visibility may be expressed as

$$V = 1 - 2\left(\frac{\pi}{s_f}\right)^2 \left(\overline{x_A^2} - \overline{x}_A^2\right),$$ (12)

where

$$\overline{x_A^2} = \frac{\int\limits_{x_{A\min}}^{x_{A\max}} w(x_A)x_A^2 dx_A}{\int\limits_{x_{A\min}}^{x_{A\max}} w(x_A)dx_A}.$$ (13)

Clearly, visibility approaches 1 with decreasing particle diameter (i.e. increasing s_f) and narrowing aperture width along x_A-axis.

Comparing Eq. (10) with (2), the phase shift for the finite aperture is identical to that of a point detector at the centroid of the aperture. For small particles, the phase centroid of a finite aperture is represented by the geometrical centroid of the aperture regardless of its shape.

The phase-diameter curve is initially a straight line and is represented by a fixed point detector. However, according to Eqs. (6) & (7), location of the effective point detector may vary with the particle diameter for larger values of particle diameter. Equations (6)–(8) for phase and visibility are solved in the following sections for certain standard aperture shapes.

3 Rectangular Apertures

A rectangular aperture is characterized by the constant value of the width $w(x_A)$, so that the integrals in Eqs. (6)–(8) are easily evaluated. The results can be expressed as follows:

$$\sin\overline{\Delta\Phi} \propto \sin\left(2\pi\overline{x}_A/s_f\right)\sin\left(\pi\Delta x_A/s_f\right);$$ (14)
$$\cos\overline{\Delta\Phi} \propto \cos\left(2\pi\overline{x}_A/s_f\right)\sin\left(\pi\Delta x_A/s_f\right);$$ (15)
and

$$V = \frac{s_f}{\pi\Delta x_A}\left|\sin\left(\frac{\pi\Delta x_A}{s_f}\right)\right| = \left|\mathrm{sinc}\left(\frac{\pi\Delta x_A}{s_f}\right)\right|;$$ (16)

where
$$\Delta x_A = x_{A\max} - x_{A\min}.$$ (17)

According to Eqs. (14) & (15), phase shift can be given by an equation of the form (10), provided that the second factor, i.e. the common factor, on the right-hand side of Eqs. (14) & (15) has a positive sign. Otherwise, the phase is shifted by 180°; i.e.

$$\overline{\Delta\Phi} = 2\pi\overline{x}_A/s_f \qquad \text{for} \quad 2ns_f \leq \Delta x_A < (2n+1)s_f$$ (18)

and

$$\overline{\Delta\Phi} = \frac{2\pi\overline{x}_A}{s_f} - \pi = \frac{2\pi}{s_f}\left(\overline{x}_A - \frac{s_f}{2}\right) \quad \text{for} \quad (2n+1)s_f \le \Delta x_A < 2(n+1)s_f , \quad (19)$$

where $n = 0, 1, 2, \cdots$.

A *phase jump*, as pointed out by Naqwi & Durst (1990), occurs at the integer values of $\Delta x_A/s_f$. According to Eq. (16), the signal visibility vanishes at the point of each phase jump. For very small particles (or large s_f), the phase centroid is fixed and coincides with the geometrical centroid. It is shifted down abruptly by $s_f/2$ at $(2n+1)$th phase jump and shifted up at $2(n+1)$th jump.

With increasing particle diameter (i.e. decreasing s_f) the first phase jump occurs, when the scattered fringe spacing s_f reaches the thickness of the aperture Δx_A . Under this condition, the receiving aperture is continuously exposed to a complete scattered bright fringe and a complete dark fringe. Hence, the total amount of light collected by the receiver is unaffected by the motion of the scattered fringes. Consequently, the collected scattered light does not exhibit any oscillations, which is manifested as zero visibility.

As the scattered fringe spacing reduces below the aperture height, the receiving aperture may be considered as being composed of two segments; i.e. the effective and the ineffective segments. The height of the ineffective segment equals the scattered fringe spacing, so that it does not contribute to the oscillations in the signal. The remaining aperture constitutes the effective segment, which is responsible for modulation of the signal. The geometrical centroid of the effective segment may be regarded as the phase centroid of the aperture, as shown in Fig. 3. Equation (19) is based on the assumption that the ineffective segment occupies the upper portion of the aperture, so that the phase centroid is shifted down after the phase jump. Alternatively, the ineffective segment could be considered as occupying the lower part of aperture. This would result in an upward shift in the phase centroid. The two alternative descriptions lead to phase-diameter relations that differ by a complete cycle, i.e. 2π radians.

After the second phase jump, a second ineffective segment may be identified. According to the formulation of Eqs. (18) & (19), an ineffective segment appears in the upper part of the aperture at each odd numbered phase jump and in the lower part at the even numbered phase jumps.

With decreasing s_f , a portion of the receiving aperture is "blinded" as it collects a fixed amount of light despite the motion of the scattered fringes. Such ineffective segments appear abruptly in the case of a rectangular aperture, because the shape of the aperture agrees with the shape of the scattered fringes. In the case of a non-rectangular aperture, the ineffective segments emerge gradually; i.e., the phase centroid is shifted gradually. Consequently, the signal visibility does not vanish completely.

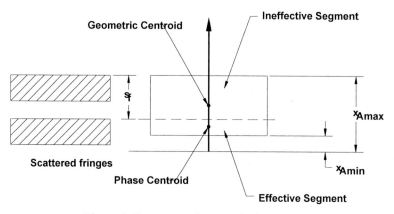

Figure 3: Response of a rectangular aperture

In the case of a rectangular aperture, it seems most appropriate to design the aperture, such that the entire size range of interest lies in the first lobe of the visibility curve, so as to avoid the zero visibility condition. As a consequence, it suffices to treat the geometrical centroid as the phase centroid. In the earliest phase Doppler systems, the receiving apertures were nearly rectangular. This has been the reason for ignoring the shifts in the phase centroid in the early works on the technique.

As shown in the following section, the shifts in the phase centroid — in the case of non-rectangular apertures — may be substantial and may occur gradually with increasing particle diameter.

4 Trapezoidal Apertures

A trapezoidal aperture has the simplest shape that offers a variable width $w(x_A)$. Response of this aperture is examined to illustrate the gradual shifts in the phase centroid that may occur with increasing particle size.

The width of a trapezoidal aperture, as shown in Fig. 4, is given by

$$w(x_A) = w_0 - s_w x_A , \quad x_{A\min} \leq x_A \leq x_{A\max} , \tag{20}$$

where $x_{A\min} \geq 0$, $w_0 \geq 0$ and in the case of positive s_w, $x_{A\max} \leq w_0/s_w$. In limiting cases, this aperture may be reduced to a triangle or an inverted triangle with its vertex in the plane of symmetry.

Substituting Eq. (20) into Eq. (6), the following expression is obtained for the sine of the phase shift:

$$\frac{2\pi F}{C_p s_f}\sin\overline{\Delta\Phi} = w_0\Big[\cos\big(2\pi x_{A\min}/s_f\big) - \cos\big(2\pi x_{A\max}/s_f\big)\Big]$$
$$+ s_w\Big[x_{A\max}\cos\big(2\pi x_{A\max}/s_f\big) - x_{A\min}\cos\big(2\pi x_{A\min}/s_f\big)\Big] \tag{21}$$
$$- \frac{s_w s_f}{2\pi}\Big[\sin\big(2\pi x_{A\max}/s_f\big) - \sin\big(2\pi x_{A\min}/s_f\big)\Big]$$

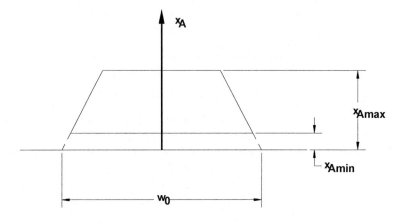

Figure 4: Trapezoidal aperture

Similarly, by substituting Eq. (20) into Eq. (7), the cosine of the phase shift is expressed as

$$\frac{2\pi F}{C_p s_f}\cos\overline{\Delta\Phi} = w_0\Big[\sin\big(2\pi x_{A\max}/s_f\big) - \sin\big(2\pi x_{A\min}/s_f\big)\Big]$$
$$- s_w\Big[x_{A\max}\sin\big(2\pi x_{A\max}/s_f\big) - x_{A\min}\sin\big(2\pi x_{A\min}/s_f\big)\Big] \tag{22}$$
$$- \frac{s_w s_f}{2\pi}\Big[\cos\big(2\pi x_{A\max}/s_f\big) - \cos\big(2\pi x_{A\min}/s_f\big)\Big]$$

The above relations can be reduced to Eqs. (14) & (15) for $s_w = 0$; i.e. a rectangular aperture.

For a trapezoidal aperture, the signal pedestal as given by Eq. (4) reduces to

$$P = C_p\big(x_{A\max} - x_{A\min}\big)\Big[w_0 - s_w\big(x_{A\max} + x_{A\min}\big)/2\Big]. \tag{23}$$

It may be noticed that the response of a trapezoidal aperture may be expressed in terms of three independent parameters; i.e. $2\pi x_{A\max}/s_f$, $x_{A\min}/x_{A\max}$ and $s_w x_{A\max}/w_0$. A triangular aperture may be defined by assigning the values of 0 and 1 to the second and third parameter respectively. The phase shift for this triangular aperture is examined below.

For the triangular aperture, Eqs. (21)–(23) are reduced to the following relations:

$$\frac{2\pi F}{C_p s_f w_0} \sin \overline{\Delta\Phi} = 1 - \mathrm{sinc}\left(3\pi d_p^*\right);$$ (24)

$$\frac{2\pi F}{C_p s_f w_0} \cos \overline{\Delta\Phi} = \frac{1 - \cos\left(3\pi d_p^*\right)}{3\pi d_p^*};$$ (25)

$$P = C_p \, w_0 x_{A\max}/2 \, .$$ (26)

Using Eqs. (24)–(26), the signal visibility, as defined by Eq. (8), is expressed as

$$V = \frac{2\sqrt{9\pi^2 d_p^{*2} + 2\left(1 - \cos 3\pi d_p^*\right) - 6\pi d_p^* \sin 3\pi d_p^*}}{9\pi^2 d_p^{*2}},$$ (27)

where the non-dimensional particle diameter

$$d_p^* = \frac{2\overline{x}_A}{s_f} \, .$$ (28)

The symbol $\overline{x}_A$ represents the geometrical centroid of the triangular aperture, so that $\overline{x}_A = x_{A\max}/3$.

The non-dimensional diameter would be 1 for $\Delta\Phi_{12} = 360°$, provided that the phase centroid is fixed at the geometrical centroid, i.e. the initial slope of the phase-diameter curve is maintained. The size range $0 < d_p^* \leq 1$ will be hereafter referred to as the *nominal size range*. As shown later, shaped apertures allow the $360°$ limit to be deferred to a larger value of d_p^*, such as ~3.

According to Eq. (2), location of the phase centroid $\tilde{x}_A$ is given by

$$\tilde{x}_A = \frac{\overline{\Delta\Phi}}{2\pi} s_f = \frac{\overline{\Delta\Phi}}{3\pi d_p^*} x_{A\max} \, .$$ (29)

Since $\overline{x}_A = x_{A\max}/3$, the relationship between the phase centroid and the geometrical centroid may be expressed as

$$\frac{\tilde{x}_A}{\overline{x}_A} = \frac{\overline{\Delta\Phi}}{\pi d_p^*} \, .$$ (30)

Figure 5 shows the phase shift, visibility and the centroid location for the triangular aperture.

The phase shift between a pair of triangular apertures saturates within the nominal size range and oscillates below $180°$. As obvious from Eq. (25), $\overline{\Delta\Phi}$ is $90°$ for $d_p^* = 2/3, 4/3, 2, \cdots$, as the cosine of $\overline{\Delta\Phi}$ vanishes at these points. Consequently, $\Delta\Phi_{12}$ is $180°$ at the above values of d_p^*. It is also clear from Fig. 4 as well as Eqs. (24) & (25) that the limiting value of phase shift for very large diameters is $180°$.

According to Fig. 5, signal visibility decreases gradually for a triangular aperture but does not vanish completely. The phase centroid is initially located at the geometrical centroid, i.e. $\tilde{x}_A/\bar{x}_A = 1$. It shifts towards the plane of symmetry with the increasing particle diameter.

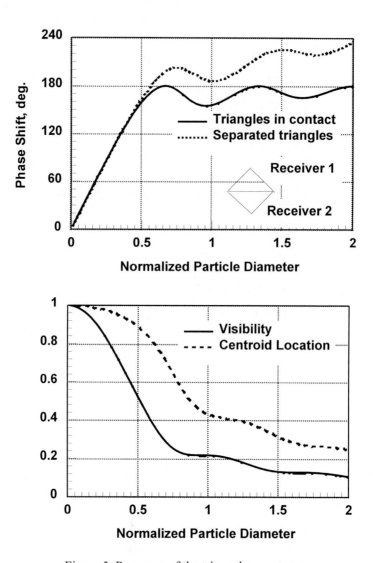

Figure 5: Response of the triangular apertures

The upper plot in Fig. 5 also includes the case of triangular apertures that are separated, so that $x_{A\min}/x_{A\max} = 0.03$, which corresponds to the experimental setup discussed in Sec. 7 below.

5 Aperture Shape for an Arbitrary Response Curve

The above discussion clearly shows that nonlinear phase-diameter relationships can be obtained using aperture shapes that are significantly different from a rectangle. Nonlinear response curves are desirable for the following reasons:

(i)   To extend the size range, while maintaining a high sensitivity to small particles;

(ii) To obtain a uniform percent-of-the-reading sensitivity, i.e. an invariant value of

$$S_{por} = \frac{d\left(\overline{2\Delta\Phi}\right)}{100 d(d_p)/d_p} \tag{31}$$

for a pair of symmetric receivers;

(iii) To eliminate the 2π *ambiguity* by ensuring that the phase shift does not exceed 360°.

In order to meet the above requirements, one needs to solve the inverse problem, i.e. to calculate the aperture shape $w(x_A)$ for a given phase-diameter relationship $\overline{\Delta\Phi}(d_p)$. To this end, a complex integral expression is obtained by multiplying Eq. (6) with $\sqrt{-1}$ and adding to Eq. (7), i.e.

$$\exp\left[i\,\overline{\Delta\Phi}\left(d_p^*\right)\right] = \int_{-\infty}^{+\infty} w^*(x_A^*)\exp\left(2\pi i d_p^* x_A^*\right)dx_A^* , \tag{32}$$

where the non-dimensional particle diameter is defined by Eq. (28). Furthermore, non-dimensional aperture width and height are introduced that are defined as below:

$$w^* = \frac{2\overline{x}_A C_P}{F} w , \tag{33}$$

$$x_A^* = \frac{x_A}{2\overline{x}_A} . \tag{34}$$

Note that the distance along x_A-axis is normalized with the spacing between the geometrical centroids of the two symmetrically located apertures. The above formulation allows one to treat the relationship between the aperture and the response curve as a *Fourier transform*.

In order for w^* to be real, the left-hand side of Eq. (32) must exist for both the positive and negative values of d_p^* and must satisfy the following relationship (see Press et al. (1986), Chap. 12):

$$\overline{\Delta\Phi}\left(-d_p^*\right) = -\overline{\Delta\Phi}\left(d_p^*\right) . \tag{35}$$

Negative particle diameter, appearing in Eq. (35), is a mathematical artifact that allows one to make use of the Fourier transform for solving the inverse problem. Inverting the transform in Eq. (32),

$$w^*(x_A^*) = \int_{-\infty}^{\infty} \exp\left[i\overline{\Delta\Phi}(d_p^*)\right]\exp\left[-2\pi i d_p^* x_A^*\right] d(d_p^*).$$

(36)

It may be noticed that $w^*(x_A^*)$ is a delta function if the phase-diameter relationship is linear; i.e., the above formulation readily allows to recover the point detector arrangement.

For some phase-diameter relations, the corresponding function $w^*(x_A^*)$ may become negative for certain values of x_A^*, indicating that a realizable aperture shape does not exist. However, the exponential response curve, discussed below relates to a realizable aperture.

Various desirable features of a non-linear response curve discussed above are met closely by an exponential function of the following form:

$$\overline{\Delta\Phi} = \pi\left(1 - e^{-d_p^*}\right), \qquad \text{for} \qquad d_p^* \geq 0.$$

(37)

This function eliminates the 2π ambiguity by confining the phase shift to π for a single detector; i.e. the phase shift between two symmetric receivers would be restricted to 2π.

The percent-of-the-reading sensitivity for the above response curve is obtained by substituting Eq. (37) in Eq. (31), so that

$$S_{por}[\text{deg.}/\%] = 3.6 d_p^* e^{-d_p^*}.$$

(38)

Although S_{por} is not constant, it varies slowly and remains above $0.5°$ per % in the range $0.2 \leq d_p^* \leq 3$. The largest value of S_{por} is about $1.32°$ per % and occurs at $d_p^* = 1$.

By extending the phase-diameter relationship of Eq. (37) to the negative diameters, in accordance with Eq. (35), a fast Fourier transform (FFT) algorithm could be used to compute the corresponding width function. The results are shown in Fig. 6. The aperture shape is fairly simple and may be approximated by a lognormal function.

Based on the above considerations, pairs of symmetrically located *lognormal apertures* are analyzed in detail in the following section.

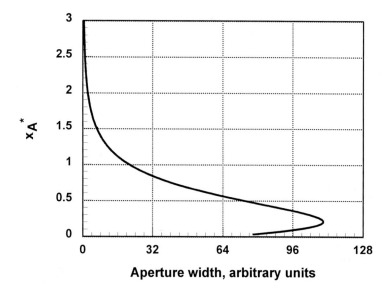

Figure 6: Aperture for an exponential response curve

6 Lognormal Apertures

It is understood that the response of the aperture in Fig. 6 would be unaffected if it is made symmetric by mirror imaging its contour about x_A-axis. The resulting aperture is shaped like an "onion" or a "tear drop", such as the shape depicted in Fig. 7. Such shapes can be implemented conveniently in practice. The lognormal apertures considered hereafter have symmetric onion-like shape.

A lognormal function has two independent variables. Four independent variables are needed if the coordinates of the lognormal function are shifted arbitrarily. Such a function may be expressed as

$$w = \frac{A_{\lgn}}{\left(x_A + x_{A0}\right)} \exp\left\{-\frac{1}{2\sigma_{\lgn}^2}\left[\ln\left(x_A + x_{A0}\right) - \mu_{\lgn}\right]^2\right\} - \Delta w,$$

$$\text{for} \qquad x_{A\min} \le x_A \le 1. \tag{39}$$

The above aperture can be specified in terms of four parameters, i.e.
(i) $x_{A\min}$: the minimum value of normalized x_A, so that the maximum value is 1;
(ii) x_{A0} : the zero-shift of lognormal distribution along x_A-axis;
(iii) x_{Ap} : location of the maxima of the aperture width;
(iv) $\sigma_{\lgn}$: width of the lognormal distribution.

By normalizing the maximum width of the aperture to 1, the parameters $A_{\lgn}$, $\mu_{\lgn}$ and Δw may be obtained from the relations:

$$\mu_{lgn} = \sigma_{lgn}^2 + \ln\left(x_{Ap} + x_{A0}\right); \tag{40}$$

$$\frac{1}{\Delta w} = \left(1 + x_{A0}\right)\exp\left\{\frac{1}{2\sigma_{lgn}^2}\left[\ln\left(\frac{1 + x_{A0}}{x_{Ap} + x_{A0}}\right) - \sigma_{lgn}^2\right]^2 - \ln\left(x_{Ap} + x_{A0}\right) - \frac{\sigma_{lgn}^2}{2}\right\} - 1; \tag{41}$$

$$A_{lgn} = \left(1 + \Delta w\right)\exp\left[\ln\left(x_{Ap} + x_{A0}\right) - \frac{\sigma_{lgn}^2}{2}\right]. \tag{42}$$

Equations (6)–(8) were solved numerically for the above aperture shape. This numerical problem involves solution of integrals whose integrands are oscillating and may have many oscillations within the range of integration. For solution of these integrals efficient numerical schemes, provided by Stamnes (1986), were used.

Simulations show that it is important to minimize $x_{A\min}$ in order to significantly extend the nominal size range beyond 1. However, there are practical limits on the minimum gap between two symmetrically located apertures. Using a practically achievable value for the minimum gap, an optimal lognormal aperture can be specified by the following combination of the shape parameters:

$$x_{A\min} = 0.04, \quad x_{A0} = 0.1, \quad x_{Ap} = 0.16, \quad \sigma_{lgn} = 0.65.$$

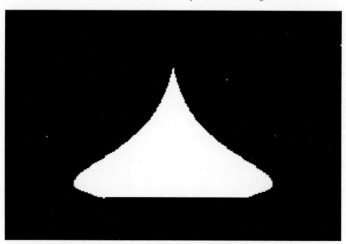

Figure 7: An optimal lognormal aperture

This shape is depicted in Fig. 7. The geometrical centroid of this aperture is located at $x_A = 0.3097$. The corresponding response curve, signal visibility and percent-of-the-reading sensitivity are given in Fig. 8. It can be seen that the size range is extended by a factor of three as compared to the nominal size range. Furthermore, the visibility is always non-zero and large enough to produce measurable signals. The sensitivity is such that over most of the size range, about 1° phase shift is obtained for 1% variation in the size.

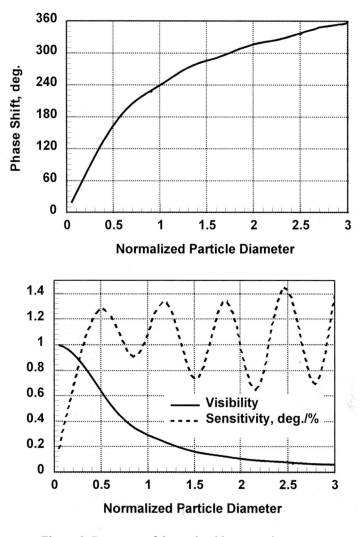

Figure 8: Response of the optimal lognormal aperture

Reduction in the signal visibility with increasing particle diameter is, in fact, a desirable feature of PDV. It is known that the signal strength increases with the square of particle diameter, if the diameter is significantly larger than the wavelength of light. However, due to decreasing visibility, the amplitude of fluctuation F would not increase so excessively. As explained in Sec. 2, parameter F represents the signal strength sensed by the signal processor. Uniform signal strength allows to obtain a uniform bit-resolution for the digitized signals, so that all the signals are processed with about the same precision. In the case of analog processors, the output of analog components varies with the signal amplitude. This source of uncertainty is suppressed if the signal amplitude is rather uniform.

Another adverse effect of large variations in the signal strength is discussed in Sec. 7, where effects of non-uniform illumination of the particle are taken into consideration. It is shown that the measurements are less prone to error if the signal strength is rather uniform over a wide range of particle diameters. The experimental results in the following section confirm that this objective can be achieved using lognormal apertures.

7 Experimental Results pertaining to Aperture Shape

The receiving optics of Adaptive Phase/Doppler Velocimeter (APV of TSI Inc.) allows to use external masks on the two receiving apertures, so that triangular apertures, similar to those described above, could be implemented. Measurements were taken with a spray of water, in which drop velocity correlates strongly with the drop size, i.e. larger drops move faster.

The measured phase shift versus velocity correlations are given in Fig. 9 for two sets of apertures, i.e. rectangular and triangular apertures. The triangular apertures were 36 mm high with $x_{A\min}/x_{A\max} \approx 0.03$ so that their response is expected to follow the curve for the separated receivers shown in Fig. 5. The rectangular receivers were matched with the triangular receivers in the location of the centroid as well as the cross-sectional area.

Figure 9 clearly shows that the phase response can be very different for two aperture sets with the same geometrical centroids. The rectangular receivers show an increase in the average velocity with increasing phase. Since phase increases linearly with drop size for rectangular apertures, this result illustrates the correlation between the drop size and drop velocity.

The triangular apertures exhibit a saturation of the phase response slightly above 180°, which verifies the theoretical result shown in Fig. 5.

In another set of experiments, the optimal aperture of Fig. 7 was used to measure a similar spray and the results were compared with the response of a pair of narrow rectangular apertures. Both the aperture shapes had the same geometrical centroids.

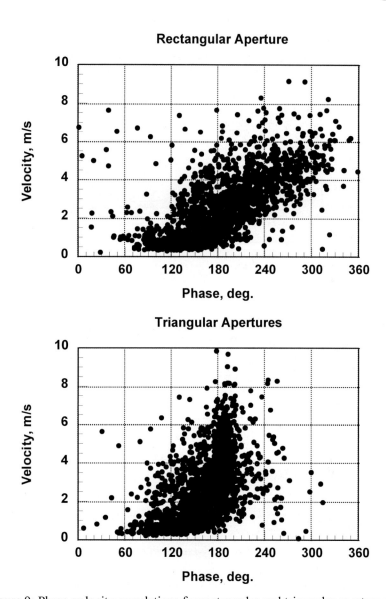

Figure 9: Phase-velocity correlations for rectangular and triangular aperture sets

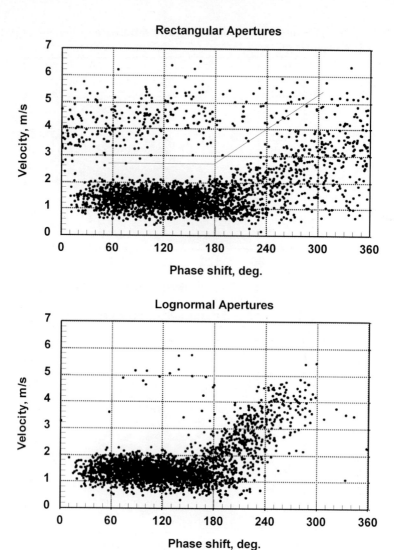

Figure 10: Phase-velocity correlations for rectangular and lognormal apertures

The phase-velocity correlations for the two aperture shapes are shown in Fig. 10. Clearly, the phase shift exceeds 360° in the case of rectangular apertures. The true phase shift for the data points above the dividing solid line in the upper plot of Fig. 10 is expected to be 360° in excess of the measured phase shift. After shifting the phases accordingly, the phase data were converted into the size distributions.

For the optical setup under consideration, the phase-to-diameter conversion factor of the rectangular apertures was 3.67°/µm; i.e. the nominal size range was 98.2 µm. This factor pertains to pure refraction (see Naqwi & Durst, 1991) and is based on the following optical parameters: beam half-angle, 3.94°; elevation angle, 2.11°; Wavelength of laser, 0.5145 μm; refractive index of the drops, 1.33 and off-axis angle, 74.74°. In the case of lognormal apertures, the correlation shown in Fig. 8 is used to convert the measured phase shifts into the normalized diameters, which are multiplied by the nominal size range to obtain the actual diameters.

The resulting size distributions are given in Fig. 11. The nominal size range is divided into 36 bins, so that each size bin is 2.73 μm wide. The vertical axis that represents the particle count in the corresponding bin is normalized, so that the total area under each curve is 100%.

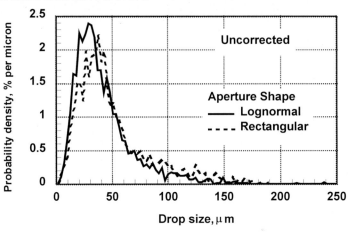

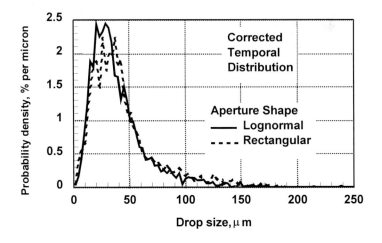

Figure 11: Size distributions measured by the lognormal and rectangular apertures

According to Fig. 11, large drops are measured more frequently by the rectangular apertures, as the visibility of the corresponding signals is higher for the rectangular apertures. In the case of rectangular apertures, large drops generate measurable signals from a large portion of the fringe volume, including the outer regions where the intensity of illumination is low. Hence, the effective measuring volume is larger for large drops, provided that rectangular apertures are used. As shown by Saffman (1987), the r.m.s value of the burst length (which is defined as the product of particle velocity and the signal duration; both of these parameters were recorded) for a given size bin is a measure of the effective measuring volume for the corresponding drop size.

The r.m.s. burst lengths are plotted in Fig. 12 as a function of the drop size. For sizes exceeding 100 µm, there are very few drops in each size bin. Hence, large fluctuation are found in the r.m.s. burst length versus drop size relationship. Curve fitting with simple functions (using TableCurve 2D of Jandel Scientific) is used to estimate the stationary values of the r.m.s burst lengths.

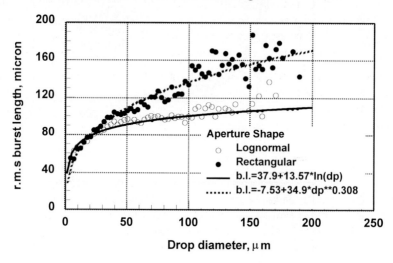

Figure 12: Root-mean-square burst lengths measured by lognormal and rectangular apertures

According to Fig. 12, the effective measuring volume increases more significantly with the particle diameter for the rectangular apertures, as opposed to the lognormal apertures. In the case of lognormal apertures, the signal visibility decreases with the increasing particle diameter and hence, limits the strength of the filtered signals. As a result, the effective measuring volume diameter ceases to increase with the drop diameter.

The size distributions of Fig. 11 are corrected using the r.m.s burst length information from Fig. 12. The corrected temporal size distributions measured by the two aperture shapes are included in Fig. 11 and appear to agree well with each other.

It is known that signals from the outer regions of the measuring volume are undesirable as they may become dependent on the particle trajectory besides the size (see Gréhan et al., 1994 and the references therein). By eliminating such signals, the lognormal apertures also suppress the trajectory ambiguity associated with large drops.

8 Conclusions

Effects of the shape of receiving apertures on the response of a phase Doppler system are examined assuming uniform spacing between the scattered fringes. It is shown that specially shaped apertures can be used to (i) extend the size range without sacrificing measurement sensitivity for small particles, (ii) obtain a uniform percent-of-the-reading sensitivity, (iii) suppress the 2π-ambiguity; i.e. keep the phase-shift smaller than 360° for a very wide range of particle diameters, (iv) obtain relatively uniform amplitude of filtered signals, by ensuring that the signal visibility decreases consistently with increasing particle diameters, (v) suppress the trajectory ambiguity associated with large particles.

The above effects have been verified experimentally. Future work would involve rigorous simulations of the shaped apertures, using Mie scattering theory and the generalized Lorenz-Mie theory (see Gréhan et al., 1994 and the references therein).

References

Gréhan, G.; Gouesbet, G.; Naqwi, A. & Durst, F.: "Trajectory ambiguities in phase Doppler systems: study of a near-forward and a near-backward geometry", Particle and Particle System Characterization 11 (1994) 133–144.

Naqwi, A. & Durst, F.: "Constraints on the size and shape of receiving aperture in a phase Doppler system", Particle and Particle System Characterization 7 (1990) 113–115.

Naqwi, A. & Durst, F.: "Light scattering applied to LDA and PDA measurements. Part 1: Theory and numerical treatments'', Particle and Particle System Characterization 8 (1991) 245–258.

Saffman, M.: "Automatic calibration of LDA measurement volume size", Applied Optics 26 (1987) 2592–2597.

Stamnes, J. J.: "Waves in Focal Regions", Hilger, Bristol, 1986.

I.8 The Differential Speckle Strophometry - A Light Scattering Technique for the Measurement of Velocity Gradients in Turbulent Fluid Flow

R. Schulz and W. Staude

Universität Bremen, Fachbereich 1 - Physik, D - 2800 Bremen 33,
Federal Republic of Germany

Abstract. A laser light scattering technique is described which allows the measurement of velocity gradients and their temporal correlation functions in fluid flow. This non-invasive and gauge-free technique was applied to a turbulent channel flow. Results of the measurements, including the turbulent energy dissipation and the enstrophy are presented.

1.Introduction. It is a well known fact that the light scattered by moving particles carries information on the velocity of these particles. The wide spread Laser Doppler Anemometry is a technique which in an ingenious way exploits this fact for the determination of flow velocities. But there is even more information contained in the scattered light. Thus we could show [1] iin what way a velocity gradient of the flow influences the light scattered by immersed particles: The random intensity pattern (speckle pattern) produced by randomly distributed particles within the fluid exhibits a *motion* if certain components of the velocity gradient tensor are different from zero. A theoretical treatment of this effect shows that each velocity component of the speckle motion is uniquely determined by a particular linear combination of the components of the velocity gradient tensor, the coefficients being dependent only on geometrical parameters describing the optical set-up. It is therefore possible to determine velocity gradient components of a flow by measuring components of the velocity pattern.

It is worth noting that the mentioned effect only occurs if *many* light scattering particles are in the region where the scattered light emerges, because the light pattern produced by only one scattering particle exhibits no motion. It is therefore not possible to understand the method described in this paper on the basis of the LDA measurement scheme.

A method to measure velocity gradients, which is based on essentially the same physical effect as the one just mentioned, was proposed by de Gennes [2], but it was not developed to a level, such that measurements comparable to the ones presented here can be performed.

It is obvious that a measurement scheme which is non-invasive and gaugefree has advantages over the hot wire technique, which was used by Balint et al. [3] to perform for the first time a systematic investigation of velocity gradient properties in a flow.

2. Theory. The velocity field $\vec{u}(\vec{r},t)$ within a flowing fluid is a smooth function in space and time. One can therefore perform a Taylor series expansion with respect to the spatial variables. If only a small volume within the fluid is considered, one can neglect all powers in the spatial variables higher than the first ones an obtains

$$\vec{u}(\vec{r},t) = \vec{u_0}(t) + \hat{\Gamma}(t).\vec{r}. \tag{1}$$

$\hat{\Gamma}(t)$ is the 3x3 velocity gradient tensor. $\vec{u_0}(t)$ is the mean value of the velocity within the considered volume if the origin of the coordinate system is taken to lie in the centre of mass of the considered volume. In turbulent flow both quantities are random functions of time.

Let us now consider a light scattering experiment as depicted schematically in figure 1. A laser beam is focussed into a small region within a flow seeded with small light scattering particles. Light emerging from the so called scattering volume passes a lens and can then be observed on a screen. We will assume that the scattering volume, i.e. the region from which scattered light does reach the screen, is small enough such that the velocity field within this volume can be expressed by equation (1).

Within the scattering volume the illuminating light wave is assumed to be plane and can therefore be characterised by its wave vector $\vec{K_e}$..The direction of this vector is given by the propagation direction of the light wave, and its length depends on the wavelength λ of the light awave through the relation $|\vec{K}| = 2\pi/\lambda$.

The screen is placed in the focal plane of the lens. This assures that all light wave components emerging in the scattering volume, which travel in the same direction (i.e. have the same wave vector $\vec{K_a}$) are focussed on one particular point P_a on the screen. The relative phases of the just mentioned components determine the brightness at this point. If all particles move with the same velocity, all phase differences of the scattered waves remain unchanged and therefore the pattern on the screen doesn't change. The situation is different, if the particles within the scattering volume move with different velocities. Then apparently the relative phases change and consequently also brightness at any point on the screen.

In the case of a velocity field of the form given in equation (1) with a non-vanishing velocity gradient the phase changes are such that generally a motion of the pattern results. This is due to the fact that the relative phases of the scattered waves at an arbitrary point P_a at time t are identical to those of a point P'_a at a later time t'. An illustration of this behaviour is the following: Let us assume that the motion of the particles is just a rotation around the laser beam axis. Then the particles move in a velocity field of type/refGGradient. If we thenconsider the scattered light in a direction rotatated in the same way as the particles, obviously all phase relations of the scattered waves remain the same. Due to the properties of the lens the motion on the screen will be in the opposite direction, but the pattern will just move and no other changes occur.

A quantitative relation between the velocity of the pattern and the velocity gradient was deduced in [1] . This relation turns out to be especially simple, if the pattern velocity is expressed via the temporal change $\vec{\dot{k}_a}\perp$ of the wave vector $\vec{K_a}$ wich "points" towards the arbitrary point P_a:

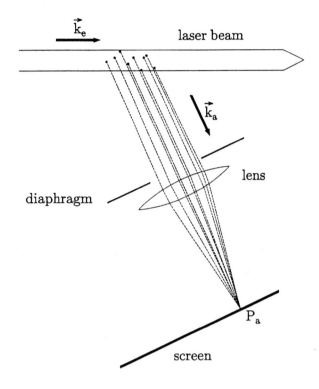

Figure 1: Schematic representation of a light scattering experiment. Light scattering particles which are hit by the laser light emit secondary waves. The dashed lines denote the light paths to a point on the focal plane. Their optical lengths determine the phases of the scattered light components at this point. The diameters of the laser beam and the diaphragm, respectively, determine the scattering volume.

$$\vec{K_a}\perp=(\hat{\Gamma}^T.(\vec{K_e}-\vec{K_a}))\perp \qquad (2)$$

Γ^T is the transposed gradient tensor. The index $\perp$ denotes components perpendicular to $\vec{K_a}$. This is due to the fact that pattern motion is described only by changing direction of the vector $\vec{K_a}$ which is tantamount to a rotation of it. No change in the length of the wave vector of light can occur in a light scattering experiment.

With the help of geometrical optics one can relate the pattern velocity $\vec{u}$ on the screen to $\vec{K_a}\perp$ and obtains

$$\vec{v}\perp=-\frac{(\hat{\Gamma}^{T.}(\vec{K_e}-\vec{K_a}))\perp}{|\vec{K_e}-\vec{K_a}|}f \qquad (3)$$

where f is the focal length of the lens.

Unforunately one observes not only a motion of the speckle pattern but superimposed random fluctuations which are known as speckle boiling. This is caused by the stochastic particles through the scattering volume, and by the parallel component of the vectors appearing on both sides of of equation (2). this speckle boiling generally leads to very fast decorrelations of the pattern, which renders the measurement of the velocity a rather difficult task. Nevertheless we were able to develop several measurement schemes [4] for a reliable determination of this velocity. One of these methods is based on differential properties of the pattern, it is the topic od this paper. We could show [5] that the following relation holds between the pattern velocity component v_x say and derivatives of the pattern intensity $I(\vec{r}, t)$ on the screen:

$$<v_x>=-\frac{<\dot{I}(\vec{r},t).\text{sign}(I_x(\vec{r},t))>}{<I_x(\vec{r},t).\text{sign}(I_x(\vec{r},t))>} \qquad (4)$$

I_x denotes in this equation the spatial derivative of the intensity with respect to x, $\dot{I}$ is the temporal derivative, and sign is the sign-function with the properties:

$$\text{sign}(x)=\begin{cases} 1 & \text{for } x>0 \\ 0 & \text{for } x=0 \\ -1 & \text{for } x<0 \end{cases} \qquad (5)$$

The brackets in equation (4) denote averaging, which for our measurement scheme will be temporal averaging.

Based on this equation it is possible to determine average values of the velocity gradient components as we showed in [4].

An important extension of the briefly scetched method is obtained, if the light scattered in two different directions is used to measure correlations of velocity gradient components. Let us denote the light intensities on the two screens by $I^1(\vec{r}_1,t)$ and $I^2(\vec{r}_2,t)$, where the vectors $\vec{r}_i$ indicate points in these screen planes. One then has the following relation:

$$
\langle v_i^1(t) v_j^2(t+\tau)\rangle = \tag{6}
$$

$$
\frac{\langle \dot{I}^1(\vec{r}_1 t).\mathrm{Sign}(I_i^1(\vec{r}_1,t))\rangle}{\langle I_i^1(\vec{r}_1,t).\mathrm{Sign}(I_i^1(\vec{r}_1,t))\rangle} \times
$$

$$
\frac{\dot{I}^2(\vec{r}_2 t).\mathrm{sign}\left(I_j^2(\vec{r}_2,t)\right)\rangle}{\langle I_j^2(\vec{r}_2,t).\mathrm{sign}\left(I_j^2(\vec{r}_2,t)\right)\rangle}
$$

The "speckle-velocity crosscorrelation function"

$\langle v_i^1(t) v_j^2(t+\tau)\rangle$ is according to the theory scetched above directly proportional to a temporal correlation function of two different linear combinations of components of the velocity gradient tensor. If a sufficient number of these correlation functions for different scattering geometries is measured one can calculate any desired auto- or cross-correlation function of gradient tensor elements.

For a general flow there are 9 auto- and 36 independent crosscorrelation functions. These numbers decrease if the flow under consideration has symmetry properties. For an incompressible fluid the numbers reduce to 8 and 28, respectively. This is due to the relation

$$
\Gamma_{xx} + \Gamma_{yy} + \Gamma_{zz} = 0 \tag{7}
$$

A channel flow which can approximately be viewed as two dimensional in the x-y-plane (the mean flow velocity is in the x-direction) has furthermore the following symmetries:

.invariance against reflexion at the x-y-plane

.invariance against translation in x-direction

.invariance against translation in z-direction

These symmetry properties reduce the number of independent correlation functions to 8 and 10, respectively.

3. The measurement scheme. The derivatives of the light intensity which appear in equations (4) and (6) cannot be measured directly. Consequently one is forced to measure spatial and temporal intensity differences, which requires two detectors at different locations which perform sucessive intensity measurements.

The time interval between sucessive intensity measuremens (the so called sample time) must be chosen in a way that intensity changes during that time are only small. This requires the possibility to adjust the sample time according to the pattern velocity. In our experiments we used sample times from 5 - 50 μs depending on the flow properties. The mutual distance of the two detectors was about one tenth of the speckle size in the screen plane, this assures sufficiently small spatial intensity differences.

Gauge measurements with these parameters on the flow within a rotating cuvette show a most 5% deviations of the values obtained with our measurement scheme from the true gradient values.

The intensity measurements were done with photomultipliers in the photon count mode. A measure for the average intensity within a sample time is then the number of detected photons during this time. These numbers were counted electronically and after each sample time fed to the memory of a PC. After a certain time the data transfer to the PC was interrupted, and the calculation of the correlation functions according to expressions derived from equations (4) and (6) was started. This procedure was repeated until the scatter in the calculated correlation functions was considered sufficiently small.

The resulting measurement time varied between 10 and 40 minutes. This amounts to an averaging over a number of sample times of the order of magnitude of 10^8. This number can be reduced considerably if the light intensity is increased. The measurements cited in this paper were performed with an average of detected photons per sample time of not more than 5.

4. The flow. We did the measurements on a flow in an open water channel with the dimensions of 500x18x3 cm^3. All results cited in this paper refer to a Reynolds number R_e of 4200, where R_e is defined by

$$R_e = \frac{u.d}{v} \qquad (8)$$

Here d is the channel with, u the centre line velocity, and v the Kinematic viscosity. The measurements were done at 3.5 m from the inlet and 16 cm above the channel bottom. The critical Reynolds number of this channel was determined to be 2400. The temperature of the water was controlled to better then 0.5 degrees. The seeding was done with latex particles with a particle size of ≈200 nm.

5. The optical set-up. An argon laser of maximum power of 3 W served as the light source. The laser was run in the TEM_{00} mode at wave lengh of 488nm and a power of 300 mW.

In order to obtain a sufficiently large number of independent correlation functions we built a set-up where the illuminating light wave could be switched by use of a Pockels cell between two antiparallel directions. Both light beams were directed with mirrors and lenses into the flow in a way that the waists of the two beams had equal size and coincided.

The scattered light was observed in two perpendicular directions, which were also perpendicular to the directions of the illuminating light beams. The resulting scattering volume was almost spherical of a diameter of 150 πm.

With this set-up we could measure 8 auto- and 16 crosscorrelation functions of the speckle velocities in the two observation planes. This was sufficient to calculate all non-vanishing correlations functions of the components of the velocity gradient tensor.

For measurements at different distances from the channel wall we could translate the whole optical set-up with respect to the flow channel.

6. Results. Examples of measured correlation functions of the speckle velocity are shown in figures 2 and 3.

From a set of 24 correlation functions measured for one definite position of the scattering volume within the flow we calculated the correlation functions of the components of the velocity gradient tensor. It should be noted that this calculation is not straight forward since not all of the measured correlation functions are linear independent, and therefore, the data are not compatible with each other. We did the calculation by use of least mean square fit procedure.

Examples of these correlation functions are shown in the figure 4, where the autocorrelation functions of the z-component of the vorticity for different distances y^+ from the channel wall are plotted.

In this figure as well as in the following ones we used scaled variables, which can be calculated from the kinematic viscosity v of the fluid and the so called wall gradient γw. This is the derivative of the mean flow velocity at the wall in the direction normal to the wall. One then has the following scaled variables:

$$\vec{u}^+ = \frac{\vec{u}}{\sqrt{v \cdot \gamma w}} \tag{9}$$

$$\tau^+ = \tau \cdot \gamma w$$

$$y^+ = y \cdot \sqrt{\frac{\gamma w}{v}}$$

From the zeroth channels of the correlation functions we calculated the turbulent energy dissipation ε^+,

$$\varepsilon^+ = \sum_{i,j=1}^{3} \left\langle \left(\frac{du_i^+}{dx_j^+} + \frac{du_j^+}{dx_i^+} \right)^2 \right\rangle \tag{10}$$

and the enstrophy ω^+,

$$\omega^+ = \frac{1}{2} \cdot \sum_{i,j=1}^{3} \left\langle \left(\frac{du_i^+}{dx_j^+} - \frac{du_j^+}{dx_i^+} \right)^2 \right\rangle. \tag{11}$$

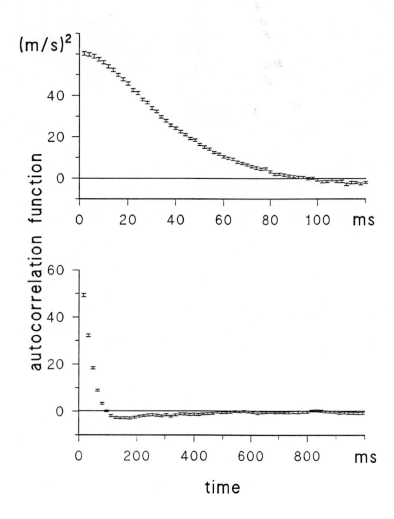

Figure 2: Example of a measured autocorrelation function of the speckle velocity. Note the different scales!

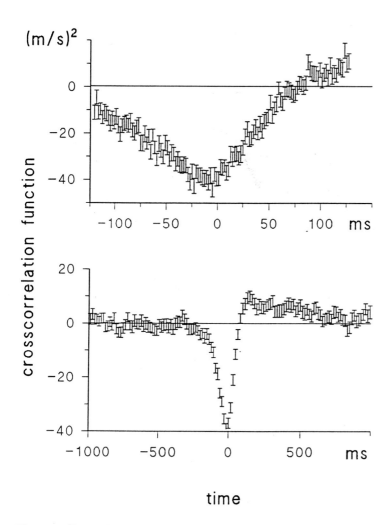

Figure 3: Example of a measured crosscorrelation function of the speckle velocity. Note the different scales!

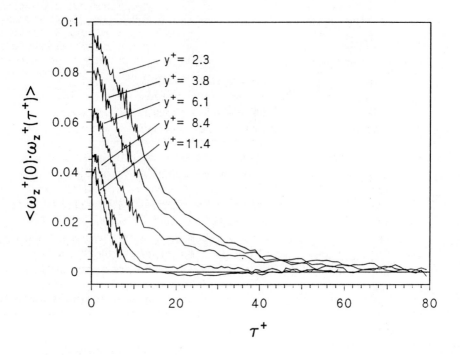

Figure 4: Autocorrelation functions of the z-component of the vorticity for different wall distances y^+.

In these two equations the symbols u_i^+ denote the flutuating parts of the velocities according to the Reynolds decomposition.

The results are shown in the figures 5 and 6.

The examples shown in this paper clearly demonstate the power of the differential speckle strophometry for measurements of correlations and correlation function of velocity gradient components. It is also possible to use the effect of speckle motion for measurements of momentaneous velocity gradient values, as was shown in [6]. It is furthermore worth noting that our method is especially suited for measurements at locations in flows where the velocity is low, as for instance near walls or stagnation points. This is due to the fact that in these cases the speckle boiling is very slow. The method can also be extended to measure two-point correlation functions, measurements of this type are presently being performed.

References

[1] Keveloh, C. and Staude, W. 1988, he measurement of velocity gradients in laminar and turbulent flow, J. Phys. D: Appl. Phys. vol.21, pp.237-245

[2] de Gennes, P.G. 1977, Principe de nouvelle mesure sur les écoulements par échauffements optique localisés, J. Physique Lett. vol.38, pp.L-1 - L-3

[3] Vukoslavcevic, P., Wallace, J.M., and Balint, J.-L. 1991, The velocity and vorticity vector fields of a turbulent boundary layer. Part 1: Simultaneous measurement by hot wire anemometry, Part 2: Statistical properties, J. Fluid. Mech., vol. 282, pp.25-86

[4] Breyer, H., Kriegs, H., Schulz, R., and Staude, W.1993, The measurement of velocity gradients in fluid flow by laser light scattering, Part2: Statistical properties of gradients, Exp. in Fluids, vol.15, pp.240-246

[5] Staude, W.1996, The velocity of random time-dependent patterns and its experimental determinations, J. Phys. D, vol.29, pp.307-314

[6] Kriegs, H. and Staude, W.1995, A laser pulse technique for the measurement of time-resolved velocity gradients in fluid flow, Meas.Sci.Technol vol.6, pp.653-662

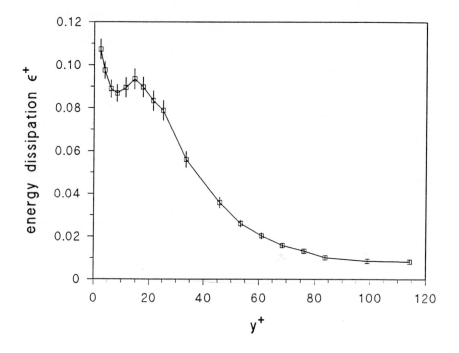

Figure 5: The turbulent energy dissipation in a channel flow as a function of the wall distance y^+. All quantities scaled according to equations (9)

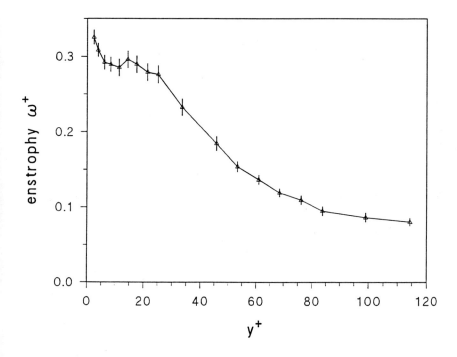

Figure 6: The enstrophy in a channel flow as a function of the wall distance y^+. All quantities scaled according to equations (9)

Chapter II

Instrumentation Developments

II.1 Molecular Tagging Diagnostics for the Study of Kinematics and Mixing in Liquid Phase Flows

M. M. Koochesfahani[1], R. K. Cohn[1], C. P. Gendrich[1], and D G. Nocera[2]

Departments of Mechanical Engineering[1] and Chemistry[2]
Michigan State University, East Lansing, MI48824 USA

Abstract. This work reports our recent developments of novel techniques for velocimetry and studies of the Lagrangian evolution of mixing interfaces based on molecular tagging approaches. These developments take advantage of a class of newly engineered phosphorescent supramolecules that are water soluble. Previous implementations using photochromic molecules and caged fluorescein are briefly discussed and compared. The application of molecular tagging velocimetry is demonstrated in several flow fields, including examples which illustrate its capability in flows with significant out-of-plane motion and its, potential for simultaneous passive scalar and velocity measurements.

Keywords. molecular tagging, optical diagnostics, supramolecules, velocimetry

1. Introduction

The capability to tag a portion of a flow non-intrusively and observe its subsequent evolution offers new possibilities for velocimetry and the study of entrainment and mixing in fluid flows. Chemical compounds with long luminescence lifetimes are needed for this approach. Current laser induced fluorescence (LIF) techniques are generally not suitable for this purpose because of the short lifetime of the fluorescence process (τ = a few nanoseconds).

The majority of past efforts in the use of molecular tagging have concentrated on flow velocimetry. Regardless of the details of the photophysics in each molecular design, all of these efforts can be characterized under a broader common heading of **M**olecular **T**agging **V**elocimetry (MTV). In this method of velocimetry, the flowing medium is premixed with molecules that can be turned into long lifetime tracers upon excitation by photons. Typically a pulsed laser is used to „tag" small regions of interest. The tagged regions are imaged at two successive times within the lifetime on the tracer. The measured vector provides the estimate of the velocity vector.

Past implementations of MTV in liquid phase flows primarily include the use of photochromic molecules (Popovich & Hummel 1967; Falco & Chu 1987; Ojha, et al. 1989; and Chu & Liao 1992, among others), caged fluorescein (Lempert, et al. 1995; Harris, et al. 1996a & 1996b; Biage, et al. 1996), and specially engineered phosphorescent supramolecules (Koochesfahani, et al. 1993; Stier, 1994; Gendrich, et al. 1994; Hill & Klewicki, 1995, Cohn, et al. 1995, and Gendrich & Koochesfahani 1996). In gas phase flows, the fluorescence of excited-state oxygen (Miles, et al. 1987; 1989; 1993) and the phosphorescence of biacetyl (Hilbert & Falco 1991, Stier, et al. 1995) have typically been utilized. The review by Falco & Nocera (1993) discusses some of these and other efforts known at that time.

The purpose of this work is to give a brief summary of various MTV implementations and their chemical mechanisms, followed by our recent contributions to improve and extend this technique. This includes a discussion of the ideas behind the design of the new phosphorescent complexes and our approach to imaging and tracking the tagged regions. Examples of the application of MTV to different flow fields will be provided, one of which highlights the ability to make measurements when strong out-of-plane motions are present. Other examples will illustrate the potential for simultaneous passive scalar and velocity measurements.

2. Chemical Mechanisms

A molecular complex is suitable for molecular tagging applications if its lifetime as a tracer is long enough relative to the flow convection time scale to allow sufficient displacement of the tagged regions. The discussion in this work focuses primarily on liquid phase applications, however it is worthwhile to briefly note some highlights of work in gas phase. The use of excited-state oxygen fluorescence, pioneered by Miles, et al. (1987, 1989, 1993) under the acronym RELIEF, is the only tagging method currently available in an oxygen environment (e.g. air). The lifetime of the tracer (vibrationally excited O_2) is of order 100 μs, making it suitable for high speed flows. The phosphorescence of biacetyl, which has a reported lifetime of order 1 ms, has also been used for velocimetry (Hilbert & Falco 1991, Stier, et al. 1995). However, its use is limited to oxygen-free environments due to the phosphorescence quenching by oxygen.

2.1. Photochromics

In a photochromic process a molecule M is excited to produce a high energy form of M, designated M', which has a different absorption spectrum giving rise to a color change (e.g. from clear to dark blue). The long lifetime tracer is the newly produced M', which persists for several seconds to minutes. The nonradiative conversion from M to M', i.e. the tagging process, occurs rapidly (within the duration of a few nanosecond long laser pulse). The photochromic process is reversible; M' thermally converts back to M over time, therefore the chemical is reusable. Photochromic dyes are generally insoluble in water, so organic liquids such as kerosene are typically used as the flowing medium. The use of photochromic chemicals requires two photon sources; typically a UV laser (e.g. $\lambda = 351$ nm from an excimer laser) to induce the color change and a white light source to interrogate the tagged regions.

Hummel and his group (e.g. Popovich & Hummel 1967, Ojha, et al. 1989) originated the use of photochromic chemicals as a velocity measurement tool by tagging the flow along single lines. Significant improvements were made in the pioneering work of Falco & Chu (1987), who used a laser grid to tag the flow and coined the acronym LIPA (**L**aser **I**nduced **P**hotochemical **A**nemometry). The distinction between these different tagging approaches is a very important one, which will be discussed in Section 3.

Some of the advantages of photochromics (long lifetime, reusable) are offset by the need to use special fluids such as kerosene. It is possible to ease this restriction by making chemical modifications to enable many photochromic dyes to dissolve in water. Such examples have been reported by Yurechko & Ryazantsev (1991), and Douglas, et al. (1991). The most significant drawback in using photochromic chemicals is that the image is produced by a change in absorbence, thereby requiring a measurement of the difference between incident and transmitted light. Emitted light (against a black background) is more easily and accurately detected than transmitted light; consequently, images based on luminescence are better suited to MTV applications. Despite some of these difficulties, photochromics are being used very effectively to advance the understanding of flow physics (Chu & Liao, 1992; Chu, et al. 1993).

2.2. Caged Fluorescein

In this compound a chemical group is attached to fluorescein in order to render it non-fluorescent. The caging group is removed upon absorption of UV photons ($\lambda \cong 350$ nm), thereby creating regular fluorescein which fluoresces with very high quantum efficiency. In this case, the long lifetime tracer is the uncaged fluorescein, which persists for a very long time and can be interrogated at the time of interest through its luminescence upon re-irradiation. Two sources of photons are therefore needed, one to break the cage and the other to excite fluorescence. The use of caged fluorescein and similar compounds for molecular tagging velocimetry was first introduced by Lempert, et al. (1995) under the acronym PHANTOMM (**PH**oto-**A**ctivated **N**on-intrusive **T**racking **O**f **M**olecular **M**otion). In both this work and several recent novel applications of it (Harris, et al. 1996a & 1996b; Biage, et al. 1996), laser line tagging is used (see Section 3 for further discussion).

There are two aspects of caged fluorescein that one must be aware of in designing an experiment. The cage-breaking process is irreversible; each caged molecule can be used only once. The cage-breaking process is not rapid and occurs with a time constant of the order of a few milliseconds. This delay between laser tagging and generation of enough fluorescein to obtain an image with sufficient signal/ noise will dictate the fastest flow speeds that can be accommodated. Gendrich, et al. (1996) have recently presented a technique for overcoming this problem based on a two-detector approach. On the other hand, very slow speed flows can be handled with ease considering the very long lifetime (practically infinite) of the uncaged fluorescein. It is anticipated that future research will lead to the design of new generations of caged fluorescent molecules without these drawbacks.

2.3. Phosphorescent Supramolecules

In using a phosphorescent compound for molecular tagging, excitation by photons is used to produce a long-lived excited state which is interrogated through its phosphorescence emission as it radiatively returns to its ground state. The long lifetime tracer is the excited state molecule itself. In this case only one source of photons is needed, and the excitation/emission process is reversible. The difficulty is that the long-lived excited state (i.e. phosphorescence) suffers from O_2 and H_2O quenching, and as a result, suitable molecular complexes have not been available until recently.

New findings by Nocera and his group (Ponce, et al. 1993, Mortellaro & Nocera 1996, Hartmann, et al. 1996) show that supramolecules may be designed that exhibit long-lived phosphorescence, which is not quenched. A successful design by Ponce, et al. (1993) indicates that the quenching of a lumophore can be stopped, and the phosphorescence emission recovered, by mixing certain alcohols with an aqueous solution of a cyclodextrin (CD) cup that contains the lumophore. Cyclodextrins are molecules constructed from sugars connected in a head-to-tail arrangement. The molecule is cup-shaped with its size determined by the number of sugars in the structure. The CD used in our application is Gß-CD, which is constructed of 7 glucose subunits, resulting in an outer cup dimension of 15.3 Å and an inner cup cavity dimension of 7.8 Å. The size of the cavity is important for effective binding to the guest molecule, in our case 1-bromonaphthalene (1-BrNp), which is the lumophore.

The long-lived, green phosphorescence (lifetime $\tau \approx 5$ ms) of 1-BrNp is efficiently quenched by oxygen, even when the lumophore is inside a CD cup. The phosphorescence can be recovered upon adding an alcohol (ROH) to the solution as shown in Figure 1. Detailed studies of Ponce, et al. (1993) and Hartmann, et al. (1996) show that a ternary complex (1-BrNp · Gß-CD · ROH) is formed, where the alcohol hydrogen bonds to the rim of the CD cup and acts as its lid, thereby shielding 1-BrNp from oxygen. The phosphorescence enhancement can be very large, approaching 10^4-10^5, depending on the fit of the alcohol lid to the CD cup. For the studies described in this work, cyclohexanol was used as the alcohol.

When using this phosphorescent complex, it is useful to recall that luminescence lifetime refers to the time when the emission has decayed to 37% (e^{-1}) of its initial intensity. The actual usable delay time between laser tagging and interrogation can be considerably longer and is dictated by the type of detection used. We have, for example, measured flows with delays of up to 30 ms using image-intensified cameras. Nevertheless, the lifetime of this molecular design is not suitable for very low speed flows. Such cases are best investigated using caged fluorescein or photochromics.

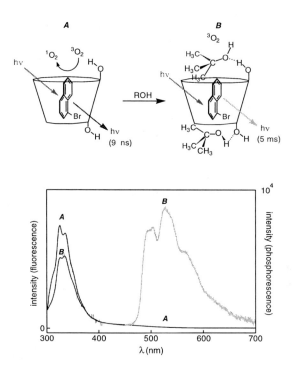

Fig. 1. Emission from 1-BrNp in Gβ-CD (A) Only blue fluorescence is exhibited in the absence of an appropriate alcohol (ROH); (B) A bright green phosphorescence plus the initial fluorescence is seen upon the addition of ROH, which prevents the quenching of 1-BrNp phosphorescence by O_2.

Depending on the placement of the three components of the phosphorescent complex in the flow, we have devised methods for (1) purely velocimetry, (2) molecularly tagging a passive scalar mixing region and monitoring its Lagrangian evolution, and (3) molecularly tagging a chemical reaction interface between two streams and observing its Lagrangian evolution. The first two of these applications are described in this work. Details of the experimental implementations for all three applications can be found in Gendrich, et al. (1996).

3. Tagging, Detection, and Processing Schemes

3.1. Tagging

Tagging the flow along single lines was originally used by Hummel and his group (e.g. Popovich & Hummel 1967, Ojha, et al. 1989). The velocity is determined from the displacement of the tagged lines in much the same manner as using hydrogen bubble lines generated by a wire (for example, see Lu & Smith, 1985). Laser line tagging is still the only method used to date in the works utilizing caged fluo-

130

rescein (Lempert, et al. 1995; Harris, et al. 1996 & 1996b; Biage, et al. 1996) and excited-state oxygen fluorescence (Miles, et al. 1987, 1989, 1993). It is very important to recognize that line tagging allows the measurement of only one component of velocity, that normal to the tagged line. In addition, the estimate of this velocity component has an inherent error associated with it. Following the analysis of Hill & Klewicki (1995), and referring to Figure 2, this error can be cast in the form

$$\frac{\Delta u}{u} = \tan \theta \; \frac{\partial u}{\partial y} \; \Delta t$$

In this expression, u is the estimated velocity component normal to the tagged line, $\Delta u = u_{actual} - u$ is the error in the estimated velocity, θ is the local flow angle given by $\tan \theta = v / u$ with v being the flow velocity parallel to the tagged line, and Δt is the time delay between tagging and interrogation. Clearly an *a priori* knowledge of the flow field is necessary in order to provide an estimate of the error. It can be observed, however, that this inherent error is identically zero only in flows where the velocity component v along the tagged line is zero (i.e. unidirectional flows) or where the velocity gradient $\partial u/\partial y = 0$. In a general flow field where these constraints are not met, the error can be reduced by decreasing the delay time Δt, but it cannot be made arbitrarily small, since Δt has to be large enough for the resulting displacement of the tagged line to be measured with adequate accuracy. While keeping these issues in mind, it is sometimes possible to take advantage of an *a priori* knowledge of the flow field under investigation to design the experimental parameters such that the inherent error discussed here becomes minimal compared to other measurement errors. In cases where this can be done (see Hill & Klewicki 1995, for example) the line tagging approach can provide reliable information on one component of the velocity vector.

In order to measure two components of the velocity in a plane unambiguously, the luminescence intensity field from a tagged region must have spatial gradients along two, preferably orthogonal, directions. For single-point velocimetry, this is

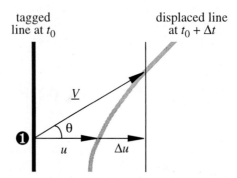

Fig. 2. The velocity at point ❶ is $\underline{V}$. Using the line center displacement gives a lateral velocity estimate u, with error Δu. The vertical velocity component cannot be estimated with the flow tagged in this manner.

easily achieved using a pair of crossing laser beams; a grid of intersecting laser lines allows multi-point velocity measurements. Use of this tagging scheme was first suggested by D'Arco, et al. (1982), and it was later improved upon and utilized by Falco & Chu (1987). An example of a region in a flow tagged in this manner using our new phosphorescent supramolecules, and the same region interrogated subsequently at two different delay times, is depicted in Figure 3. The flow field is that generated by a vortex ring impacting a solid wall at normal incidence, which will be discussed as one of the examples in Section 4. The superposition of the image at the longer time delay (Figure 3c) onto the initially tagged image (Figure 3a), shown in Figure 4, serves to highlight the velocity error inherent in the line tagging approach discussed earlier. The image at the shorter time delay in Figure 3b is more typical of that used in our velocimetry applications.

As will be seen later, laser grid tagging is only a special case of a more generalized approach to induce a spatially non-uniform luminescence intensity in a tagged region. For example, the non-uniform passive scalar concentration field typical of most turbulent flows can sometimes be used as a natural source of luminescence non-uniformity without the need for grid illumination.

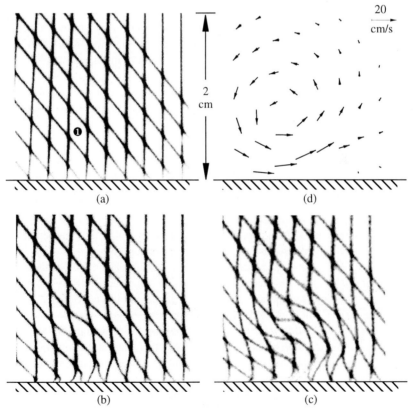

Fig. 3. (a) Undistorted grid. Distorted grid after two different delays: (b) $\Delta t = 5$ ms. (c) $\Delta t = 19$ ms. (d) The velocity field resulting from correlating (a) with (b).

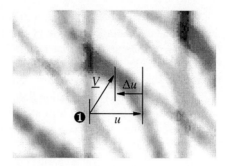

Fig. 4. The superposition of Figure 3c (dark grey) over Figure 3a (light grey) in the neighborhood of the point marked ❶ in Figure 3a. The notation is the same as in Figure 2.

3.2. Detection

Depending on the specific requirements of a particular application, a variety of image acquisition methods have been used in the past including cameras recording on film, CCD cameras, and gated image-intensified cameras. The common element among all previous studies is that a single detector is used; the initial tagging pattern is recorded once, usually at the beginning of the experiment, and then the delayed images are acquired. The implicit assumption in this approach is that the initial tagging pattern remains spatially invariant throughout the experiment. Because current processing schemes (see Section 3.3) measure the displacement of the tagged regions with sub-pixel accuracy, small variations in the initial pattern will be misinterpreted as flow velocity fluctuations. We have improved the accuracy of the MTV technique by employing a two-detector imaging scheme.

The experimental arrangement, shown in Figure 5, involves a link between the pulsed excimer laser ($\lambda = 308$ nm) and two image detectors through a digital delay generator. Immediately after the laser fires, the first detector records an image of the tagged flow, and after a prescribed time delay Δt, the second detector records a delayed image. These images are digitally acquired in real time by two separate acquisition systems. The advantage of this arrangement is that any spatial "wandering" of the tagged regions (e.g. due to laser beam pointing instability, vibration of the optics, etc.) does not contribute to error in the measurement of the displacement of these regions. Furthermore, the two-detector setup is clearly indispensable when the intensity field in the initial tagging pattern cannot be assumed, for example in the case of a non-uniform scalar mixing field.

The data shown in this work were acquired using the arrangement shown in Figure 5 in conjunction with a variety of CCD detectors (interlaced cameras, electronically-shuttered frame transfer cameras, and gated image-intensified cameras) depending on the optical/imaging requirements unique to the particular experiment. Each image was digitized to 8 bits into a 512 x 512 pixel array. Time series data were acquired at a rate of 30 image pairs per second directly onto a 10-GB hard disk array.

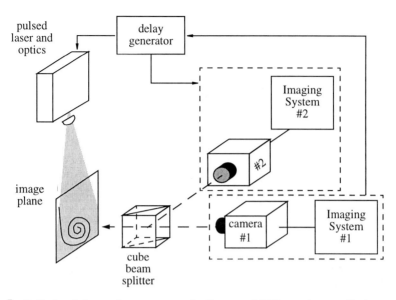

Fig. 5. Optical and electronic arrangement for 2-camera MTV experiments. Both cameras view the same image plane through the cube beam splitter. Synchronization between the two cameras and laser is provided by a digital delay generator.

3.3. Processing

The traditional method for finding the displacement of tagged lines or grids has been to locate the center of each line through various techniques. Most of the recent techniques use the best fit to an assumed laser line shape, for example, a gaussian intensity distribution. We are not aware of a systematic statistical study of the performance of this approach while considering the effects of experimental parameters such as image contrast, signal to noise ratio, etc. The recent study of Hill & Klewicki (1995) reports the accuracy in determining the displacement vector to be ±0.35 pixel rms. Even though efforts are under way to improve the accuracy further using better algorithms, the basic premise behind this approach is the use of a known line shape for the tagged line, which may not be known *a priori* in some situations, due to a variety of reasons associated with laser beam transmission through a flowing medium, bleaching effects, etc.

We have taken a different approach in an attempt to implement a generalized scheme that is independent of the details within a tagged region, and can accommodate arbitrary tagging patterns including those due to non-uniform scalar mixing fields. The displacement of the tagged regions is determined using a direct digital spatial correlation technique. The example provided in Figure 6 illustrates a region in an actual experiment tagged by a laser grid, the tagged region at a later time, and the corresponding spatial correlation coefficient field for one of the grid crossings. A well-defined correlation peak occurs at the location corresponding to the displacement of the tagged region by the flow; the displacement peak is located to sub-

134

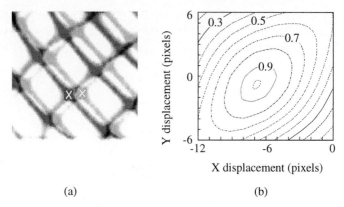

Y displacement (pixels)

X displacement (pixels)

(a)   (b)

Fig. 6. Experimental MTV grid. (a) ▨ lines are the grid at t = 0; ▇ lines are the grid after a 6 ms delay. (b) Correlation coefficient contours for the indicated intersection in (a)

pixel accuracy using a multi-dimensional polynomial fit. This procedure is similar to what could be used in DPIV processing of particle image pairs. One advantage of our processing technique over traditional line-center methods is robustness to the presence of noise due to the averaging process inherent in the correlation procedure. Based on both experiments and an extensive statistical study on the performance of this correlation approach, we have found that we can typically measure the displacement of the tagged regions with a 95% confidence limit of ±0.1 subpixel accuracy (i.e. 95% of the displacement measurements are accurate to better than 0.1 pixel). This corresponds to an rms accuracy of ±0.05 pixel, assuming a Gaussian distribution for error. The details can be found in Gendrich & Koochesfahani (1996).

It should be noted that the spatial correlation method just described, though of more general utility than the line-center methods, may lead to a somewhat degraded spatial resolution. The contrast is analogous to that between PIV (using a group of seed particles) versus individual particle tracking.

4. Examples of MTV Measurements

When the three components of the phosphorescent complex (1-BrNp · Gß-CD · ROH) are premixed with water in the entire flow facility, we obtain a technique intended solely for velocity measurements. Since the flowing medium is homogeneously mixed, a pulsed laser grid pattern is used to molecularly tag the regions of interest.

An example of the type of time series data obtained from a single MTV grid intersection is shown in Figure 7. This figure illustrates the time evolution of two components of the velocity vector at a point near the axis of symmetry of a passing vortex ring. We note that this represents raw data (i.e. no filtering or smoothing) from one of many grid points where simultaneous velocity data were obtained. For an ideal vortex ring, at this location in the flow the v-component is expected to remain at zero and the u-component to smoothly increase to a maximum and subsequently drop to zero as the ring passes by. This behavior is captured by the data in Figure 7.

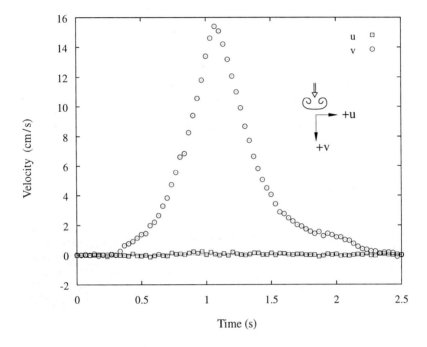

Fig. 7. The time evolution of (u, v) velocity components measured by MTV near the center of a passing vortex ring.

The next example provides a sample of the instantaneous whole-field measurements of two components of the velocity vector over a plane in a study of unsteady boundary layer separation in a vortex-wall interaction. A vortex ring approaching a wall at normal incidence generates an unsteady adverse pressure gradient on the wall, which results in boundary layer separation and the formation of a secondary vortex. An LIF visualization of this flow right after the formation of the secondary vortex is depicted in Figure 8(a) (only the left half of the flow is included). Figure 8(b) shows the corresponding instantaneous MTV velocity data (1 mm grid spacing with first grid 0.5 mm away from the wall) and the computed vorticity field. The data sequence in Figure 9 is selected from a much longer time series showing an enlarged view of the details of the velocity and vorticity fields during the boundary layer separation process. Except for the inherent smoothing involved in mapping the velocity data from the irregular MTV grid onto a regular grid, the data shown in Figures 8 and 9 represent raw data, i.e. no attempt was made to remove a "bad" vector, replace it, or apply spatial filtering.

136

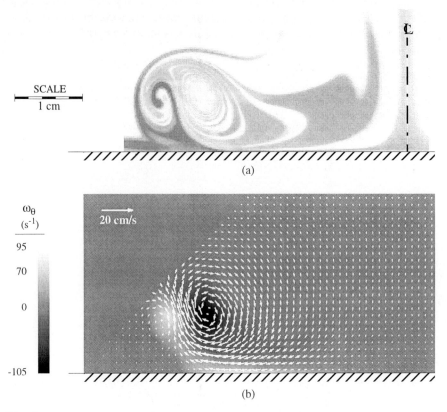

SCALE

1 cm

(a)

ω_θ
(s^{-1})

95

70

0

-105

20 cm/s

(b)

Fig. 8. (a) LIF image of the left-half of a downward moving vortex ring. The ring and wall-layer fluids are marked by green- and red-emitting laser dyes (the light grey and dark grey colors above), respectively. (b) Velocity and vorticity fields measured using MTV.

At this stage, a comparison between the MTV and PIV methods is warranted. The MTV technique can be thought of as the *molecular* version of PIV with two major advantages. First, since particles are replaced by molecules dissolved in the flowing medium, the problems related to particles tracking the flow, density mismatch between particles and the fluid, particle seeding density, etc. are all absent from MTV. Second, the performance and accuracy of PIV in measuring the in-plane velocity vector can degrade in highly 3-D velocity fields due to particle motion in/out of the plane of the laser sheet. In MTV, the measurement of in-plane velocity vectors is quite insensitive to the out-of-plane velocity component. The reason is that only the tagged region is luminescent; the motion of any molecules located outside of this tagged region into or out of the image field of view does not contaminate the spatial correlation process, since those molecules are not luminescent.

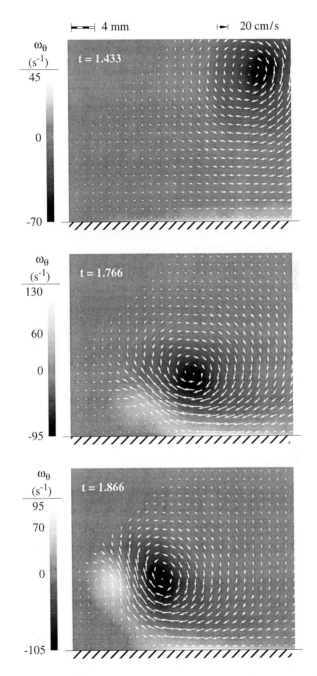

Fig. 9. Velocity and vorticity details of the ring/wall interaction measured using MTV. Only the left-half of the flow field is shown.

138

The latter advantage of MTV offers a unique capability for velocimetry in highly
three-dimensional flows, which we demonstrate in an application to a forced wake.
We have previously reported that forcing a low Reynolds number 2-D wake downstream of a splitter plate can lead to a highly three-dimensional flow and a large
increase in mixing (MacKinnon & Koochesfahani 1993, Koochesfahani, et al.
1994). Figure 10 illustrates preliminary instantaneous velocity vectors (v, w components) in the cross-stream (y-z) plane at three different instances during the per-
turbation period at a downstream station x=13cm from the splitter plate. The
estimated streamwise vorticity fields at two of those instances are included in Figure 11. The region shown in these figures covers an area 1.8 cm × 4 cm. As before,
except for remapping onto a regular grid, these are raw data. It is important to recognize that the mean streamwise flow direction is actually out of the page in this
flow field. The high degree of three-dimensionality and the spatial structure of the
streamwise vorticity are apparent. The maximum cross-stream velocity in Figure
10 is about 40% of the wake freestream velocity.

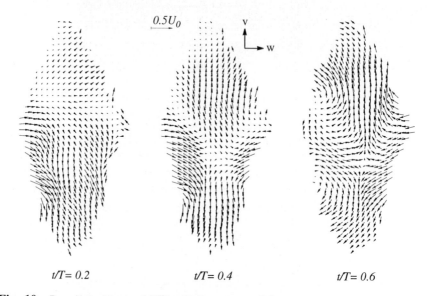

Fig. 10. Raw instantaneous MTV velocity vectors of the cross-stream flow in a forced
wake at three different instances in the forcing cycle (T is the forcing period). Note that the
mean flow is out of the page.

5. Passive Scalar Mixing Dynamics

In the non-reacting (passive scalar) implementation, the alcohol, CD, and lumophore are premixed in one stream, and the other mixing stream is for example pure
water or contains an alcohol solution. The Lagrangian evolution of the scalar mixing field is then monitored over the luminescence lifetime. An example of this

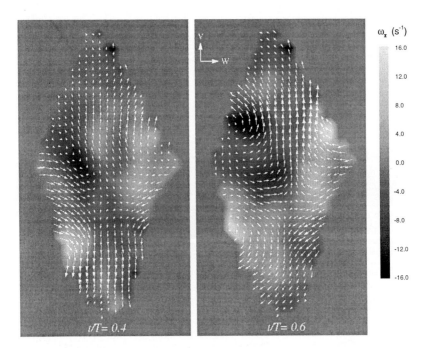

Fig. 11. Instantaneous MTV velocity vectors and streamwise vorticity of the cross-stream flow in a forced wake.

implementation is shown in Figure 12. A vortex ring's fluid is premixed with the alcohol/CD/lumophore solution, and the ambient fluid contains only the alcohol solution. The UV laser is arranged to illuminate a series of parallel "bands" in the flow. Figure 12 shows the molecularly-tagged patches of the vortex ring fluid at the initial time (20 μs after laser firing), marked light grey, and the evolution of the same patches 9 ms later (marked dark grey). The velocity vectors at the corners of the patches are determined using the same direct spatial correlation technique described earlier. Figures 13(a,b) show the velocity vectors in the laboratory frame and the vortex frame, respectively.

Even though the vortex ring example just shown has a rather simple concentration field (i.e. uniform within the tagged patches), it does highlight the potential of a molecular tagging approach for simultaneous concentration (from the first image) and velocity (using image pairs as in Figure 12) measurements in more complex and turbulent flows. In this case the spatially non-uniform scalar concentration field typical of most turbulent flows can be used as a natural source of the required luminescence intensity variation described earlier. Where this can be accomplished, a much simpler optical arrangement with a laser sheet can replace the more involved technique of "writing" a prescribed laser pattern into the flow (e.g. the usual grid pattern). An example is given in Figure 14, illustrating the vortex ring of a starting jet illuminated by a pair of crossing laser sheets. The placement of the chemical components is the same as that in the example of Figure 12.

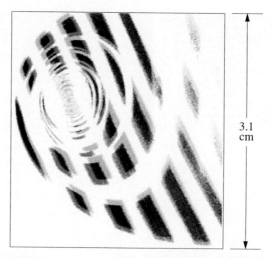

Fig. 12. Lagrangian evolution of molecularly tagged fluid patches in a vortex ring. Only the left half of a downward-moving ring is shown. The ▨ patches are regions imaged 20 µs after laser firing; the ■ patches are the same regions 9 ms later.

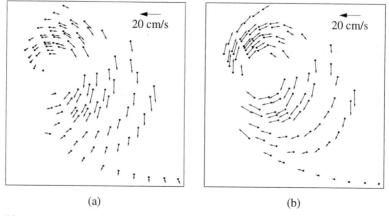

(a)   (b)

Fig. 13. (a) Velocity vectors estimated using the corners of the patches in Figure 12. (b) The velocity field of (a) with the core convection velocity subtracted.

Figure 14(a) is the first image of the tagged flow acquired 20 µs after the laser firing; the second image, not shown here, is acquired 6 ms later. In a manner similar to that already described, the Lagrangian displacement of small regions (source windows) in the first image can be obtained using spatial correlation with "roam windows" from the second image. One example of a suitable source window is illustrated in Figure 14(b) along with the roam window from the second image inFigure 14(c); the dashed black square in the roam window indicates the initial location of the source window. The resulting correlation coefficient field indicates

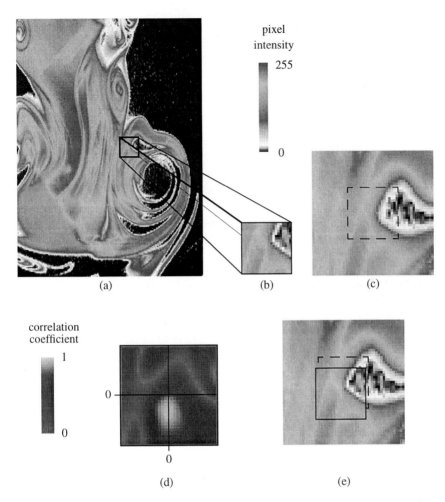

Fig. 14. Correlation technique applied to experimental data from a mixing jet flow field. (a) Earlier image. (b) 21 x 21 pixel source window from (a). (c) 51 x 51 pixel roam window from the second image (not shown here). (d) correlation coefficient field. (e) superposition of source window over roam window.

a well-defined peak whose pixel coordinates are determined to be $(\Delta x, \Delta y) = (-0.9, -5.3)$. Figure 14(e) shows the superposition of (b) over (c), after (b) has been moved to the location of the peak correlation (the solid black square).

It is implicitly understood that the correlation procedure applied to continuously distributed luminescence fields such as Figure 14(a) will yield accurate displacement information in two directions only if sufficient intensity gradients are present within the selected source windows. In the areas of the flow where this requirement is not satisfied, a non-uniform laser illumination can be used instead. The example of Figure 14 again highlights the potential for simultaneous concentration (from the first image) and velocity (using image pairs) measurements. Similar to the discussion earlier in Section 2.3, in cases where monitoring the evolution of the passive scalar field over very long time periods is desired, a much longer lifetime tracer such as caged fluorescein should be used.

6. Conclusions

New developments in molecular tagging diagnostics for velocimetry (MTV) and the study of the Lagrangian evolution of scalar mixing fields have been described. The MTV approach is the molecular counterpart of PIV and as such offers certain advantages. In addition MTV tends to have much reduced sensitivity to out-of-plane motion contaminating in-plane velocity data. This aspect has been demonstrated in an application of MTV to the highly three-dimensional velocity field of a forced wake. The potential of the molecular tagging approach for studying the dynamic evolution of mixing interfaces and combined scalar/velocity measurements has also been demonstrated.

Acknowledgments

This work was supported in part by the Air Force Office of Scientific Research under grants F49620-92-J-0338 and F49620-95-1-0391, and the MRSEC Program of the National Science Foundation Award Number DMR-9400417.

7. References

Biage, M., Harris, S. R., Lempert, W. R., and Smits, A. J. 1996, Quantitative velocity measurements in turbulent Taylor-Couette flow by PHANTOMM flow tagging, Proceedings of the *Eighth International Symposium on Applications of Laser Techniques to Fluid Mechanics*, Lisbon, Portugal, July 8-11, 1996, 15.4.1-15.4.8.

Chu, C. C. and Liao, Y. Y. 1992, A quantitative study of the flow around an impulsively started circular cylinder, *Exp. Fluids*, vol. 13, 137-146.

Chu, C. C., Wang, C. T., and Hsieh, C. H. 1993, An experimental investigation of vortex motions near surfaces, *Phys. Fluids A.*, vol. 5, no. 3, 662-676.

Cohn, R. K., Gendrich, C. P., Mackinnon, C. G., and Koochesfahani, M. M. 1995, Crossflow velocimetry measurements in a wake flow, *Bull. Am. Phys. Soc.*, vol. 40, no. 12, 1962.

D'Arco, A., Charmet, J. C., and Cloitre, M. 1982, Nouvelle technique de marquage d'ecoulement par utilisation de molecules photochromes, *Revue Phys. Appl.*, vol. 17, 89-93.

Douglas, P., Enos, R. D., Azzopardi, B., and Hope, C. B. 1991, Characterisation of a photochromic triarylmethane dye sulphite and its applications to the visualisation of water flows, *7th Int. Topical Meeting on Photoacoustic and Photothermal Phenomena*, The Netherlands, August, 1991.

Falco, R. E. and Chu, C. C. 1987, Measurement of two-dimensional fluid dynamic quantities using a photochromic grid tracing technique, *SPIE*, vol. 814, 706-710.

Falco, R. E. and Nocera, D. G. 1993, Quantitative multipoint measurements and visualization of dense solid-liquid flows using laser induced photochemical anemometry (LIPA), in *Particulate Two-Phase Flow*, Ed. M. C. Rocco; Butterworth-Heinemann, 59-126.

Gendrich, C. P. and Koochesfahani, M. M. 1996, A spatial correlation technique for estimating velocity fields using Molecular Tagging Velocimetry (MTV), *Exp. Fluids*, vol. 22, no. 1, 67-77.

Gendrich, C. P., Koochesfahani, M. M., and Nocera, D. G. 1994, Analysis of molecular tagging velocimetry images for obtaining simultaneous multi-point velocity vectors, *Bull. Am. Phys. Soc.*, vol. 39, no. 9, 1980.

Gendrich, C. P., Koochesfahani, M. M., and Nocera, D. G. 1996, Molecular tagging velocimetry and other novel applications of a new phosphorescent supramolecule, submitted to *Exp. Fluids*.

Harris, S. R., Lempert, W. R., Hersh, L., Burcham, C. L., Saville, A., Miles, R. B., Gee, K., and Haughland, R. P. 1996a, Quantitative measurements on internal circulation in droplets using flow tagging velocimetry, *AIAA J.*, vol. 34, no. 3, 449-454.

Harris, S. R., Miles, R. B., and Lempert, W. R. 1996b, Observations of fluid flow produced in a closed cylinder by a rotating lid using the PHANTOMM (Photo-Activated Non Intrusive Tracking of Molecular Motion) flow tagging technique, Proceedings of the *Eighth International Symposium on Applications of Laser Techniques to Fluid Mechanics*, Lisbon, Portugal, July 8-11, 1996, 15.3.1-15.3.9.

Hartmann, W. K., Gray, M. H. B., Ponce, A., and Nocera, D. G. 1996, Substrate induced phosphorescence from cyclodextrin · lumophore host-guest complexes, *Inorg. Chim. Acta*, vol. 243, 239.

Hilbert, H. S. and Falco, R. E. 1991, Measurements of flows during scavenging in a two-stroke engine, SAE Technical Paper 910671.

Hill, R. B. and Klewicki, J. C. 1995, Data reduction methods for flow tagging velocity measurements, *Exp. Fluids*, vol. 20, no. 3, 142-152.

Koochesfahani, M. M., Gendrich, C. P., and Nocera, D. G. 1993, A new technique for studying the Lagrangian evolution of mixing interfaces in water flows, *Bull. Am. Phys. Soc.*, vol. 38, no. 12, 2287.

Koochesfahani, M. M., Beresh, S. J., and MacKinnon, C. G. 1994, Volumetric visualization of 3-D flow structure in a low Reynolds number forced wake, *Bull. Am. Phys. Soc.*, vol. 39, no. 9, 1947.

Douglas, P., Enos, R. D., Azzopardi, B., and Hope, C. B. 1991, Characterisation of a photochromic triarylmethane dye sulphite and its applications to the visualisation of water flows, *7th Int. Topical Meeting on Photoacoustic and Photothermal Phenomena*, The Netherlands, August, 1991.

Falco, R. E. and Chu, C. C. 1987, Measurement of two-dimensional fluid dynamic quantities using a photochromic grid tracing technique, *SPIE*, vol. 814, 706-710.

Falco, R. E. and Nocera, D. G. 1993, Quantitative multipoint measurements and visualization of dense solid-liquid flows using laser induced photochemical anemometry (LIPA), in *Particulate Two-Phase Flow*, Ed. M. C. Rocco; Butterworth-Heinemann, 59-126.

Gendrich, C. P. and Koochesfahani, M. M. 1996, A spatial correlation technique for estimating velocity fields using Molecular Tagging Velocimetry (MTV), *Exp. Fluids*, vol. 22, no. 1, 67-77.

Gendrich, C. P., Koochesfahani, M. M., and Nocera, D. G. 1994, Analysis of molecular tagging velocimetry images for obtaining simultaneous multi-point velocity vectors, *Bull. Am. Phys. Soc.*, vol. 39, no. 9, 1980.

Gendrich, C. P., Koochesfahani, M. M., and Nocera, D. G. 1996, Molecular tagging velocimetry and other novel applications of a new phosphorescent supramolecule, submitted to *Exp. Fluids*.

Harris, S. R., Lempert, W. R., Hersh, L., Burcham, C. L., Saville, A., Miles, R. B., Gee, K., and Haughland, R. P. 1996a, Quantitative measurements on internal circulation in droplets using flow tagging velocimetry, *AIAA J.*, vol. 34, no. 3, 449-454.

Harris, S. R., Miles, R. B., and Lempert, W. R. 1996b, Observations of fluid flow produced in a closed cylinder by a rotating lid using the PHANTOMM (Photo-Activated Non Intrusive Tracking of Molecular Motion) flow tagging technique, Proceedings of the *Eighth International Symposium on Applications of Laser Techniques to Fluid Mechanics*, Lisbon, Portugal, July 8-11, 1996, 15.3.1-15.3.9.

Hartmann, W. K., Gray, M. H. B., Ponce, A., and Nocera, D. G. 1996, Substrate induced phosphorescence from cyclodextrin · lumophore host-guest complexes, *Inorg. Chim. Acta*, vol. 243, 239.

Hilbert, H. S. and Falco, R. E. 1991, Measurements of flows during scavenging in a two-stroke engine, SAE Technical Paper 910671.

Hill, R. B. and Klewicki, J. C. 1995, Data reduction methods for flow tagging velocity measurements, *Exp. Fluids*, vol. 20, no. 3, 142-152.

Koochesfahani, M. M., Gendrich, C. P., and Nocera, D. G. 1993, A new technique for studying the Lagrangian evolution of mixing interfaces in water flows, *Bull. Am. Phys. Soc.*, vol. 38, no. 12, 2287.

Koochesfahani, M. M., Beresh, S. J., and MacKinnon, C. G. 1994, Volumetric visualization of 3-D flow structure in a low Reynolds number forced wake, *Bull. Am. Phys. Soc.*, vol. 39, no. 9, 1947.

II.2   Advantages of UV Lasers in Laser and Phase Doppler Anemometers for Submicrometer Particles

F. Durst, A. Melling and P. Volkholz

Lehrstuhl für Strömungsmechanik, University of Erlangen-Nürnberg Cauerstr. 4, 91058 Erlangen, Germany

Abstract. For laser Doppler (LDA) and phase Doppler anemometry (PDA) measurements with submicrometer particles ultraviolet lasers with wavelengths between 240nm and 280nm are compared with lasers emitting visible radiation. UV lasers offer a main advantage in the higher signal-to-noise ratio (SNR) obtainable, since the scattered power is significantly increased at shorter wavelengths. Furthermore, for PDA the capability of making size class distinctions in the submicrometer size range is improved, as shorter wavelengths lead to steeper phase difference-to-diameter relationships. In particular, the use of ultraviolet (UV) laser light enables the size class distinction of 0.2μm from even smaller particles, that is not possible when using visible radiation. The feasibility of performing laser Doppler and phase Doppler anemometry measurements with ultraviolet laser light is confirmed by experiments.

1   Introduction

Laser Doppler and phase Doppler anemometers typically use He-Ne or argon ion lasers, i.e the wavelengths 633nm, 514nm, or 488nm. Laser diodes emitting at infrared wavelengths, e.g 820nm, are also frequently used. These wavelengths are well-suited for typical LDA and PDA applications that generally use particles larger than the wavelength, i.e. significantly larger than 1μm. For such particles, the scattering cross section C_{sca}, defined as the ratio of total scattered power P_s^{tot} to laser intensity in the probe volume I_0

$$C_{sca} = \frac{P_s^{tot}}{I_0} \qquad (1)$$

is in approximation proportional to the square of the particle diameter d_p.

$$C_{sca} \propto d_p^2 \qquad (2)$$

Therefore, for particles larger than about 1μm, only the output power, and not the wavelength, of a laser is relevant to the total scattered power. For

146

particles smaller than the wavelength λ, however, the scattering cross section becomes a function also of the wavelength and can be approximated by [2]

$$C_{sca} \propto \left(\frac{d_p}{\lambda}\right)^4 \qquad (3)$$

This abrupt transition from $C_{sca} \propto (d_p/\lambda)^4$ to $C_{sca} \propto d_p^2$ at $d_p \approx \lambda$ is shown in fig. 1, where the scattering cross section is plotted logarithmically as a function of the ratio d_p/λ.

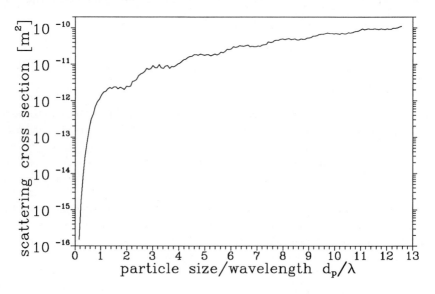

Figure 1: Scattering cross section C_{sca} as a function of the ratio particle size to wavelength d_p/λ (plotted for refractive index $m=1.6$)

In the present work particles as small as $0.1\mu m$ are of interest. Fig. 1 makes clear that even when using a powerful argon ion laser at $\lambda = 514nm$, the scattered power from a $0.1\mu m$ $(d_p/\lambda = 0.2)$ particle becomes very small, making a reduction of the wavelength necessary. This paper shows that the use of ultraviolet (UV) laser light is a logical and viable alternative.

The rest of the paper discusses the theory of UV-based LDA and PDA-systems and presents experimental results that demonstrate the feasibility of LDA and PDA with ultraviolet light. The theory presented in section 2 compares the SNR of Doppler signals from $0.1\mu m$ particles when using green and UV light. Also a comparison of the phase-diameter relationships for green and UV light in the submicrometer range is made, particularly with regard to size discrimination between 0.1 and $0.2\mu m$ particles. Section 3 discusses the experimental setup that was used to perform verification measurements alternately with green and UV-light. Measured SNR values for Doppler signals from $0.1\mu m$ particles are presented and compared with the theoretically expected values. Finally, in section 4, preliminary PDA-measurements of 0.1 and 0.5μm particles show that PDA using UV-light is possible in the submicrometer size range, but also indicate where improvements are necessary.

2 Theory of LDA and PDA for submicrometer particles; comparison green-UV

In this section a detailed analysis shows how using UV (257nm) light for submicrometer particles will improve

1. the SNR of the Doppler signals.

2. the phase difference-to-diameter relationship, referred to as the phase-diameter relationship in the following.

3. the maximum particle concentration at which PDA measurements can still be performed.

2.1 Improvement of the SNR of Doppler signals obtained from submicrometer particles using UV-light

For submicrometer particles down to 0.1μm, which are of interest in this paper, the scattered power and the SNR become so small, that it is very difficult to obtain processable Doppler signals.

The aim in this section is, therefore, to estimate theoretically the SNR of Doppler signals as a function of scattered power and signal frequency only, when using a photomultiplier (PM)[1]. This is done by calculating the combined SNR due to the photomultiplier and to quantum fluctuations in the light scattering process.

Theoretical SNR estimate for photomultipliers

The SNR of a photomultiplier (PM) can be estimated by dividing the mean signal current $\bar{i}_s$ by the root of the sum of mean square shot noise $\overline{i_n^2}$, mean dark count noise $\overline{i_d^2}$, and mean square thermal noise $\overline{i_T^2}$.

$$SNR_{PM} = \frac{\bar{i}_s}{\sqrt{\overline{i_n^2} + \overline{i_d^2} + \overline{i_T^2}}} \qquad (4)$$

The mean signal current $\bar{i}_s$ produced by a Doppler burst is

$$\bar{i}_s = \frac{G}{\sqrt{2}} \cdot V \cdot \bar{i}_c \qquad (5)$$

---

[1] Avalanche Photodiodes (APD) are not considered here, since they cannot detect UV-light.

G: gain of PM

V: signal visibility, defined as $V = \frac{I_{max} - I_{min}}{I_{max} + I_{min}}$, where I_{max} and I_{min} are the maximum and minimum intensities detected by the photomultiplier, respectively.

The factor $1/\sqrt{2}$ accounts for the fact that the Doppler signal is an oscillating signal.

$\bar{i}_c$ is the mean photocathode current and is given by

$$\bar{i}_c = \frac{P_s}{h \cdot \nu} \cdot \eta_Q \cdot e \tag{6}$$

P_s: power scattered on to photocathode
h: Planck's constant
ν: laser light frequency
η_Q: quantum efficiency of photocathode
e: electron charge

The term $P_s \cdot \eta_Q / (h \cdot \nu)$ represents the number of photons detected per unit time.

The mean square shot noise $\bar{i_n^2}$ of a PM is given by

$$\bar{i_n^2} = G^2 \cdot 2e \cdot \bar{i}_c \cdot \Delta f \tag{7}$$

Δf is the bandwidth of the electronics and is usually equivalent to the Doppler frequency f_D of the maximum velocity to be measured. Similarly, the mean dark count noise $\bar{i_d^2}$ is given by

$$\bar{i_d^2} = G^2 \cdot 2e \cdot \dot{n}_d \cdot e \cdot \Delta f \tag{8}$$

$\dot{n}_d$: dark counts per second

The mean square thermal noise is given by

$$\bar{i_T^2} = \frac{4k_B T}{R_L} \cdot \Delta f \tag{9}$$

k_B: Boltzmann's constant
T: temperature
R_L: load resistance.

Putting equations 5, 6, 7, 8, and 9 together, the SNR of a PM can be given as

$$SNR_{PM} = \frac{\frac{G}{\sqrt{2}} \cdot \frac{V \cdot P_s \eta_Q e}{h\nu}}{\sqrt{G^2 2e \left(\frac{P_s \eta_Q e}{h\nu} + \dot{n}_d \cdot e \right) \cdot \Delta f + \frac{4k_B T}{R_L} \cdot \Delta f}} \tag{10}$$

In eq. 10 most values are the same for green and UV light, the exceptions being $P_s, \eta_Q, \nu, \Delta f$. How these four values differ for green and UV light will be discussed below.

Factors influencing the comparison green–UV

The main advantage of using UV-light is the increase in the scattered power P_s. According to eq. 3, at 257nm wavelength, the scattered power is about 16 times larger than at 514nm up to $d_p \approx 0.5\mu m$, assuming equal light intensities in the probe volume. While this is a convenient assumption with which to start, it is unrealistic when comparing light scattering and detection of UV and green light relevant to LDA and PDA. The following factors also have to be taken into account:

1. Continuous wave (cw) laser sources generally deliver less power when emitting ultraviolet rather than visible radiation.
   The UV laser source considered in this paper, for example, was an argon ion laser (Coherent Innova FReD 300) with a built-in frequency doubling crystal to convert radiation in the green line to UV. It produced about 10 times less output power at 257nm than at 514nm at equal plasma tube currents.

2. According to the photocathode material, the quantum efficiency of photomultipliers optimized for UV-light is generally 5 to 10 percent points higher than that of the best photomultipliers for green light.

3. Since the fringe spacing Δx in the probe volume is given by

$$\Delta x = \frac{\lambda}{2 \cdot \sin(\alpha)} \tag{11}$$

   halving the wavelength also halves the fringe spacing. Hence, when converting from green to UV-light, the diameter of the probe volume can also be reduced by a factor 2 without reducing the number of fringes. At a particular laser output power P_0, the laser intensity in the probe volume I_0 can, therefore, be increased by a factor of 4 for a given number of interference fringes.

4. The number of photons arriving at the detector per signal cycle is given by

$$n_{ph} = \frac{P_s}{h \cdot \nu} \cdot \frac{\eta_Q}{f_D} = \frac{P_s \cdot \lambda}{h \cdot c} \cdot \frac{\eta_Q}{f_D} \tag{12}$$

Since the Doppler signal frequency f_D is given by

$$f_D = \frac{1}{\lambda} \cdot v_\perp \cdot 2\sin(\alpha) \tag{13}$$

$v_\perp$: velocity component perpendicular to the probe volume fringes
eq. 12 predicts the following dependence:

$$n_{ph} \propto \lambda^2 \cdot P_s \tag{14}$$

This means that at equal scattered powers, using laser light at a wavelength of 257nm instead of 514nm reduces the number of scattered photons/cycle by a factor 4. Since the reduction of the probe volume size, referred to in point 3 above, increases the scattered light intensity by the same factor four, the number of photons/cycle remains unchanged by such a reduction of the probe volume diameter. However the SNR is increased, since the SNR $\propto 1/\sqrt{\Delta f}$ for the PM and the connected electronics.

These four points are taken into account in figure 2, where the number of detected photons per signal cycle is plotted versus particle size.

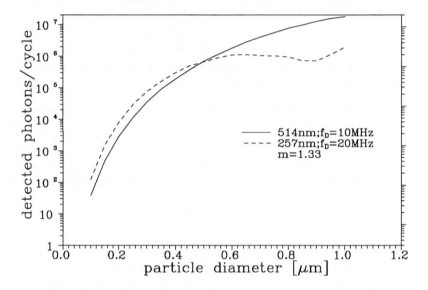

Figure 2: Detected number of photons per Doppler signal cycle versus particle size at the wavelengths 257nm and 514nm (output power at 257nm ten times less than at 514nm) at constant velocity.

Despite the reduction in the photon rate with the wavelength and the 10 times lower power of the UV laser relative to the green laser, fig. 2 shows that more photons are detected per signal cycle with UV-light than with green light for particles in the size range 0.1 to 0.6 μm. In this calculated example, there are 121 detected photons/cycle at 257nm and only 37 at 514nm for a 0.1μm particle.

SNR due to quantum noise

The scattering process is a statistical process, i.e. the number of photons/cycle plotted in fig. 2 has to be seen as the expectation value. Hence, low numbers of photons/cycle lead to quantum noise, which can be quantified by the standard deviation σ_{ph} of the number of photons/cycle, approximately given by

$$\sigma_{ph} = \sqrt{n_{ph}} \tag{15}$$

The resulting relative error, i.e. the noise-to-signal ratio, is given by

$$\frac{\sqrt{n_{ph}}}{n_{ph}} = \frac{1}{\sqrt{n_{ph}}} \tag{16}$$

The signal-to-noise ratio due to quantum noise SNR_{QN} is therefore $\sqrt{n_{ph}}$. Returning to the example given above, 37 photons/cycle correspond to a SNR of 6, whereas 121 photons/cycle correspond to a SNR of 11.

Derivation of the combined SNR

In order to give a final estimate of the expected SNR of a Doppler signal produced by a 0.1 μm particle, as shown in figure 3, the combined SNR of the quantum noise and of the PM has to be obtained.

Relative random errors $\Delta l_i / l_i$ of measured quantities l_i add up to a combined relative error $\Delta R / R$ according to:

$$\frac{\Delta R}{R} = \sqrt{\sum_i \left(\frac{\Delta l_i}{l_i}\right)^2} \tag{17}$$

In our case, considering the noise-to-signal ratios, this translates to:

$$SNR_{tot} = \frac{1}{\sqrt{\left(\frac{1}{SNR_{QN}}\right)^2 + \left(\frac{1}{SNR_{PM}}\right)^2}} \tag{18}$$

Comparison green-UV of the combined SNR

With eq. 10, 16, and 18 a comparison of the SNR versus bandwidth for green and UV light is possible. A comparison for a 0.1μm latex particle in air is shown in fig. 3, again assuming that the UV laser is a factor 10 times weaker than the green laser. This plot shows that the total SNR (in decibel) using UV light is about 9 to 12dB higher than using green light, even when considering the fact that the Doppler frequency is doubled when the wavelength is halved. As an example, in fig. 3 one sees that the SNR using UV-light at 100MHz equals the SNR using green light at about 10MHz, i.e. it is possible to measure 10 times higher frequencies using UV-light, which, according to eq. 13, converts to 5 times higher velocities, without a loss in SNR. Using a UV-based LDA-system can therefore be advantageous when using submicrometer seeding particles in high speed flows with high turbulence frequencies.

Further advantages from using UV laser light can be expected when using compact and portable lasers. Argon ion lasers are increasingly being replaced by frequency-doubled Nd:YAG lasers, which emit laser light at a wavelength of 532nm. The conversion efficiencies are about 50-60%, resulting in output powers of about 500mW. These Nd:YAG lasers can also be twice frequency-doubled producing a wavelength of 266nm. The conversion efficiencies from green to UV can be as high as 22% [5], which results in output powers of about 100mW. The estimated SNR's at 266nm and 532nm are shown in figure 4. Whereas the SNR at 266nm and 100mW is approximately the same as for 257nm in figure 3, the SNR at 532nm and 500 mW quickly drops to very low levels.

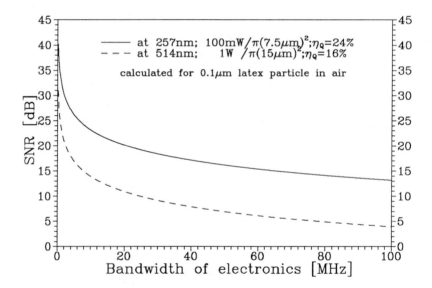

Figure 3: Estimated SNR of a Doppler signal produced by a 0.1 μm latex particle in air versus bandwidth

2.2 Phase difference-to-diameter relationship

In PDA, particle sizes are determined from the measured phase difference between the Doppler signals detected at two spatially separated detectors with the help of the so-called phase factor F:

$$\Phi_m = F(m, \alpha, ...) \cdot d_p \tag{19}$$

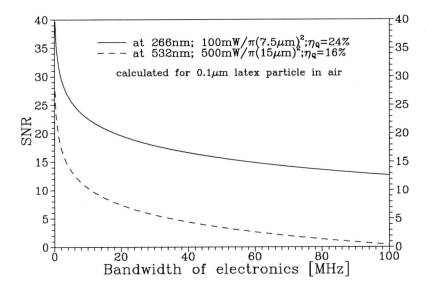

Figure 4: Estimated SNR of a Doppler signal produced by a 0.1 μm latex particle in air versus bandwidth

For particles that are large compared to the wavelength, geometrical optics applies, and the phase factor is a constant for a fixed optical geometry [1]. For particles that are smaller than about 10 times the wavelength, geometrical optics no longer applies, and the exact Mie-theory [2] has to be used to predict the phase-diameter relationship. Figure 5 shows the phase-diameter relationship in the size range from 0.1 to 1μm using green and UV light with identical optical geometries.

First, it is important to note that the phase-diameter relationship is non-linear and non-monotonic. Second, the phase-diameter relationship is steeper when using UV light, especially for particles smaller than 0.3μm. In the example shown in fig. 5 the phase difference at $\lambda = 514$nm for a 0.2μm particle is only -0.4°, whereas at $\lambda = 257$nm the phase difference is -7.43°. Since the phase resolution in PDA is limited to about 4° [3], a discrimination of 0.2μm particles from 0.1μm (and smaller), for which the phase difference can be considered to be 0°, is not possible at $\lambda = 514$nm. It is also not possible to attain a measurable phase difference by increasing the beam intersection angle or the elevation angle as is shown in fig. 6, where the calculated phase difference is plotted for a 0.2μm latex particle as a function of the beam intersection half angle for several elevation angles. Using a beam intersection half angle where the slope of phase difference curve is steep (see fig. 6) is not feasible, since the required alignment accuracy cannot be obtained in practice. Therefore, to differentiate the size classes 0.1 and 0.2μm, the utilization of UV-light becomes essential.

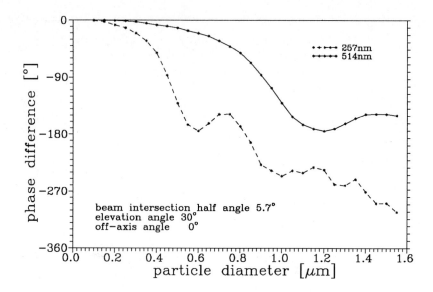

Figure 5: Comparison of the phase-diameter relationship using identical optical geometries

2.3   Probe volume and particle concentration

When considering submicrometer particles much higher number concentrations than in the 10 to $100\mu m$ size range may be encountered. For example, a volume fraction of only $6.5 \cdot 10^{-7}$ corresponds to a particle concentration C of $10^7 cm^{-3}$ for $0.5\mu m$ particles. Therefore, a submicrometer PDA should be capable of performing measurements at very high particle concentrations. A prerequisite for valid PDA measurements is that only one particle is in the probe volume at a time. Assuming that particles in a fluid or spray are uniformly distributed in space, the probability p_k that k particles are in the probe volume V_{PV} simultaneously is given by the Poisson-distribution:

$$p_k = \frac{(C \cdot V_{PV})^k}{k!} \cdot e^{-C \cdot V_{PV}} \tag{20}$$

The probability P_2 that there are at least 2 particles in the probe volume simultaneously is then

$$P_2 = 1 - p_0 - p_1 \tag{21}$$

Reducing the wavelength from green to UV by a factor of 2 allows the probe volume V_{PV} to be reduced by a factor of 8. Hence, according to eq. 20 and 21, the utilization of UV-light enables PDA measurements at 8 times higher particle concentrations without increasing the probability of there being more than one particle in the probe volume.

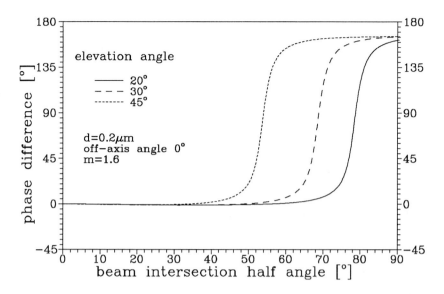

Figure 6: Phase difference of a 0.2μm latex particle at $\lambda = 514$nm as a function of the beam intersection half angle. For beam intersection half angles less than 45° the phase difference remains less than 4°.

3 Realization of the UVLDA-System

3.1 Experimental setup and its components

The experimental setup used for the LDA and PDA verification measurements is shown in figure 7. The cw UV laser light, redirected by mirror 1, hit the beam splitter plate at an angle of 60°. The transmitted and reflected beams were made parallel by mirrors 2, 3 and 4 and focused into a probe volume by lens 1 with a focal length of $f_1 = 100$mm. The additional reflections at mirrors 3 and 4 were necessary to avoid a path length difference between the two beams, since the coherence length was only about 40mm. At a distance of 200mm, lens 2 (focal length $f_2 = 150$mm) focused the scattered light on to a 100 μm pinhole covering the photocathode of photomultiplier PM_2. The signal from the PM was first bandpass filtered, then digitized by a transient recorder and finally sent to a PC for processing. PM_1 was only used for the PDA measurements. This setup was also used for the experiments using green laser light, except that the beam splitter plate and mirrors 3 and 4 were replaced by a path length compensated beam splitter.

The following system components were used:

1. Laser: Coherent Innova FReD 300
   The Coherent Innova FReD 300 is a water-cooled argon ion laser, that

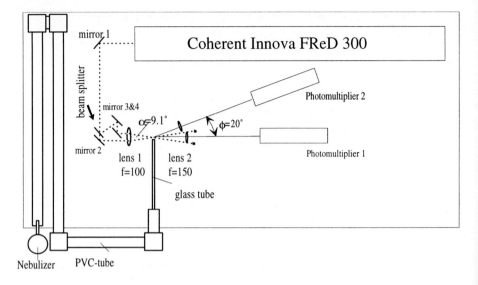

Figure 7: Experimental setup: overview

uses a frequency doubling BBO-crystal to produce up to 100mW of UV-light. All the visible lines of the argon ion laser can thereby be frequency doubled. In our case only the frequency doubling from $\lambda = 514$nm to $\lambda = 257$nm was used. The laser can be operated in the visible range simply by removing the frequency doubling crystal.

2. Optics:

- mirrors:
  257nm: UV-mirrors
  514nm: standard metal-coated mirrors

- lenses:
  257nm: UV-lenses (quarz)
  514nm: same lenses as for 257nm

- beam splitter:
  257nm: beam splitter plate for 250nm – 320nm (Melles Griot)
  514nm: path length compensated beamsplitter (Carl Zeiss Jena)

3. Photomultipliers:

Photomultipliers selected for visible light are usually not suited for the detection of UV-light [2]. A PM with a bialkali photocathode and a quantum efficiency $\eta_Q = 24\%$ at $\lambda = 257$nm, Thorn/EMI 9893QB100, was chosen instead. This PM has a very high gain of about max. $80 \cdot 10^6$

---

[2] The quantum efficiencies at wavelengths below ≈ 350nm are virtually zero, since the windows absorb UV.

at its operating voltage of about $U_{PM} \approx 1.6$kV and a very low dark count of 20 Hz. Its suitability for the detection of UV-light is ensured by the use of a quarz-glass window. This PM was also used for the experiments with green light, the quantum efficiency being $\eta_Q = 16\%$ at $\lambda = 514$nm.

A special feature of this PM is that, due to internal focusing, only a 2.5mm diameter area on the photocathode is active. This reduces noise due to stray light, but makes the optical alignment very difficult.

4. Electronics:

   For the amplification and filtering of the raw signals, a PDA-Extension Box (Invent GmbH) was used. The signals were then digitized by a transient recorder (LeCroy 9424) and sent to a PC for further processing.

5. Particle generation:

   In order to generate an air flow seeded with submicrometer particles, monodisperse latex particles supplied by Bangs Laboratories were dispersed in highly pure (99.8%) ethanol. This dispersion was then nebulized and passed through a 2.5m long PVC-tube with a diameter of 50mm, so that the ethanol evaporates. On the last 40cm, the ethanol/latex mixture was passed through a glass tube with a diameter of 10mm, which could be heated to several hundred °C, in order to ensure that the ethanol was completely evaporated, leaving only the latex particles. In such a way, seeded air flows with 0.1μm, 0.5μm, or with both 0.1μm and 0.5μm latex particles were obtained.

The main optical parameters of the LDA-system are listed in table 1.

Table 1: Optical parameters of the implemented LDA-system for both $\lambda = 257$nm and $\lambda = 514$nm

wavelength	257nm	514nm
beam spacing a	32mm	30mm
focal length transmitting lens f	100mm	110mm
beam-intersection half-angle α	9.1°	7.8°
fringe spacing Δx	0.81μm	1.89μm
number of fringes N	61	32
diameter of probe volume b_{MV}	50μm	60μm
focal length of receiving lens	150mm	170mm
distance probe volume – receiving lens	200mm	220mm
receiving-cone half-angle	6.8°	6.2°
magnification of receiving lens	1:3	1:3
pinhole diameter	100μm	100μm

3.2 Verification experiments

Verification experiments were performed to check that

1. LDA- and PDA-measurements are possible with UV-light

2. improved SNR's can be obtained through the use of UV-light.

The LDA-measurements were performed in the seeded air flow described in section 3.1. In order to compare SNR's, LDA-measurements were performed with equal output powers for UV and green. Table 2 lists the measured SNR's, which were obtained for the following parameters:

- scattering angle: $\Theta = 20^0$

- particle size: d_p nominally 0.096μm; width of the size distribution about 20%

- velocity: $v \approx 1.3$ m/s

- operating voltage at the PM: $U_{PM} = 1550$V

Measurement of SNR value:

The SNR of each Doppler burst was determined by the signal processing software, which measures signal frequency and the phase difference between the two channels by computing the cross spectral density function (CSD) from the digitized data. The phase difference Φ between channels x and y is determined from the real and imaginary parts of the cross spectral density function $G_{xy}(f)$ according to

$$\Phi_{xy}(f) = \arctan\left(\frac{\mathcal{R}e(G_{xy}(f))}{\mathcal{I}m(G_{xy}(f))}\right) \tag{22}$$

With this CSD-method, the program can also determine the average signal and the noise powers over the number of Doppler signal periods processed (usually 10–12 periods) and calculate the SNR in dB according to:

$$SNR[dB] = 10 \cdot log_{10}\left(\frac{signal\ power}{noise\ power}\right) \tag{23}$$

The SNR values listed in table 2 were obtained by evaluating several hundred Doppler bursts and forming the median SNR value of all bursts. Since a meaningful comparison is only possible between the maximum obtainable SNR with each wavelength, the trigger level was set to such a high level that only the signals from the largest particles passing through the middle of the probe volume were considered for the measurement. The corresponding theoretical SNR values listed in table 2 were obtained with eq. 18 using the scattered power, calculated using Mie-Theory. Since the width of the size

Table 2: SNR comparison between UV and green

wavelength	257nm	514nm
output power	20mW	20mW
signal amplitude	150mV	20mV
signal frequency	0.94 MHz	0.63 MHz
fringe contrast	0.67 [3]	1.0
SNR calculated	17.7dB	8.3dB
SNR measured	**17.1dB $\pm$ 0.5dB**	**9.4dB $\pm$ 0.5dB**
output power	20mW	–
signal amplitude	150mV	–
signal frequency	0.73 MHz	–
fringe contrast	0.67	–
SNR calculated	18.8	–
SNR measured	**18.5dB $\pm$ 0.5dB**	–

distribution was about 20%, the maximum latex particle size expected was about 0.12μm, and this diameter was used in the calculation.

The data in table 2 show that the SNR with UV-light is significantly larger than with green light, even using probe volumes of almost equal diameter (see table 1). By exploiting the possibility with UV-light of halving the probe volume diameter, without reducing the number of fringes, the same result could have been obtained with only 5mW of UV output power. Furthermore, the fringe contrast at 257nm was only ≈ 0.7 due to the poor intensity match between the two beams. A 50/50 beam splitter for UV-light would help to increase the SNR value at 257nm even more.

The calculated SNR values according to eq. 18 are very close to the measured values, confirming that the expected advantages in using UV-light as plotted in figures 3 and 4 can actually be achieved.

4   Extension to PDA measurements

4.1   Layout of the PDA-system

As shown in fig. 7, the setup was extended to a PDA by including a second PM to pick up a phase difference signal. To overcome the ambiguity of the phase-diameter relationship shown in fig. 5 three detectors would be necessary [6, 4]. In the presented series of measurements, however, a two detector system sufficed for the size range from 0.1μm to 0.6μm that was of interest.

---

[3] The fringe contrast is taken to be 0.67 for the 60/40 UV-beam splitter. The best visibility that can therefore be obtained is 0.67.

The phase-diameter relationship for this PDA-system of fig. 7 is shown in fig. 8 for refractive index values ranging from 1.6 to 1.7.

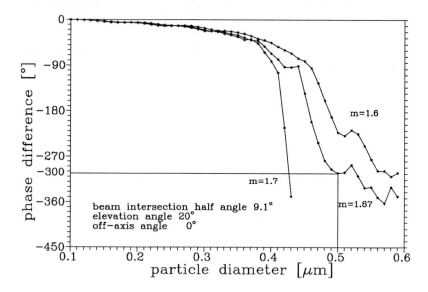

Figure 8: Phase-diameter relationship for the implemented PDA, plotted for several refractive indices.

4.2 Verification experiments

Experiments were made

1. to check whether PDA-measurements can be performed in the submicrometer range with UV-light, and

2. to quantify the precision of the phase determination at low SNR's and at phase values near 0°.

Figure 9 shows an example of a measured phase difference distribution after seeding the air flow with 0.1μm and 0.5μm latex particles. The peaks near phase differences of about 15° and 300° correspond to the 0.1μm and 0.5μm particles respectively. To determine the resolution with which these known sizes could be deduced from the phase difference distribution, it was necessary to calculate the corresponding size distribution from fig. 9. Since the refractive index of latex at $\lambda = 257$nm could not be found in the literature, this had to be estimated. Phase-diameter relationships were computed for several refractive index values between $m = 1.6$ (the refractive index of latex at $\lambda = 514$nm) and $m = 1.7$ to determine the theoretical phase difference for 0.5μm particles, as shown in fig. 8. Comparing these results with the observed peak of the phase difference distribution at 300° in fig. 9, the best phase difference match for a diameter of 0.5μm was found for a refractive index value of $m = 1.67$.

Using $m = 1.67$, the size distribution corresponding to the phase-diameter relationship was found, as shown in fig. 10. The widths of the size intervals in this figure are inversely proportional to the local gradient of the phase-diameter curve and, hence, vary greatly.

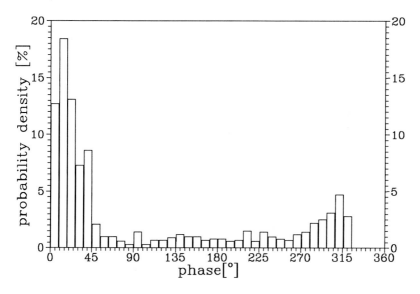

Figure 9: Measured phase difference distribution of the air-flow seeded with 0.1μm and 0.5μm latex particles

According to the phase-diameter relationship in fig. 8, the phase distribution of the first peak in fig. 9 corresponds to the size range from 0.1 to 0.35μm as shown in fig. 10. The poor size resolution arises because 0.1μm particles correspond to a phase value near 0°, where the accuracy of the phase difference measurement is lowest (at least with the CSD-method, since $\Phi_{xy}(f)$ depends on an arctan function, as described by eq. 22, which has its steepest slope at the origin). The SNR-values are also lowest near 0.1μm, further reducing the measurement accuracy. Many particles in this size range were therefore evaluated as too large, although on the oscilloscope one could easily obtain signal pairs with no visible phase shift, i.e. corresponding to the 0.1μm particles.

Improvements in the amplification and filtering of the signals could increase the measurement accuracy. However, for the reliable measurement of the phase difference value 0°, which is vital in submicrometer PDA, changes in the phase difference determination software near 0° will have to be made.

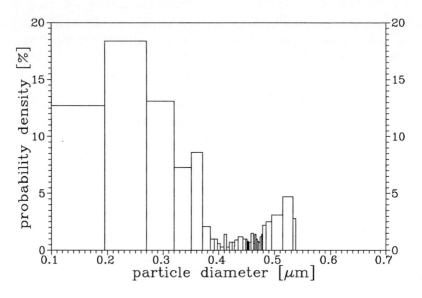

Figure 10: Size distribution determined by comparing fig. 9 and fig. 8

5   Summary

In this paper it has been shown that UV lasers can be used for LDA and PDA measurements in the submicrometer range and that they offer several advantages compared with visible lasers. Theoretical analyses of the scattered light intensity and the phase-diameter relationship for submicrometer particles over a realistic range of design parameters of LDA and PDA systems have shown the following advantages:

1. significant increases in signal-to-noise ratio, especially towards higher Doppler frequencies

2. steeper phase-diameter relationships in the 0.1 μm to 0.2 μm range, permitting particles as small as 0.2 μm to be distinguished from even smaller particles

3. measurements at up to 8 times higher particle concentrations.

Experiments using nominally monodisperse latex particles of two sizes have confirmed the theoretical improvement in signal-to-noise ratio. The detection of particles down to 0.1 μm diameter has also been demonstrated, but the size discrimination below 0.2 μm diameter is unsatisfactory. Improvements will be achieved principally by changes in the data evaluation software, and to a lesser extent through refinement of the signal processing. The extension to higher concentrations is readily achievable by a reduction of the probe volume dimensions at the shorter wavelength, so that measurements at concentrations up to 10^7 cm^{-3} should be possible.

6 Acknowledgements

The work presented in this paper is part of a research project funded by the Deutsche Forschungsgemeinschaft (grant Du 101/35-1).

References

[1] Bauckhage, K. The phase-Doppler-difference-method, a new laser-Doppler technique for simulataneous size and velocity measurements, Parts 1&2. *Part. Part. Syst. Charact.*, 5:16–22 and 66–71, 1988.

[2] Bohren, C. and Huffman, D. *Absorption and Scattering of Light by Small Particles.* John Wiley & Sons, Inc, 1983.

[3] Domnick, J., Ertel, H., and Tropea, C. Processing of phase-Doppler signals using the cross spectral density function. In *Proc. 4th Int. Symp. on Applications of Laser Anemometry to Fluid Mechanics*, page 3.8, Lisbon, 1988.

[4] Durst, F., Melling, A., and Volkholz, P. The extension of phase-Doppler anemometry to small particles. In *Proc. 4th Int. Congress on Optical Particle Sizing*, page 185, Nürnberg, 1995.

[5] Liu, L.Y., Oka,M., Wiechmann, W., and Kubota, S. Longitudinally diode-pumped continuous wave 3.5-W green lasers. *Optics Letters*, 19:189 – 191, 1994.

[6] Naqwi, A. and Ziema, M. Extended phase Doppler anemometer for sizing particles smaller than 10μm. *J. Aerosol Sci.*, 23(6):613–621, 1992.

II.3 Measurement Uncertainties of Phase Doppler Technique due to Effect of Slit Location, Control Volume Size Effects

Yuji IKEDA, Toshiaki HIROHATA and Tsuyoshi NAKAJIMA

Department of Mechanical Engineering
Kobe University
Rokkodai, Nada, Kobe 657 JAPAN

Abstract . Measurement uncertainties of phase Doppler technique in application for practical combusting spray was investigated. There are many error sources reported as Gaussian beam effect, trajectory effect and slit effect and so on. But further measurement uncertainties may be yielded such as slit location shift effect, optical measurement volume size choice and flame front pressure effect. In this study, two dominant factors have been examined in measurements of spray burner of 0.1MW.

It is found that the slit location shift effect can hardly be avoided in combustion and numerous errors may be caused. Quantitative uncertainty value will be examined in this study. The control volume size should be determined in consideration of velocity gradient in shear flow region. But this matter is not considered in the practical measurement. This error sources were also discussed here. The flame front will change the optical path like slit location shift so as to have measurement error in the small diameter droplet.

Keywords . Phase Doppler technique, Spray measurement, Measurement uncertainties, Slit location effect, Measurement volume size effect.

1 Introduction

Over the past decade, phase Doppler measurement techniques (PDA / PDPA) have been developed and employed for practical spray studies and researches (Bachalo (1980), Bauckhage (1985), McDonell et al. (1986), Durst et al. (1988), Edwards & Rudoff (1990), Taylor (1993), Lai et al. (1994), and so on). Numerous researches have been carried out using this phase Doppler technique in order to provide sophisticated data as each droplet velocity and diameter without disturbing the flow. Currently, many researches have been carried out by focusing on the next generation , that is, not just measurements of velocity and diameter but also investigations of droplet dynamics, drag coefficients, turbulence interaction, agglomeration, evaporation, group combustion (Mizutani et al. (1994)), mass flux in practical combusting spray systems and so on (Hardalupas et al. (1994), Edwards (1994)).

The error estimation (Grehan et al. (1991) , Sanker et al. (1992), Xu et al. (1994), Aizu et al. (1993)), affecting factor analysis (Bachalo& Sanker (1994)) and measurement uncertainties of this technique have been highly required to be quantified both in experiment and theoretical analysis. But these conventional

166

researches have been carried out mainly in a steady state and very simple experiment. The affecting factors considered are Gaussian beam effect (Grehan et al.(1991), Sanker & Bachalo (1992)), scattering light analysis to reduce noisy phase information, refraction / reflection analysis, planer optical alignment analysis (Tropea et al. (1995), Sanker & Bachalo (1995)), Rainbow effect (Sanker et al. (1994)), trajectory effect (Scott et al. (1994), Sanker et al. (1995)), slit effect (Tropea et al. (1994), Maeda et al. (1996)), optimization of optical system (Naqwi (1996), Ikeda et al. (1990)), set-up of processor, data number and data rate, liquid refractive index variation (Brenn et al. (1994, 1996), Durst & Brenn (1994), Naqwi et al. (1991)), and so on.

In application of this phase Doppler measurement system for practical combusting spray, additional uncertainty sources are raised up, which should be taken into account in a dynamics state.

The purpose of this study is to investigate these additional factors in dynamic state such as slit location effect, measurement volume size effect. Two kinds of experiment were curried out using two sets of PDA. A practical oil burner was used in the measurement under non-combusting condition.

2 Sources of measurement uncertainties

Besides of the conventional measurement uncertainty sources, there are mainly seven sources of measurement uncertainty in application of PDA systems for practical combusting spray measurements as illustrated in Fig. 1.

The first one is a phase shift wave front distortion at a flame front at the receiving optics due to refractive index changing. This error was examined in LDV

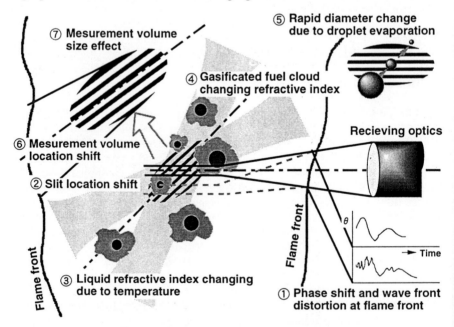

Fig. 1 *Measurement uncertainty sources*

(Ancimer & Fraser (1994)) well, but the distortion effect has not been investigated.

The second one is the slit location shift effect which is related to the first one. If there is a flame front and the refractive index changes, the optical path of receiving optics will shift. At the result, the slit location will shift. This slit location is normally adjusted empirically by manual at the center of the measurement volume before the measurement. If this slit is not locating at the center, the measured data will be dramatically influenced and affected by Gaussian beam effect. This will be a big problem in obtaining mass flux. Furthermore, the slit aperture was defined as 100μm in an atmospheric pressure condition and room temperature. If the flame front is curved or not smooth such as distributed reaction zone (Kuo (1986)), the slit width will differ from the defined one. This cause slit location shift and very much influence in calculating volume flux.

The third one is the measurement error associated with liquid refractive index changing due to temperature variation. In combusting spray measurements, the refractive index of the fuel was changing due to combustion so that the assumption of PDA measurement, that is, the refractive index is constant will fail. Otherwise, the PDA can not measure the droplet diameter just from phase information without refractive index information. Because it is obvious that the refractive index of fuel changes with the temperature in combusting spray. The refractive index measurements were examined in order to show these effects but not so well quantified (Brenn et al. (1994), Naqwi et al. (1991)). Then, we have to use the refractive index data of the fuel at standard condition. The temperature in combusting spray flame, the local temperature is not constant because the droplets travelling. Evaporation, gasified flow, high temperature surrounding gas are existing. The temperature is also varying. Although we can make measurement of the refractive index of the fuel, the required temporal resolution for the PDA should be very high which can be hardly achieved. This means that there are some uncertainties in diameter measurement, which can not be avoided.

The fourth one is the refractive index changing around the droplet. The gasificated fuel forms a cloud in which the refractive index of this cloud also very. Theoretical evaluation of this effect will be discussed later paper.

The fifth one is the rapid droplet diameter changing like small droplet evaporation within a measurement volume. The PDA measures the phase when the droplet travels through 5-8 fringes, so that the measurement can not be performed if the droplet diameter changing time is faster than that measurement time. The diameter measured is an averaged within a measurement volume. This fact will cause error.

The sixth one is the measurement volume location shift effect, which is the same as the second one. The measurement volume will be formed at a different location due to the refractive index changing at the flame front and in the gasificated fuel cloud.

The seventh one is the measurement size effect. It is desirable to have long focal length both in transmitting and receiving optics for combusting spray measurements to prevent from fuel oil adhesion, soot and heat radiation. The long focal length makes larger measurement volume in which the fringe spacing is enlarged and the light intensity decreases (low SNR). The fringe spacing should be optimized in consideration of measurable ranges of velocity and diameter. The large measurement volume has to average the velocity and diameter within the large measurement volume, so that the characteristics of the large velocity gradient flow or strong shear flow region cannot be demonstrated well. This effect is an essential for LDV and PDA. If the measurement volume is too much small, which causes small fringe spacing, large fringe number and very wide frequency dynamic range requirement. In PDA, the measurable diameter range is determined by these value, so that the optimization of the optics to the flow should be done before measuring the final data.

The evaluation of the system is very important factor, but we have to say that this has been performed by experimentally and empirically.

These seven uncertainty sources can be classified into four groups in this story; flame-front effect (1), slit location shift effect (2,6,7), measurement volume size effects (5,7) and refractive index changing effect (2,4). In this study, these two measurement uncertainty sources will be investigated, that is, slit location shift, measurement volume size and flame front presence.

3 Experimental apparatus

A gun-type oil burner was used in this experiment as shown Fig. 2 as well as the flame picture (Ikeda et al. (1995), Kawahara et al. (1996, 1997)). This oil burner is for 0.1 MW class boiler and furnace. An A-type fuel oil was pressurized at 0.7MPa and atomized by the hollow-cone nozzle (Danfos : H type) of 60 degree. The fuel rate used is 9.45×10^{-3} m³/h. The PDA setup is illustrated in Fig. 3. As a signal processor, Dantec processor was used.

4 Results

4.1 Slit Location Shift Effect

As explained in above section, the slit location shift will be happened due to flame front effects both on the transmitting and receiving optical paths. But we have to

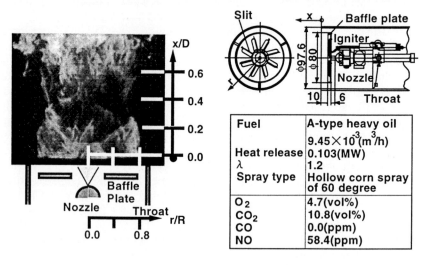

Fig. 2 *Burner and flame photograph*

pay more attention to the measurement set-up at the initial stage. The slit location is normally adjusted by same one, which is an empirical based set up.

In order to examine the slit location shift effect, two measurement points were chosen. One (point A) is the dense spray region of large velocity gradient, the other (point B) is on the center axis of high data rate region as shown in Fig. 4 (Kawahara et al. (1995)). Two dimensional velocities and the droplet diameter were measured. The sample number was 10,000. The size-classified (Edwards & Rudoff (1990), Presser et al. (1995), Seay et al. (1995), Ikeda et al. (1996))

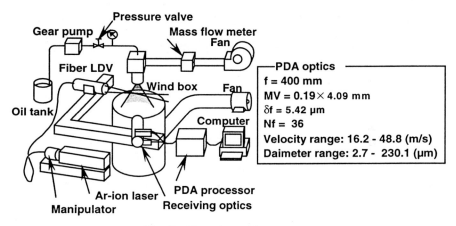

Fig. 3 *Experimental apparatus*

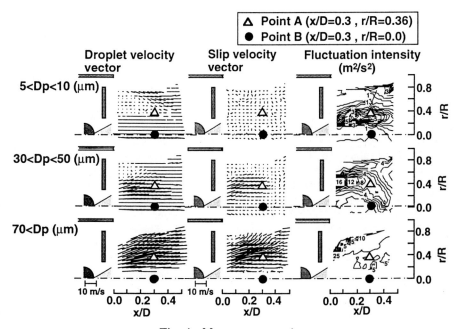

Fig. 4 *Measurement points*

droplet velocity vectors are shown as well as slip velocity and fluctuation intensity. At point A, this point locates at the tail of recirculating flow region. This figure shows that the small droplet can follow to the reversing flow, so that the small droplet velocity was negative, while the large droplet penetrated into this recirculating flow region in which the larger droplet velocity was positive. The slip velocity vectors show the large shear stress region and their magnitude. The detailed flow characteristics were explained in the previous paper (Kawahara et al. (1995)).

Five slit locations were examined in the optical setup by manually. The measurement volume size was 0.19×4.09 mm and the slit width was about 100 μm. As shown in Fig. 5, five sliced slit volumes are not the same, and the light intensity differences should be taken into account.

Figure 6 shows the raw data at two slit locations, #3 (center) and #5 (side end). As seen clearly, there are many discrepancies. For evaluating these data, the measured axial and radial mean velocities (U and V), and Sauter and Volume mean diameters (D_{32}) are shown in Fig. 7. The ratio of velocity and diameter were defined the ratio to that at the slit #3. The ratio 0.25 means that the velocity (diameter) is 25% larger than that measured at slit #3.

The point A is the recirculation flow tail. As shown in Fig. 6, these two measured data are not the same. The negative velocity was measured at slit #5 and the axial velocity profile was almost flat, while there was no negative velocity at slit #3 and the maximum velocity peak was about 4 m/s. Even in the diameter, the slit #5 could not measure the droplet less than 50μm. But the slit #3 data is almost under 50μm. These two diameter are extremely different from each other. This point A is the strong shear flow region so that the difference of the slit location shows entirely different measurement point of different droplet dynamics. The velocity difference at these two locations was over 30%.

The point B is on the center axis. The fuel was atomized at hollow-cone angle of 60 degree so that there was residual and reversed droplet on the center axis. This point is less shear than point A. In the axial velocity distribution, the peak velocities are not so much different each other. But the diameter distribution are totally different.

In order to understand the difference and the error of the slit location in detail, Fig. 7 was demonstrated.

In Fig. 7 at point A, the axial velocity at slit #2 to #4 are not so much different. The velocity ratio is under 10%. But the data at slit #1 and #5 are very much different, the velocity at slit #5 is 70% higher than that at slit #3 (center). At point B, the axial velocity at different locations are gradually increasing, but the discrepancies are less than 20%. The radial velocity was almost 0.3 m/s but the radial velocity changes are very large as shown in this figure.

For SMD at point A, the slit #3 shows 60μm. The range of the measured diameter was almost 30μm, which was about 30 %. At point B, the diameter ratio at different location was over 100%.

Here, we have to determine which the measurement results were right or wrong. Those results are derived from the system set-up or due to different droplet dynamics. Then, these measurement status were evaluated with data rate, spherical rate and accepted rate as shown in Fig. 8 in order to find the optimal slit location.

Both in these points, the measurement at slit location #3 should be the best,

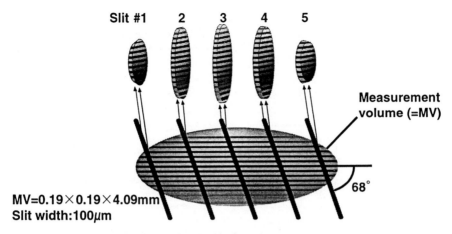

Fig. 5 *Slit location*

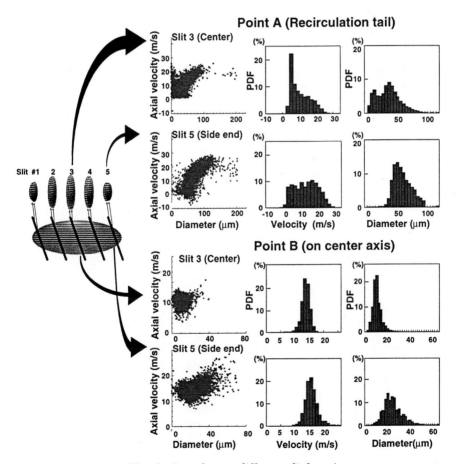

Fig. 6 *Raw data at different slit location*

172

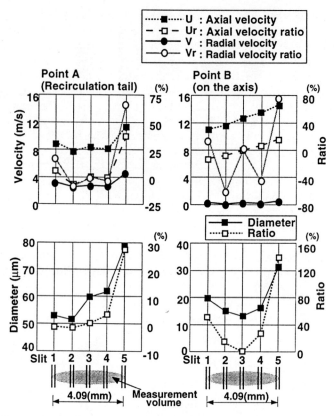

Fig. 7 *Comparison of measured data coaxial mean velocity (U), radial mean velocity (V) and Sauter mean diameter D_{32}*

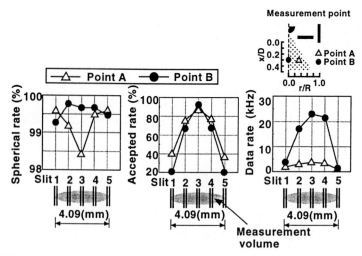

Fig. 8 *Comparison of spherical rate, accepted rate and data rete*

which can be demonstrated in these three factors. At the side end of the measurement volume, #1 and #5, the accepted rate are lower both in two points, although the high spherical rate measured was over 98%. It is very difficult to say that the measured results at #1, 2 and 3 are right or wrong. Finally, we have to say that the discrepancies in velocity and diameter at central three slit location do not derive from the measurement error, the measured results are accurate and appropriate in terms of measurement technique and system set-up. It should be emphasized the measurement uncertainty called "slit location shift effect", this effect can not be avoided, essentially happen, then how to correct or reduce this error.

When the measurement is carried out in a combusting spray, there is a flame front. The refractive indices at the flame fronts will cause the measurement volume location shift both on transmitting and receiving optical paths, which will also induce this effect. Then, the practical measurement results should be evaluated in this factor effected and show the uncertainty in quantitative value.

For further examination of the measured results, the PDF of size-classified droplet sample number and those velocities are shown in Fig. 9. It is found that this slit location shift effect can be seen in the larger droplet diameter. The axial velocities of small droplet have less difference than those of the larger droplet. The PDF of

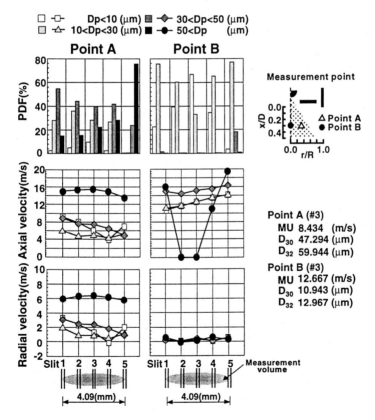

Fig. 9 *Size-classified droplet distributions at points A and B*

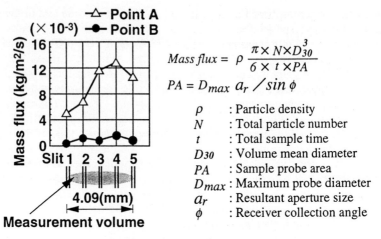

$$Mass\ flux = \rho\ \frac{\pi \times N \times D_{30}^{3}}{6 \times t \times PA}$$

$$PA = D_{max}\ a_r \Big/ \sin \phi$$

ρ : Particle density
N : Total particle number
t : Total sample time
D_{30} : Volume mean diameter
PA : Sample probe area
D_{max} : Maximum probe diameter
a_r : Resultant aperture size
ϕ : Receiver collection angle

Fig. 10   *Mass flux*

droplet number show the same effect.

The one of the big advantage of the PDA technique is to be able to estimate a mass flux. The mass flux measurement errors have been evaluated in many researches (Edwards et al. (1990), Sanker & Bachalo (1995), Bachalo (1994)), but those are carrying out in the steady state. It should be noticed that the measurement of mass flux contains of large uncertainty due to the slit location effect in combusting spray.

As shown in Fig. 10, the mass flux at slit #5 is almost three times higher than that at slit #3 at point A.   Considering that the large velocity gradient exists and droplet movements of large and small are different from each other, it is easy to assume that there is some uncertainties.   But the value of three times higher is out of question.   We have to take care of the initial optical set-up and uncertainty due to Gaussian effect (Grehan et al. (1991), Sanker et al. (1992)) and slit effect (Tropea et al. (1995), Sanker et al. (1995)), however the measured data might contain these error.   There are many reports to obtain mass or volume flux by the phase Doppler technique.   The key point is the slit aperture and its size.   In combustion, the error due to slit location shift may not be such large value.   But it is highly required to consider the uncertainty and error in the measurements.   The evaluation of the data as well as the system set-up is the only way to the eliminate these uncertainties.

4.2 Measurement Volume Size

For practical combusting spray measurements, it is desirable to have long focal length both in transmitting and receiving optics to prevent from oil adhesion, soot and heat radiation.   But, the long focal length causes the larger measurement volume which will have weak light intensity (low SNR) and volume averaging effect at a large velocity gradient region and large droplet diameter distribution region. In order to prevent from this volume averaging effect and above-mentioned slit location effect, a small measurement volume is the one of the best solution.   This is the trade-off relation.   The smaller the measurement volume becomes, the higher

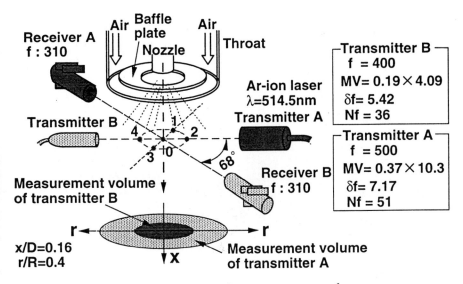

Fig. 11 *Two coincident measurement volume*

the measurement accuracy becomes and the smaller the uncertainty becomes. But the measurable range of velocity and diameter decreases. An optimization of this effect is the quite important task for researcher before choosing the system set-up specification.

In this study, two measurement volumes were formed at the same location and overlapping each other and the measurements were carried out to show this measurement volume size effect as shown in Fig. 11. The larger measurement volume was 0.37 × 10.3 mm and the smaller was 0.19 × 4.09 mm. The measurement set-up was the same as the above-mentioned. Five measurement points were examined to show this effect at different conditions. One is on the axis #0, the others are four points in radius with 20 mm around the axis. The spray is a hollow-cone of 60 degree but an absolute axisymmetry was not achieved well. Figure 12 shows the comparisons of the measured data of axial mean velocity, D_{30} and D_{32} both in two measurement volume sizes. At the point #0, which is on the axial axis. The larger measurement volume results indicate the lower velocity and the larger diameter than those in the small measurement volume. This discrepancies came from the volume averaging effect. At the four radial locations, the discrepancies of D_{30} and D_{32} are not so large, but the velocity differences are not negligible.

Here, as evaluated in the above-mentioned as slit location shift effect, the measurement status was examined by data rate, spherical rate and accepted rate as shown in Fig. 12. At the point #0, the larger measurement volume can show the high data rate of 25 kHz but lower accepted rate of 73%. On the contrary, the small measurement volume cannot provide such high data rate but very high accepted rate of 97%. The spherical rate for both measurement volumes is over 99.7%. It is found that these two measurements were conducted at high level status having less error. For the status of the other four points, there are the same level and possible to identify that the measurement was very accurate.

The influence of measurement volume size on the droplet diameter distribution

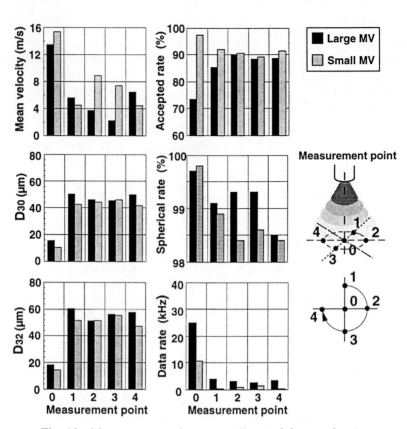

Fig. 12 *Measurement volume size effect and data evaluation*

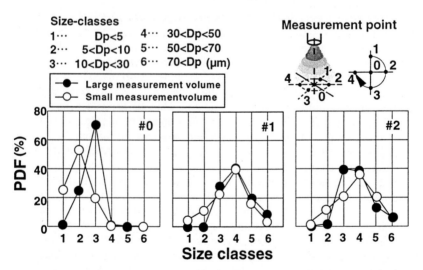

Fig. 13 *Measurement volume size effect on size-classified droplet*

were examined as shown in Fig. 13. At point #0 on the axis, the small measurement volume can detect large sample number of small droplet than those by the larger measurement volume. There are no differences for over 30μm droplet. But the difference of PDF of 10-30μm droplet is very remarkable. At the radial two points #1 and #2, the droplet diameter distributions are compared with each other. These two data show almost no significant difference. Then, it is found that the measurement volume size effect can be caused at the high data rate measurement.

5 Conclusion

Measurement uncertainties of phase Doppler technique were investigated in two experiments. The effects of slit location shift and measurement volume size were examined. The results are summarized as follows:

The slit location shift effect causes a significant measurement uncertainty, especially in high data rate. The optical alignment of the slit location at the initial stage is the most important factors, and this slit location shift effect can be improved dramatically using two PDA receiving optics. The measurement uncertainty of the mass flux at low data rate yields large value. The measurement volume size effect can be seen in detection rate of small droplet at high data rate.

6 References

Durst,F., et al., (1988) Proceedings of the Phase-Doppler Technique, *Experiments in Fluids*, vol.6.

Bachalo, W.D., (1980) Method for Measuring the Size and Velocity of Spheres by Dual-Beam Light Scatter Interferometry., *Appl. Opt.* vol.19, No.3, 363-370.

Bauckhage, K., (1985) *International Conference Laser Anemometry Advances and Application*, Manchester, 261-278 .

Edwards, C.F., and Rudoff, R. C., (1990) Structure of a Swirl-Stabilized Spray Flame by Imaging, Laser Doppler Velocimetry, and Phase Doppler Anemometry, *Proc. 23d_Symp. (Int.) on Comb.*, 1353-1359.

Taylor, A.M.K.P., (1993) *Instrumentation for Flows with Combustion*, Academic Press.

Lai, M.-C., et al., (1994) Atomization and Vaporization Characteristics of Airblast Capillary Fuel Injection inside a Venturi Tube, *Proc. IUTAM Symp. on Mechanics and Combution of Droplets and Sprays*, Taiwan, 31-40.

Mizutani, Y., et al., (1994) Optical Observation of Group Combustion Behaviors of Premixed Spray, *Proc. IUTAM Symp. on Mechanics and Combution of Droplets and Sprays*, Taiwan, 157-166.

Hardalupas, Y., et al., (1994) Coaxial Atomization and Combustion, *Proc. IUTAM Symp. on Mechanics and Combution of Droplets and Sprays*, Taiwan, 41-74.

Edwards, C.F., (1994) Application of Ideal Spray Concepts to Understanding the Stochastic Dynamics of Sprays, *Proc. IUTAM Symp. on Mechanics and Combution of Droplets and Sprays*, Taiwan, 277-290.

McDonell, V.G., et al., (1986) A Comparison of Spatially-Resolved Drop Size and Drop Velocity Measurements in an Isothermal Chamber and a Swirl-Stabilized Combustor, *Proc. 21st Symp. (Int.) on Comb.*, 685-694.

Grehan, G., et al., (1991) Evaluation of a Phase Doppler System Using Generalized Lorenz-Mie Theory, *Int. Conf. on Multiphase Flows '91*, 291-296.

Sankar, S.V., et al., (1992) Trajectory Dependent Scattering in Phase Doppler Interferometry: Minimizing and Eliminating Sizing Error, *Proc. 6th Int. Symp. of Laser Tech. to Fluid Mech.*, Lisbon., 12.2.1

Xu.T.-H., Tropea, C., (1994) Improving the Performance of Two-Component Phase Doppler Anemometers, *Meas. Sci. and Techn.*, 5, 969-975.

Aizu, Y., et al., (1993) PDA Systems without Gaussian Beam Defects., *Proc. 3rd Int. Conf. Optical Particle Sizing.*, Japan, 461-470.

Bachalo, W.D and Sankar, S.V, (1994) Factors Affecting the Measurement Resolution and Accuracy of the Phase Doppler Particle Analyzer, *The Second Int. Conf. on Fluid Dynamics Meas. and Its Appl.*.

Tropea, C., et al., (1995) Dual Mode Phase Doppler Anemometer., *Proc. 4th Int. Cong. Optical Particle Sizing*, vol.4, 287-296.

Sankar, S.V., and Bachalo, W.D., (1995) Performance Analysis of Various Phase Doppler Systems., *Proc. 4th Int. Cong. Optical Particle Sizing*, vol.4, 407.

Sankar, S.V., et al., (1994) Simultaneous Measurements of Droplet Size, Velocity, and Temperature in a Swirl-Stabilized Spray Flame., *Proc. 7th Int. Symp. of Laser Tech. to Fluid Mech.*, Lisbon., 12.3.1

Scott, A.S., et al., (1994) Theoretical analysis of the effects of particle trajectory and structural resonances on the performance of a phase-Doppler particle analyzer., *Appl. Opt.*, vol.33,No.3, 473-483.

Sankar, S.V., et al., (1995) An Adaptive Intensity Validation Technique for Minimizing Trajectory Dependent Scattering Errors in Phase Doppler Interferometry., *Proc. 4th Int. Cong. Optical Particle Sizing*, vol.4, 285.

Tropea, C., et al., (1994) Dual-Mode Phase Doppler Anemometer., *Proc. 7th Int. Symp. of Laser Tech. to Fluid Mech.*, Lisbon., 18.3.1

Maeda, T., et al., (1996) Determination of Effective Measurement Area in a Conventional Phase-Doppler Anemometry, *Proc. 8th Int. Symp. of Laser Tech. to Fluid Mech.*, Lisbon., 2.5.1

Naqwi, A., (1996) Optimization of the Shape of Receiving Aperture in a Phase Doppler Syatem, *Proc. 8th Int. Symp. of Laser Tech. to Fluid Mech.*, Lisbon., 2.2.1

Ikeda, Y., et al., (1990) A Compact Fibre LDV with a Perforated Beam Expander., *Meas. Sci. and Techn.*, 1, 260-4.

Brenn, G., et al., (1994) Investigation of Polydisperse Spray Interaction Using an Extended Phase-Doppler Anemometry, *Proc. 7th Int. Symp. of Laser Tech. to Fluid Mech.*, Lisbon., 21.1.1

Brenn, G., et al., (1996) Investigation on Accuracy and Resolution of Refractive Index Measurements with an Extended Phase-Doppler Anemometer, *Proc. 8th Int. Symp. of Laser Tech. to Fluid Mech.*, Lisbon., 9.4.1

Durst, F., and Brenn, G., (1994) ExtendedPhase-Doppler Anemometers and their Application to Measure Spray Properties, *Proc. IUTAM Symp. on Mechanics and Combution of Droplets and Sprays*, Taiwan, 241-264.

Naqwi, A., et al., (1991) Two-Optical Methods for Simultaneous Measurement of Particle Size, Velocity, and Refractive Index., *Applied Optics*, 30, 4949-4959.

Ancimer, R.J and Fraser, R.A, (1994) Flame-Induced Laser Doppler Velocimetry Velocity Bias, *Meas. Sci. and Techn.*, 5, 83-92.

Kuo, K., (1986) *Principles of Combustion*, Wiley.

Ikeda,Y., et al., (1995) Flux Measurements of O2,CO2 and NO in an Oil Furnace, *Meas. Sci. and Techn.*, 6, 826-832.

Kawahara, N., et al., (1996) Size-Classified Droplets Dynamics of Combusting Spray in 0.1MW Oil Furnace, *Proc. 8th Int. Symp. of Laser Techn. to Fluid Mech.*, Lisbon., 10.5.1

Kawahara, N., et al., (1995) Droplet Followability and Slip Velocity Analysis of Evaporating Spray on Gun-Type Oil Burner, *4th Int. Cong. Optical Particle Sizing*, vol.4, 593-602

Kawahara, N., et al., (1997) Droplet Dispersion and Turbulent Structure in a Pressure-Atomized Spray Flame, *AIAA Paper* 97-0125.

Edwards, C.F., and Rudoff, R. C., (1990) " Structure of a Swirl-Stabilized Spray Flame by Imaging, Laser Doppler Velocimetry, and Phase Doppler Anemometry", *Proc. Twenty-Third Symposium on combustion*, 1353-1359.

Presser, C., et al., (1995) Interpretation of Size-Classified Droplet Velocity Data in Swirling Spray Flames, *AIAA Paper* 95-0283.

Seay, J., et al., (1995) Atomisation and Dispersion from a Radial Airblast Injector in a Subsonic Crossflow, *AIAA Paper* 95-3001.

Ikeda, Y., et al., (1996) Spray Formation and Dispersion of Size-Classified Fuel Droplet of Air-Assist Injector, *Proc. 8th Int. Symp. of Laser Techn. to Fluid Mech.*, Lisbon., 13.6.1

Edwards, C.F., et al., (1990) Measurement of Correlated Droplet Size and Velocity Statics, Size Distribution, and Volume Flux in a Steady Spray Flame, *5th Int. Symp. of Laser Tech. to Fluid Mech.*, Lisbon., 31.5.1

Bachalo, W.D., (1994) Injection, Dispersion, and Combustion of Liquid Fuels, *25th Symp. (Int.) on Comb.*, 333-344.

II.4 Effective Size of the Measuring Cross-Section of a Phase-Doppler Anemometry

T. Maeda, H. Morikita, K. Hishida and M. Maeda

Department of Mechanical Engineering, Keio University
3-14-1 Hiyoshi, Kohoku-ku, Yokohama, 223, Japan

Abstract. The paper presents a technique for measuring particle mass flux in dispersed two-phase flows with improved accuracy of mass flux measurements by a *standard* phase-Doppler anemometer (PDA) with an A/D converter-based signal processor. The purpose was achieved by improved estimation of the effective area of the probe volume. The area of probe volume was calculated by a geometrical optics approximation after considering two major influences, namely 'trajectory ambiguity' and 'slit effect'. The calculation results showed that the effective measuring area in the present optical configuration can be determined as a simple function of the probe diameter, size of the spatial filter of the receiving optics and particle diameter by considering only refracted light when 'phase validation' is applied. In addition, careful setting of the trigger level of the signal processor resulted in reducing number of rejected signals owing to low signal to noise ratio (SNR) within the defined effective measuring area and, as a consequence, the accuracy of mass flux measurement was improved by 15% compared with our earlier work. The flux measurement of a hollow-cone spray indicated that the error did not exceed 20%, at a volume fraction of 2×10^{-4} %.

Keywords. phase Doppler anemometry, mass flux measurement, trajectory ambiguity, slit effect, geometrical optics approximation

1 Introduction

The determination of the mass flux of the particulate phase is as important as the particle size and velocity in the investigation of the flow characteristics of dispersed multi-phase flows. The phase-Doppler anemometer (PDA), which can provide size and velocity information for spheres, is a well accepted technique for this purpose and several signal processing techniques for accurate velocity and size measurement have recently been reported [e.g. Bachalo *et al.*, 1988; Kobashi *et al.*, 1992; Domnick *et al.*, 1993].

The mass flux of particles is defined as the total mass of particles passing across the measuring area per unit time. With regard to measurement accuracy, there are two major difficulties which still require attention: the first one is the accuracy in determining the effective size of the measuring volume and the second is the occurrence of Doppler signals with poor signal to noise ratio (SNR), which may reduce the data rate.

There are a number of studies treating the first problem. At an early stage, attention was focused on the particle size dependence of the probe volume [e.g. Saffman, 1987; Qiu et al., 1992; Higuchi et al., 1994, Sommerfeld and Qiu, 1995]. Recently, however, two other error sources were reported, namely trajectory ambiguity (TA) [Gouesbet et al., 1990; Sanker et al., 1992] and the slit effect (SE) [Durst et al., 1994], which yield errors for both particle size and mass flux due to the scattering of a relatively strong component of reflected rather than refracted light. Apart from the direct errors in size, these effects imply that the effective cross-section of a PDA's probe volume over which correct measurements can be made is different from that estimated by the work referenced above. Hence, minimising these error sources has become an urgent topic for accurate mass flux measurement. Although several techniques [Xu et al., 1994; Tropea et al., 1995, Aizu et al., 1993] were developed which overcome this problem, they require additional optical components which further complicate the alignment procedure and rule out the ease of use of a standard PDA configuration.

The second major difficulty is the variation of the SNR of Doppler signals which makes accurate determination of velocity and diameter difficult. Since a laser beam has a Gaussian intensity distribution, signal amplitude is dependent on the particle trajectory even if the particle distribution is mono-disperse. The usually large variation of particle size makes the measurement even more difficult. For accurate mass flux measurements, all the particles passing through the measuring volume must be considered without biasing, and an appropriate technique must be introduced to minimise the ambiguity caused by the signal amplitude and particle trajectory. In an effort to meet these requirements, a signal processor based on an A/D converter was developed and proven to be effective at processing signals with low SNR around 0 dB or less [e.g. Hishida et al., 1989].

The purpose of the present study is to improve the accuracy of mass flux measurement with a standard PDA configuration, by estimating the correct dimensions of the effective measuring area. The estimation has been carried out by calculating the intensity and phase of the scattered light by spheres with a geometrical optics approximation considering both trajectory ambiguity and the slit effect as described in the following section. For rejection of these erroneous signals due to these effects, a phase ratio validation scheme [e.g. Hardalupas & Taylor, 1994] was introduced and the effectiveness of the rejection is discussed. For future convenience in use of a standard PDA, an approximate value of the effective measuring area is expressed by a function of normalised particle size and dynamic range of the signal processor. The results of mass flux measurement in a water spray and some improvements in signal processing are described in section 3, which is followed by the conclusions of the present study in section 4.

2 Determination of the Effective Measuring Area

2.1 Calculation Method

In the standard PDA, shown in Figure 1, only refracted light is used for sizing liquid droplets because the refracted light intensity is mostly dominant compared to the other scattering modes. However, signals due to reflected light, which has a phase-diameter response curve with opposite slope to that of refraction, can have a comparable intensity to refracted light if the scatterer is on the farther side of the measuring volume relative to the receiving optics, due to the Gaussian intensity distribution of the laser. This effect causes erroneous sizing for particles passing through region 'A' in Figure 2 and this is the so called 'trajectory ambiguity' (TA). Similarly, the use of a spatial filter in the receiving optics can affect sizing through the 'slit effect' (SE) because it blocks refracted light from reaching the collecting lens if the particle is on the edge of the measuring area, schematically indicated as region 'B' in Figure 2.

The TA and SE are both due to reflected light and the *sizing* data significantly affected by these effects must be rejected by an appropriate validation scheme. At the same time, the *area* over which TA and SE occur must be excluded from the effective measuring area of the PDA. Although the effects have been investigated independently, they may also combine to cause error in the measured size. For prediction of the area where particles give correct size, a calculation based on the geometrical optics approximation was carried out. Only reflection and the first order refraction were calculated for two polarisation and for two laser beams of LDV.

The coordinate system for the calculation is illustrated in Figures 1 and 2. Two beams, with linear polarisation parallel to the x-z plane, form an ellipsoidal *probe volume* and the size of this volume is defined by three diameters, D_x, D_y and D_z, where the intensity is $1/e^2$ of the maximum intensity at the centre. The centre of the probe volume corresponds to the origin o. The scattered light was collected by the receiving optics, which was placed at an elevation angle, ϕ, with a spatial filter determined by the dimensions of slit, Z_p. The volume limited by the spatial filter is referred to as the *measuring volume*. Although a particle in the measuring volume may scatter light, not all the scattered light necessarily reaches the detector due to the finite size of the particle and the presence of the spatial filter. Light scattered by a particle outside the measuring volume may reach the detector as well. In addition, the minimum detectable amplitude of the signal is limited by the amplitude resolution of the signal processor. Hence, the area, over which the particle scatters light and the light can be recognised by the processor, does not precisely coincide with the cross section of the measuring volume and is referred as *detection area* in this work. The area over which a particle provides correct size information, after a validation procedure has been applied, is defined as the *effective measuring area*, as indicated by D_y' and D_z'.

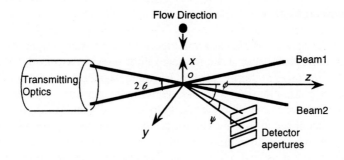

Figure 1. A standard phase-Doppler arrangement and the coordinate system used in this work.

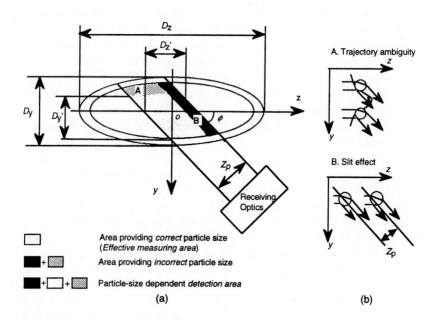

Figure 2. Definitions of the detection area and effective measuring area, and schematics of trajectory ambiguity and slit effect.

The scattered light intensity and phase difference were obtained by integrating the light intensity over the receiving aperture, using the same procedure as Hardalupas [1989]. The effect of diffraction by the slit was not taken into account in this work. For the dimensions listed in Table 1, the calculation domain for scattered light intensity and phase was limited to the plane y-o-z in Figure 2 and particles moved perpendicular to the plane y-o-z. A rectangular grid with 1 μm pitch was employed along the y-o-z plane for particle locations, and the phase differences between three detectors were calculated as functions of the particle location and size. The range of particle size was varied up to the size of the slit image, Z_p, and the particle refractive index remained constant at 1.33 for water droplets. Note that the intensity and phase were calculated for particles in the y-o-z plane and therefore the direction of the particle trajectory was not considered; this effect was eliminated by knowing the signal intensity and setting an appropriate threshold level, as described in section 3.

Table 1. Optical parameters used in the calculation.

a. Dimensions of probe volume

Condition	Slit image width [μm]	Max. diameter of particle [μm]	Probe diameter [μm]
1	200	200	180
2	400	400	180
3	400	400	350
4	200	200	250

b. Transmitting optics

Wavelength of laser [nm]	514.5
Beam intersection angle 2θ [deg]	4.66
Focal length of lens [mm]	300

c. Receiving optics

Elevation angle ϕ [deg]	30
Rotation angle ψ [deg]	1.699 (1-2ch)
	3.932 (1-3ch)
Diameter-phase coefficient [deg/μm]	0.731 (1-2ch)
	0.315 (1-3ch)
Dimensions of the aperture [mm]	3×20
Focal length of the collecting lens [mm]	300
Focal length of collimating lens [mm]	150

2.2 Size of the Effective Measuring Area

Figure 3 (a) and (b) show the measured signal amplitude contours of 20 and 180 μm diameter particles with condition No.1 in Table 1. The horizontal and angled dotted lines indicate the width of the probe volume and area limited by the slit, respectively. Since the signal processor used an A/D converter, the signal amplitude was normalised by 256 (8 bit) which corresponds to the maximum measurable signal amplitude from a particle. The maximum particle diameter was set equal to the slit width. The scattered light can be recognised only when the signal amplitude is above the minimum amplitude resolution of the converter, *i.e.* 1/256 of full scale range; thus the *detection area* was determined by this criterion and, hence, the contours correspond to the detection area and there are no contour lines outside the detection area as shown in Figure 3. The detection area is slightly displaced in the positive y and the negative z directions from the centre of the beams (probe volume) due to the geometrical relationship between the incident and scattered rays. This displacement is well known and the centre of the scattered light is quoted as the 'centre of locus' by Panidis and Sommerfeld [1996].

On the side farther from the receiving optics for a 180 μm particle, a reflection-dominated area was observed centred on negative y and positive z position due to reflected light associated with SE and TA. This effect is observed only with large particles, say above 60 μm in the present case.

Since measurement in the reflection-dominated area must be avoided because it is erroneous, a validation scheme based on the phase ratio of phases between detector channels 1-2 and 1-3 was applied [Hardalupas *et al.*, 1994; Sanker *et al.*, 1995]. The validation was done by rejecting signals with phase ratio far from the expected value, and the tolerance of the validation was defined as the Phase Acceptance Ratio, PAR,

$$\text{PAR} = 100 \cdot \left(1 - \frac{\left| d_{p1-2ch} - d_{p1-3ch} \right|}{d_{p1-3ch}} \right) \tag{1}$$

where d_{p1-2ch} and d_{p1-3ch} are the particle diameters calculated from the phase differences of two of the three detectors, respectively. This value will be close to 100 if the signal consists of refracted light only and decreases with increasing reflection component.

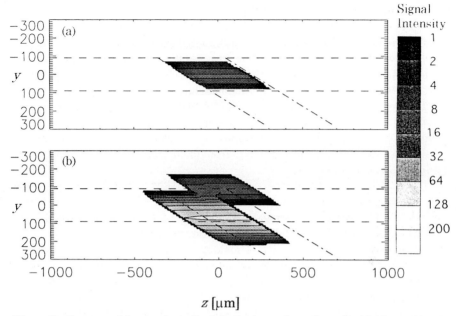

Figure 3. Contours of the signal amplitude in the measuring volume for (a) 20 μm, (b) 180 μm particles.

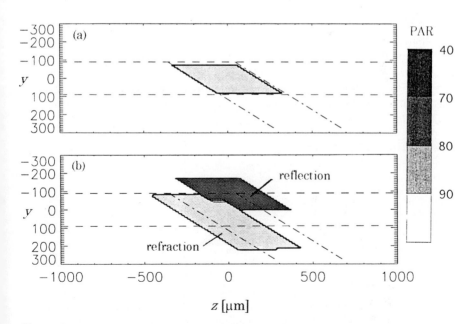

Figure 4. Contours of the phase acceptance ratio (PAR) in the measuring volume for (a) 20 μm, (b) 180 μm particles.

Figure 4 shows the PAR distribution. For a 20 μm particle, no sizing errors were observed whereas for a 180 μm particle, sizing errors exceeded 30% in the reflection-dominated area where the reflected light has relatively larger amplitude than refracted light. In this area, however, the signal amplitude was much smaller than at the centre of the refraction dominated area. The shape of the area is consistent with the result of the measurement by Willmann *et al.* [1994] and showed similar tendency over the calculated size range. These erroneous signals can be rejected by providing a limitation on the PAR tolerance and the *effective measuring area,* which is defined by the area including PAR validated particle locations, is therefore different from the detection area when the phase validation is introduced.

Figure 5 schematically illustrates the structure of the measuring cross section. The detection area (a+b+c) consists of three regions, which are refraction dominated (a), reflection dominated (b) and an intersection of these two regions (c). The dimensions of regions (a) and (b) in the z direction are limited by the slit. To the left-side of dotted line 1, only refracted light can be collected by the receiving optics due to geometrical relationship between the ray, the particle location and the slit as shown in Figure 2(b). On the contrary, only reflected light can be observed to the right side of line 2. Both reflection and refraction can be simultaneously comparable only when the particle is between lines 1 and 2 and, therefore, the effect of the trajectory ambiguity arises; in this region, reflected light affects the size measurement if the particle is above line 3. Since particles in regions (b) and (c) can be rejected by the phase validation as shown in Figure 4(b), the effective measuring area mainly consists of refraction dominated area (a) in the present optical configuration. It should be noted that the slit contributes to the reduction of the intersection area. The intersection area (c) was smaller than 4% of the refraction dominant area (a) over the calculated size range for all conditions; thus, the effective measuring area can be determined by the size of the refraction-dominated region, when the slit size and the diameter of probe volume are smaller than the maximum particle size.

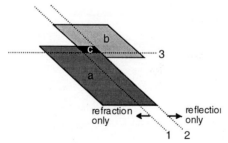

Figure 5. Schematics of the measuring cross section: (a) refraction-dominated; (b) reflection dominated; (c) intersection of (a) and (b). Areas (a) and (a+b+c) correspond to the effective measuring area and the detection area, respectively.

Figure 6 is the resultant average diameter over the *y-o-z* plane after phase validation with various PAR limitations. This figure denotes that the validation scheme is effective in rejecting erroneous measurement and that correct sizing can be achieved with a phase acceptance rate over 90%; the tolerance limit of PAR was therefore set constant at 90% in the experiments described below.

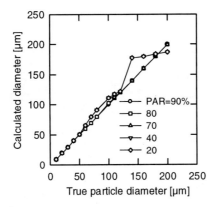

Figure 6. Accuracy of size measurement with various phase acceptance ratios (PAR) limit.

Figure 7 compares the detection area and the effective measuring area for condition No. 1. The reflection-dominated area, where sizing errors were observed, increased with particle size and the difference reached a maximum of 40% of the detection area. Hence, not all particles which scatter a detectable light intensity can be validated and the validation rate cannot be 100% if the scattered light is properly detected.

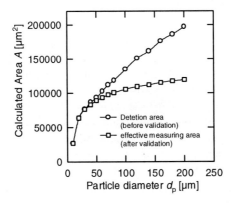

Figure 7. Comparison between the detection area due to the trajectory ambiguity and the slit effect.

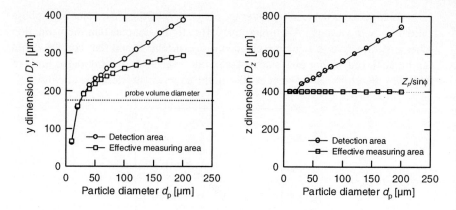

Figure 8. Dimensions of the detection area and the effective measuring area (y: along the line parallel to the slit image, z: along the line parallel to the z axis).

Detailed dimensions of the effective measuring and detection areas along the y and z axes are shown in Figure 8 (a) and (b), respectively. Figure (a) shows that the y dimension of the effective measuring area, D_y' in Figure 2, consistently increases with the particle size. On the other hand, D_z' is preserved equal to the z dimension of the slit image as shown in Figure (b), although the centre of the effective measuring area shifted toward negative z. These results suggest that the phase validation scheme can effectively eliminate the reflected light and that the sizing error due to the reflection can be neglected if the tolerance limit of PAR is set above 90%.

2.3 Derivation of an Approximate Equation for the Effective Measuring Area

The same procedure was carried out for the different optical arrangements for conditions No. 2, 3 and 4 as well as No. 1 (see Table 1). The relationship between the normalised effective measuring area (*i.e.* after validation) and the particle size showed similar tendency. Figure 9 shows the effective area, normalised by $Z_p D_p/\sin\phi$, against particle diameter normalised by the slit width. The resultant normalised areas coincided well for all conditions. Figure 10 shows the area for various dynamic ranges of the A/D converter. Thus, the size of effective measuring area can be determined as a function of the particle diameter, size of the slit and probe diameter.

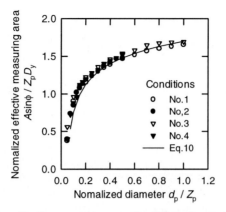

Figure 9. Size of the effective measuring area for different optical arrangements.

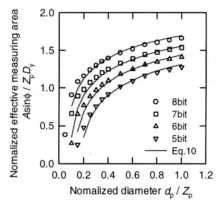

Figure 10. Size of the effective measuring area with different dynamic range / trigger level of the signal processor, in comparison with equation 10.

According to the results shown in figures 9 and 10, the size of the effective measuring volume, with the SE and TA taken into account, can be represented well by a simple function of particle size, probe volume size and dynamic range of the signal processor. The laser intensity depends on the spatial position of a particle in the measuring volume and can be given by

$$I = I_{max} \exp\left[-8\left(\frac{x^2}{D_x^2} + \frac{y^2}{D_y^2} + \frac{z^2}{D_z^2}\right)\right] \tag{2}$$

where I_{max} is the maximum intensity in the centre of the measuring volume. For the y dimension of the plane y-o-z, the incident intensity profile is given by

$$I = I_{max} \exp\left[-8\left(\frac{y}{D_y}\right)^2\right] \qquad (3)$$

As predicted by geometrical optics, the scattered light intensity is proportional to the square of the particle diameter; hence, the amplitude of the signal, S, becomes

$$S = Cd_p^2 \exp\left[-8\left(\frac{y_r}{D_y}\right)^2\right] \qquad (4)$$

where d_p and C are the particle diameter and a constant, respectively. The particle location, y_r, is measured from the location which provides the maximum intensity of the scattering light, rather than the origin of the coordinate, o. Although the location of maximum scattering intensity depends on the particle size as shown in Figure 3, it was neglected in this derivation since it does not affect the size of the effective measuring volume.

The restriction of area due to the signal processor is dependent on its amplitude dynamic range; denoting the maximum signal amplitude for particle size d_p as $S_{max}(d_p)$ and the minimum measurable amplitude as S_{min}, the diameter of the detection area in the y direction, $D_y{}'$, can be determined as follows:

$$D_y{}' = D_y \sqrt{\frac{1}{2}\ln\left(\frac{S_{max}(d_p)}{S_{min}}\right)} \qquad (5)$$

The maximum amplitude, $S_{max}(d_{p,max})$, is determined by maximum diameter of the particles, $d_{p,max}$. Then the minimum measurable amplitude, S_{min}, is determined by the dynamic range of the signal processor, R (=256 for 8 bit converter), as follows:

$$S_{min} = \frac{S_{max}(d_{p,max})}{R} \qquad (6)$$

As described by Eq.(4), the maximum amplitude scattered by particles of diameters d_p and $d_{p,max}$ are related through constant C,

$$C = \frac{S_{max}(d_p)}{d_p^2} = \frac{S_{max}(d_{p,max})}{d_{p,max}^2} \qquad (7)$$

Substituting Eqs. (6) and (7) in (5) for $D_y{}'$ yields

$$D_y' = D_y \sqrt{\frac{1}{2}\ln\left[R\left(\frac{d_p}{d_{p,max}}\right)^2\right]} \tag{8}$$

and this equation corresponds to the result in Figure 8(b).

On the other hand, the z dimension of the effective measuring area, D_z', is equal to the z dimension of the slit image:

$$D_z' = \frac{Z_p}{\sin\phi} \tag{9}$$

where ϕ and Z_p are the elevation angle and the width of the slit image, respectively. Hence, the size of the effective measuring area, A, is the product of D_y' and D_z', by neglecting the intersection area described in the previous section.

$$A \cong D_y' \cdot D_z' = \frac{Z_p D_y}{\sin\phi} \sqrt{\frac{1}{2}\ln\left[R\left(\frac{d_p}{d_{p,max}}\right)^2\right]} \tag{10}$$

Comparison of the results from geometrical optics with Eq.(10) are shown in Figure 9 and 9 by solid lines. These figures show that Eq. (10) represents the size of the effective measuring volume, although underestimation is observed at small particle size ranges. This underestimation does not affect the accuracy of mass flux measurement of sprays because most of the mass is always contained in the larger size range. The difference between the approximate equation (Eq.10) and that produced by geometrical optics was no more than 7% for $0.3 < d_p/Z_p < 1.0$.

In practice, the difficulty with using Eq. (10) is the determination of the maximum particle size, $d_{p,max}$, since there exists an error resulting from poor statistics at the largest, infrequent diameters; therefore, the methods such as the 'weighting function' [Shöne *et al.*, 1994] should be employed to overcome these difficulties. The weighting function procedure was performed by choosing a reference diameter, d_r, the size class with maximum number of particles. Once d_r and the maximum signal amplitude in the reference size class $S_{max}(d_r)$ were obtained, the maximum diameter, $d_{p,max}$, could be determined with the aid the geometrical optics as follows. Using Eq. (7),

$$d_{p,max} = d_r \sqrt{\frac{S_{max}(d_{p,max})}{S_{max}(d_r)}} \tag{11}$$

where the value of $S_{max}(d_r)$ corresponds to the maximum amplitude from a particle with a diameter of d_r. The measurement procedure should thus ensure that the signal amplitude should be adjusted, by changing laser power or detector sensitivity, to cover the whole size range. Thus, instead of Eq.(10), the size of the effective measuring area can be described as

$$A = \frac{Z_p D_y}{\sin \phi} \sqrt{\frac{1}{2} \ln\left[\frac{S_{max}(d_r)}{S_{max}(d_{p,max})} \left(\frac{d_p}{d_r}\right)^2 \right]} \qquad (12)$$

The preceding equation describes the effective measuring area based on a set of experimental data. With the aid of Eq.(12), mass flux F_n is described as follows:

$$F_n = \frac{\pi}{6T_s} \sum_{i=1}^{n} \frac{\rho N_i d_{pi}^3}{A_i} \qquad (13)$$

where, ρ, T_s and N_i are the density of the particle, total sampling time and number of particles for each particle size class i, respectively.

3 Experiment

In order to examine the accuracy in the area estimation presented in the previous section, two experiments were undertaken. Both tests were designed to provoke TA and SE because of the choice of particle size. The first test was conducted under conditions of low particle concentration spray in a low noise environment, so that the signal processing technique could be examined. The second test was to measure mass flux of a dense spray using the optimum setting of PDA established by the first experiment. The accuracy of the measurement and ability to eliminate the two effects of TA and SE were evaluated by measuring a radial profile of flux and estimating the flowrate by numerical integration and comparing it with the independently measured flow-rate.

3.1 PDA Setup

As the maximum diameter of the particle produced by the nozzle used for these two tests was known to be 250 μm from previous work, the probe diameter was set at 250 μm: the condition provoked trajectory ambiguity and the slit effect. The size of the slit was set to provide an image of 400 μm in the probe volume, which was larger than the maximum particle size. The rest of the optical parameters are presented in Table 1(b). Note that the validation tolerance of PAR was set at 90% in all the experiments, in order to provide the correct size measurement.

In order to calculate size and velocity, an FFT-based signal processor [Kobashi *et al.*, 1992; Higuchi *et al.*, 1994] was employed, as shown in Figure 11. Doppler signals were transferred into a 2M byte transient recorder to retain the time-series data. Band-pass filters were not used to avoid the influence of poor signal visibility, because the visibility can often be smaller than unity. The FFT spectrum estimation was performed by a Digital Signal Processor (Motorola 56001), selecting the centre of the burst signal, in which the SNR is high (*centre-search function* of Higuchi *et al.*, 1994).

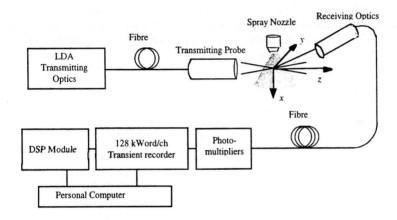

Figure 11. Block diagram of the PDA system and the signal processor.

3.2 Optimisation of the Trigger Level

For mass flux measurements of droplets with a broad size distribution, e.g. spray droplets, the variation of scattered light intensity is wide. Qiu & Sommerfeld [1992] and Higuchi *et al.* [1994] have come to the conclusion that employing an *amplitude trigger* and *centre-search function* is appropriate for the A/D converter based processors since this maximises the SNR: the SNR can be improved by 8 dB compared to that without the function [Higuchi *et al.*, 1994]. In their work, the trigger level was set as low as possible to obtain low amplitude signals from small particles.

Even with the centre-search function, however, many signals are missed due to the poor SNR of very low amplitude signals of which the maximum signal amplitude is comparable to the trigger level. Such signals are caused mainly by two unwanted conditions: one is from small particles and the other is large particles passing through the edge of the measuring volume. Because of the Gaussian intensity profile of the measuring volume, the probability of signals with low amplitude from large particles is relatively high and these missed particles yield a large underestimation of mass flux. Furthermore, at the extreme condition of setting the trigger level such that it approaches 0 volts, erroneous measurements still occur due to the small number of fringes in the signal and low SNR.

In order to overcome the problems described above, some sets of flux measurements were performed with hollow-cone spray (Delavan, type B, TN, USA) for optimisation of the trigger level in signal processing. The flow direction was kept almost normal to the interference fringes, and measurements were limited to a circular cross-section of the spray cone at 30 mm below the nozzle exit. The atomising pressure was set at 0.7 MPa, which provided a particle size range from 10 to 120 µm and a 69 µm Sauter mean diameter over the circular cross-section. The stored signals were re-processed with various trigger levels and the resultant mass flow rates were compared with the total mass of the water added to the nozzle; the true mass flow rate, obtained by mass of water supplied to the nozzle, was 0.32 mg/s. The measured flow rate was calculated by integrating the local mass fluxes measured over the cross-sectional circular plane. The mean volume fraction in the measured region was estimated at $1.8 \times 10^{-4}\%$ which is unlikely to cause frequent multiple occupancy of particles in the measuring volume.

The area-averaged mass flux downstream of the spray cone is presented as a function of the trigger level setting in Figure 12, together with the validation rate, *i.e.* the ratio of the number of validated particles to sampled particles. The results were calculated from a series of stored data for all conditions by using the effective measuring area size given by Eq.10. The local mass flux was obtained from Eq. (13). The trigger level was changed from 0.8% to 25% of full scale of the A/D converter which corresponds to the amplitude resolutions from 7 bit down to 2 bit. With increasing trigger level, the SNR of the signals increases and the effective measuring area decreases, as shown in Figure 10. Figure 12 shows that by varying the trigger level, the ratio of measured flow rate changes by up to 81% and this demonstrates that the area estimation described in section 2 is reasonably accurate. This result justifies the validity of the approximate equation of the effective measuring area (Eq. 10). Since the normalised mass flux and validation rate coincided with trigger levels below 15% and the trigger level was too high to be affected by TA, the bias error was likely to be caused by multiple scattering, signal noise and obscuration of the optical path, between the measuring volume and the receiving unit, by droplets. This result was also suggested by another independent experiment with an instrument based on imaging technique. The error was relatively large when the trigger level was set below 5% and this is due to collection of signals with poor SNR. In addition, this figure empirically justifies the compensation of the data using the validation rate to estimate true flux, although it was *not* used in this work.

Figure 13 shows the inferred size distribution. Increasing the trigger level improved the validation rate due to the reduction of the effective measuring area and high SNR; however, at the same time, small particles were missed as shown in Figure 13. Since the small particles are not dominant in flux measurement, there is a trigger level which optimises the measurement and retains good SNR. In the present test, the best condition was the 6.4% trigger level which provided the maximum mass flux in Figure 12; thus, with the aid of appropriate trigger level, the bias errors of the flux measurement due to the limitation of the performance of the processor can be minimised. The optimum trigger level might change under different conditions: however, it should be around 6% for any bit resolution of the A/D converter for sprays with typical size distribution.

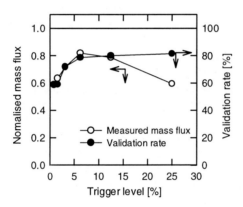

Figure 12. Mass flowrate (open circle) and validation rate (blocked) averaged over the cross section of the hollow-cone spray; the true value was determined by water mass flowing through the nozzle.

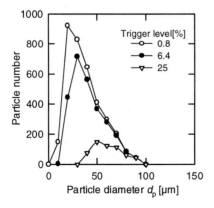

Figure 13. Particle size distribution with various trigger levels, relative to the maximum signal amplitude measured.

3.3 Mass Flux Measurement in Dense Spray

Finally, mass flux measurements of a dense spray with a wide size distribution, which could represent difficult measurement conditions for PDA, were performed with the aid of the results described in the previous section. The flow rate of the spray was about 80 times that of the former nozzle, 26 mg/s. The geometrical arrangement of nozzle was identical to the previous experiment, however, the nozzle (Ikeuchi, Japan, atomising pressure set at 0.25 MPa) provided droplet sizes with wider distribution, from 20 to 250 μm, and a Sauter mean diameter over the circular cross-section of 111 μm. The wide size distribution ensures that the large particles yield TA with the present optical arrangement. Radial profiles were measured at four cross-sections of the spray cone at 20, 30, 40 and 60 mm below the nozzle exit. For this flow condition, the appropriate trigger level was 6 %, as suggested in the previous experiment. In addition, SNR validation [Kobashi *et al.*, 1990] was introduced and the threshold value was set at -5 dB.

Figure 14 shows the total mass flow rate across each cross-sectional plane. For such a dense flow condition with large contributions from TA and SE, the accuracy of the averaged mass flowrate measurement was about 72% at x=60 mm plane in which volume fraction was 0.6×10^{-3}%. Most of the rejected particles were in the small size range (less than 50 μm) due to low SNR associated with weak scattered light, and the PAR limitation played a dominant role in removing erroneous data in the larger size range. The worst result of the flowrate was 57% at the 20 mm plane with an estimated average volume fraction of 3×10^{-3}%. The measured mass flowrate approached a constant beyond 40 mm downstream with almost the same accuracy as the dilute spray.

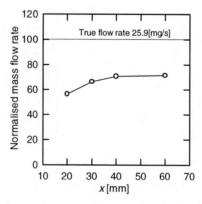

Figure 14. Integrated mass flow rate of dense spray as a function of axial location.

As a consequence, the trigger level optimisation contributed the elimination of the erroneous data due to TA and SE, and the approximate equation of effective measurement area provided the correct dimensions of the effective measuring area. The measurement error was reduced by about 15% compared with our previous work carried out under similar conditions.

4 Conclusion

In order to achieve accurate mass flux measurement, the effective measuring area for a standard configuration PDA was calculated by the geometrical optics approximation method and the results were used to derive an equation which can be use to estimate the effective area without further light-scattering calculation.

In addition, the trigger level of the A/D converter-based signal processor was carefully selected to reduce the number of particles rejected owing to low SNR. The measurement of a dense spray proved that the measurement of the mass flux using the trigger level suggested in this work can provide better accuracy than a previous work. The important findings of this work are summarised as follows:

(1) The results of calculation of scattered light intensity by the geometrical optics suggested that the phase validation scheme is an effective way of removing erroneous measurements caused by the 'trajectory ambiguity', if the validation tolerance of phase ratio (PAR) is set above 90%. This scheme can also remove the 'slit effect', when the particle size is smaller than the measuring volume.

(2) The dimensions of the *effective measuring area* can be determined by considering only refracted light, since the validation scheme can remove signals form particles in the reflection dominated area. The intersection area between refraction and reflection dominated areas in the detection area showed a maximum of about 4% of refraction dominated area with the present optical configuration and decreased with decreasing the particle size.

(3) The dimensions of the *effective measuring area* was derived as a function of the diameter of the probe volume, the size of the slit, particle diameter and the amplitude resolution of the signal processor. The approximate equation (Eq.10) showed a difference of only 7% of the effective measuring area compared to that calculated by geometrical optics. This difference does not affect the accuracy of mass flux measurement, because it decreases with increasing particle size and these large particles have the major contribution to the measured mass flux.

(4) In order to maximise the validation rate and to minimise the uncertainty of mass flux measurement, the trigger level should be optimised. The level

suggested by the present setup was 6% which is likely to be consistent for any A/D converter-based signal processor with optical configuration similar to the present work. The accuracy of mass flux measurement was evaluated in a dilute spray with a volume fraction of about 1.8×10^{-4}% and it was confirmed that the uncertainty was less than 20%, with a trigger level at 6.4% of the amplitude range of the signal processor. This uncertainty is probably due to obscuration of the optical path by droplets and multiple scattering.

(5) The selected trigger level was also applied to measure mass flux in a dense spray of volume fraction of about 2.4×10^{-3}%. Results based on the effective measuring area calculated by the approximate equation showed that the measured flux was typically between 56% and 70% of the true mass flux, which is 15% better than our earlier work carried out with the same instrument.

Acknowledgement

The authors are grateful to Dr. Y. Hardalupas for providing the original computer code and Mr. M. Watanabe for his assistance on the experiments. H M would like to thank the Japan Society for the Promotion of Science for providing the Research Fellowship (Grant No.4927) which partly supported the project.

Symbols and Abbreviations

A	effective measuring area
C_v	volume fraction
d_p	particle diameter
d_r	reference diameter
D_x, D_y, D_z	dimensions of probe volume
D_y', D_z'	dimensions of effective measuring area
ϕ	elevation angle
F_n	mass flux
I	intensity of the beam
PAR	phase acceptance rate
θ	beam intersection half angle
R	dynamic range of the signal processor
S	signal amplitude
SE	slit effect
SNR	signal to noise ratio
TA	trajectory ambiguity
x, y, z	Cartesian coordinates
Z_p	width of the image of the silt

References

Aizu, Y., Durst, F., Grehan, G., Onofri, F. and Xu, T., H.: PDA-system without Gaussian beam defects, 3rd Int. Cong. on Optical Particle Sizing, Yokohama, Japan, pp. 461-470, 1993.

Bachalo, W., D., Rudoff, R., C. and Breña de la Rosa: A mass flux measurements of a high number density spray system using the phase Doppler particle analyzer, AIAA 88-0236, 1988.

Domnick, J., Durst, F., Melling, A., Qiu, H. -H., Sommerfeld, M. and Ziema, M.: A new generation of phase-Doppler instruments for particle velocity, size and concentration measurements, 3rd Int. Cong. on Optical Particle Sizing, Yokohama, pp. 407-414,.1993.

Durst, F., Tropea, C. and Xu, T.-H.: The Slit Effect in Phase Doppler Anemometry, 2nd Int. Conf. on Fluid Dynamic Measurement and its Applications, Bejing, Chaina, pp. 38-43, Oct. 1994.

Gouesbet, G. and Gréhan, G.: Gaussian beam errors in phase-Doppler Anemometry and their elimination, 5th Int. Symp. on Applications of Laser Anemometry to Fluid Mechanics, Lisbon, Portugal, 11.6., pp. 243-259, 1990.
Hardalupas, Y., Ph.D. Thesis, Imperial College, 1989.

Hardalupas, Y. and Taylor, A. M. K. P.: Phase validation criteria of size measurements for the phase Doppler technique", Exp. in Fluids, 17, pp. 253-258, 1994.

Higuchi, M., Shirakawa, T., Morikita, M., Hishida, K. and Maeda, M.: Experimental study of multiple interacting sprays by phase Doppler anemometry, 7th Int. Symposium on Applications of Laser Techniques to Fluid Mechanics, Lisbon, paper 31.2, 1994.

Hishida, K., Kobashi, K. and Maeda, M.: Improvement of LDA/PDA using a digital signal processing systems, Third Int. Conf. on Laser Anemometry, Advances and Applications, Swansea, UK, 1989.

Kobashi, K., Hishida, K. and Maeda, M.: Measurement of Fuel Injector Spray Flow of I.C. Engine by FFT Based Phase Doppler Anemometer, in R. J. Adrian, D. F. G. Durao, F. Durst, M. Maeda, J. H. Whitelaw (Eds.) Applications of Laser Technique to Fluid Mechanics, Springer-Verlag, pp. 268-287, 1990.

Kobashi, K. Hishida, K. and Maeda, M.: Multi-purpose high speed signal processor for LDA/PDA using DSP array, Sixth International Symposium

Applications of Laser Techniques to Fluid Mechanics, paper 21.6, Lisbon, Portugal, 1992.

Panidis, Th. and Sommerfeld, M.: The locus of centres method for LDA and PDA measurements, 8th Int. Symp. on Applications of Laser Techniques to Fluid Mechanics, paper 12.5, Lisbon, Portugal, 1996.

Qiu, H. H. and Sommerfeld, M.: A reliable method for determine the measurement volume size and particle mass fluxes using phase-Doppler anemometry, Exp. in Fluids, 13, pp. 393-404, 1992.

Saffman, M.: Automatic calibration of LDA measurement volume size, Appl. Opt., Vol. 26, No. 13, pp. 2592, 1987.

Sanker, S. V., Inenaga, A. S. and Bachalo, W. D.: Trajectory Dependent Scattering in Phase Doppler Interferometry: Minimizing and Eliminating Sizing Error, in R. J. Adrian, D. F. G. Durao, F. Durst, H. V. Heitor, M. Maeda, J. H. Whitelaw (Eds.) Laser Techniques and Applications in Fluid Mechanics, Springer-Verlag, pp. 75-89, 1992.

Sanker, S. V., Bachalo, W. D. and Robart, D. A.: An adaptive intensity validation technique for minimizing trajectory dependent scattering errors in phase Doppler interferometry, 4th Int. Congress on Optical Particle Sizing, Nürnberg, Germany, pp. 1-14, 1995.

Schöne, F., Bauckhage, K. and Wriedt, T.: Size of the detection area of a phase-Doppler anemometer for Reflecting and Refracting particles, Part. Part. Syst. Charact., 11, pp. 327-338, 1994.

Sommerfeld, M. and Qiu, H. -H.: Particle concentration measurements by phase-Doppler anemometry in complex dispersed two-phase flows, Exp. in Fluids, 18, pp. 187-198, 1995.

Tropea, C. and Xu, T.-H., Onofri, F., Gréhan, G., Haugen, P. and Stieglmeier, M.: Dual mode phase Doppler anemometer, 4th Int. Congress Optical Particle Sizing, Nürnberg, Germany, pp. 287-296, 1995.

Willmann, M., Kneer, R., Eigenmann, L., Wittig., S and Hirleman, E.: Experimental investigations on the effect of trajectory dependent scattering on phase Doppler particle sizing with a standard instrument, 7th. Int. Symp. on Applications of Laser Techniques to Fluid Mechanics, Lisbon, Portugal, 18.1, 1994.

Xu. T.-H. and Tropea, C.: Improving the performance of two-component phase Doppler anemometers, Meas. Sci. Technol. 5, pp. 969-975, 1995.

II.5 The Locus of Centres Method for LDA and PDA Measurements

Th. Panidis[1] and M. Sommerfeld[2]

[1] Laboratory of Applied Thermodynamics, Mechanical Engineering and Aeronautics Department, University of Patras, GR 265 00 Patras-Rio, Greece
[2] Institute für Mechanische Verfahrenstechnik und Umweltschutztechnik, Martin-Luther-Universität Halle-Wittenberg, D-06099 Halle (Saale), Germany

Abstract. A new method is introduced for the analysis of LDA and PDA optical setups, especially suited for larger particles or bubbles. The properties of an optical setup are studied for a specific particle size in an effective control volume defined as the Locus of the Centres of the particles producing interference signals. The effect of the particle size on the size and position of such a control volume are evaluated and discussed. The convenient description using control volume fringes is adapted to suit larger particles and used to derive the equations of LDA and PDA measurements. The method sets new standards for the estimation of the active control volume as a function of the particle size, for dispersed two phase flows measurements.

Keywords. Locus of Centres method, LDA, PDA

1. Introduction

The non intrusive character of the Laser Doppler Anemometer (LDA) and its sizing counterpart Phase Doppler Anemometer (PDA) in addition to their measuring performance have established them among the most advanced techniques for fluid flow and two-phase research. A great deal of work has been done in the last two decades and significant contributions have advanced the physical understanding of the techniques. The use of geometrical optics allows a very good description of both techniques and satisfactory results for most of the applications (see e.g. Saffman, 1987; Sankar and Bachalo, 1991). Furthermore the development of more accurate tools based on Lorenz-Mie theory (LMT) has significantly improved our understanding and capability to optimise demanding optical setups (e.g. Grehan et al. 1992).

In the last years research in dispersed two phase flows has greatly benefited from the advances in the PDA technique. The ability of PDA to measure simultaneously the velocity and size of the dispersed phase has been extended to concentration and particle flux measurements (Hardalupas and Taylor, 1989; Sommerfeld and Qiu, 1995) while new capabilities seems to be on the way (Tropea et al. 1995). On the other hand the demand to understand the measurement principle for larger particles, as for example in bubbly flows, has increased. For this task geometrical optics is used with satisfactory results. Lorenz-Mie calculations are used for setup optimisation mainly for small and medium particle sizes (Saffman 1987; Grehan et al. 1992) since large particles demand significant computational power. Moreover, for large particles most of the scattering phenomena can be included in geometrical optics models. In the frame of geometrical optics analysis the present work is aiming to improve our understanding of the effect that larger particles have on the measuring parameters of LDA and PDA.

It is common practice measurements of the dispersed phase properties in two phase flows to be attributed to the individual particle present in the measuring volume. Therefore, it is reasonable to map the measuring volume to a corresponding volume that is defined by the Locus of the Centres (LoC) of the sensed particles. In the present work this methodology is applied to LDA and PDA measuring volumes revealing some aspects not readily observable in previous derivations.

Geometrical optics is used to analyse the properties of the Locus of Centres (LoC) volume in flows with reflecting or transparent spherical particles, i.e. reflection and first order refraction are considered.

In comparison with the conventional geometrical optics analysis the present work incorporates the same level of accuracy to a more convenient scheme. Of course it lacks the accuracy of LMT methods but this drawback is not considered important for larger particles for which the presented method is more suitable.

2. Locus of Centre of a Ray

In the following the principle of the LoC method is introduced for reflecting and refracting particles. Throughout this work boldface letters are used to indicate vectors. Uppercase letters, unless explicitly specified otherwise, are used for position vectors originating at point **O**, that is the centre of the co-ordinate system, while lowercase letters stand for unit vectors.

2.1 Reflecting Particles

LDA is based on the interference of Laser light scattered on a particle. In a dual beam setup two laser beams with direction unit vectors $\mathbf{k}_1$ and $\mathbf{k}_2$ are incident on the particle. For purely reflecting spherical particles only one ray for each beam, reflected on the corresponding point **X** on the particle surface, will reach the photodetector (PD) located at point **P** (figure 1). The unit vector in the direction that the PD is seeing point **X**, that is in the direction of the difference **X-P**, is **m**. The centre of the particle is located at point **Q** at distance r_p from **X** on the bisection of the transmitted and the reflected ray, that is in direction **n**.

$$\mathbf{Q} = \mathbf{X} + r_p \, \mathbf{n} \qquad (1)$$

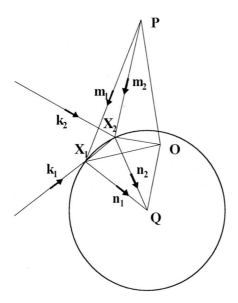

Figure 1. Reflection on a sphere

with
$$n = \frac{k + m}{|k + m|}$$
(2)

The line consisting of all points **Q** corresponding to all points **X** along a ray is the ray Locus of Centres (LoC) corresponding to this particle size.

2.2 Refracting particles

In figure 2 refraction of a ray through a sphere is studied. The upper part of the figure is referring to refraction through a particle with refractive index n_2 smaller than the refractive index n_1 of the surrounding medium, and the lower part to the inverse case. Simple geometrical considerations show that the centre of a sphere of radius r_p refracting a ray of unit vector **k** towards direction **-m**, is located at the same point as the centre of a sphere of radius r_r reflecting the same ray towards the same path. The reflection takes place either on the convex (for $n_1 > n_2$), or on the concave (for $n_1 < n_2$) surface of the sphere. If γ is the angle between direction vectors **m** and **k** then:

$$r_r = r_p \frac{\sin\left(\arctan\left(\frac{n_{rel}\cos(\gamma/2)}{1 - n_{rel}\sin(\gamma/2)}\right)\right)}{\sin(\gamma/2)}$$
(3)

where $n_{rel} = n_2/n_1$.

Positive values of r_r indicate reflection on a convex surface and negative ones on a concave. In this way the vector $\mathbf{R} = r_r \mathbf{n}$ always points from the apparent

206

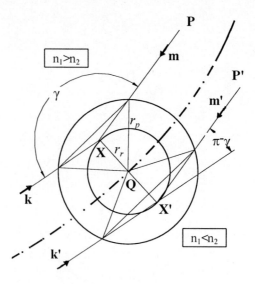

Figure 2. Refraction through a sphere

reflecting point to the centre of the sphere. According to this derivation the LoC of rays refracted through spherical particles can be obtained in the same way as with reflected ones on the bisection of the incident and the reflected or refracted ray at distances which for given angles γ will be proportional to the particle size.

In figure 3 the evolution of r_r/r_p is given for water droplets in air (n_{rel}=1.334) and for air bubbles in water (n_{rel}=0.75). For angles γ less than 97.1° total reflection either at the outer surface (n_{rel}=0.75) or at the inner surface (n_{rel}=1.334) prevents the transmission of this kind of rays. This limit corresponds to r_r=r_p for n_{rel}<1 and r_r=$r_p n_{rel}$ for n_{rel}>1. This means that the apparent reflecting sphere is larger for n_{rel}>1, or equal for n_{rel}<1 to the real one near total reflection angle. For larger angles γ, that is in the forward direction, the size of the apparent reflecting sphere is decreasing to zero.

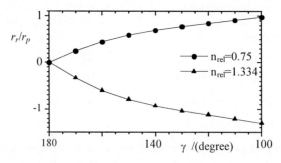

Figure 3. Radius correction for refraction

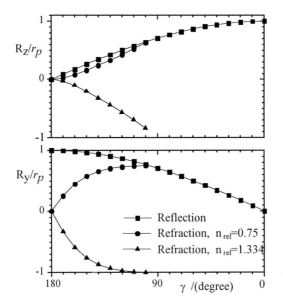

Figure 4. Non-dimensional distance of the LoC points from the ray points

2.3 Displacement of a ray's LoC

The particle centre is displaced from the real (reflection) or apparent (refraction) reflecting point by:

$$\mathbf{R} = r_r \mathbf{n} \tag{4}$$

setting $r_r = r_p$ for reflection and using equation (3) for refraction. The displacement vector $\mathbf{R}$ (it is clear that this is not a position vector) is defining in this way the location of a ray's LoC relatively to the ray.

Let us define a two-dimensional co-ordinate system y-z for the ray-receiver plane. Co-ordinate z is in the direction of ray propagation and y is normal to it. The photodetector is located on the side of negative y. Direction $\mathbf{n}$ according to equation (2) is given by:

$$\mathbf{n} = \sin(\gamma/2)\mathbf{y} + \cos(\gamma/2)\mathbf{z} \tag{5}$$

where $\mathbf{y}$, $\mathbf{z}$ are the unit vectors in the corresponding directions and γ is the intersection angle.

In figure 4 the components R_z and R_y of $\mathbf{R}$ for reflected or refracted rays non-dimensionalised with the particle radius r_p are given as functions of the intersection angle γ. For forward reflection ($\gamma=180°$) the LoC is at distance r_p normal to the ray, indicating grazing reflection on the particle surface. For backward reflection ($\gamma=0°$) the ray is reflected at normal incidence on the particle surface and the LoC is on the ray at distance r_p behind the reflecting point. For refraction the LoC is departing mostly from the apparent reflecting point on the ray just before total reflection limit. It has to be noted that since refraction for

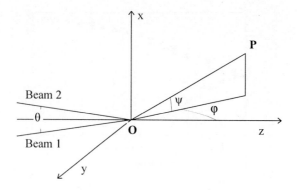

Figure 5. LDA setup

$n_{rel}>1$ is equivalent to reflection on the concave surface of a sphere the LoC for reflection and refraction depart from each other for $n_{rel}>1$ especially near the total reflection angle. On the contrary for $n_{rel}<1$ the LoC for reflection and refraction are almost the same near total reflection angle.

3. Dual Beam LDA Setup

A conventional dual beam LDA optical setup is shown in figure 5. Two laser beams, with half angle $\theta/2$, are propagating on the z-x plane towards positive z. Their waists are centred at point **O**. The photodetector is located at point **P** at an elevation angle ψ from the y-z plane. The off-axis angle φ is measured on this latter plane. The planar configuration described in the preceding section corresponds to $\theta=\psi=0°$.

3.1 LoC Control Volume

Following the same line of reasoning the LoC of the two beams can be defined. It is obvious from figure 4 that the LoC of a beam is not geometrically similar to the beam itself. Assuming though that the distance l from **O** to **P** is large, angle γ will be constant in the vicinity of the beam intersection. In that case the LoC of a beam section in the vicinity of **O** is geometrically similar to the beam section itself and parallel displaced by **R**. According to equation (2) and using the notation of the LDA setup the unit vectors $\mathbf{n_1}$ and $\mathbf{n_2}$ for beams 1 and 2 are:

$$\mathbf{n_1} = \frac{\sin(\theta/2)-\sin\psi}{f_m}\mathbf{x} + \frac{\cos\psi\cos\varphi}{f_m}\mathbf{y} + \frac{\cos(\theta/2)-\cos\psi\cos\varphi}{f_m}\mathbf{z} \qquad (6)$$

$$\mathbf{n_2} = \frac{-\sin(\theta/2)-\sin\psi}{f_p}\mathbf{x} + \frac{\cos\psi\cos\varphi}{f_p}\mathbf{y} + \frac{\cos(\theta/2)-\cos\psi\cos\varphi}{f_p}\mathbf{z} \qquad (7)$$

where **x**, **y**, **z** are the unit vectors in the corresponding directions and

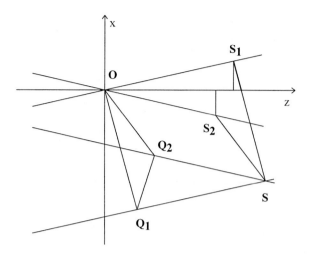

Figure 6. Intersection of rays' LoC

$$f_m = \sqrt{2}\left(1 - \sin(\theta/2)\sin\psi - \cos(\theta/2)\cos\psi\cos\varphi\right)^{1/2} \tag{8}$$

$$f_p = \sqrt{2}\left(1 + \sin(\theta/2)\sin\psi - \cos(\theta/2)\cos\psi\cos\varphi\right)^{1/2} \tag{9}$$

The LoC of the two beams are evolving in different directions as the size of the particle increases and for large particles they may depart completely from each other. In the frame of the present analysis a particle transmits only one ray from each beam to the PD, for every mode of transmission (reflection and refraction). In order to have interference from the two beams at the PD, a particle centre should be in the intersection of the LoC of the two beams. Moreover the apparent reflecting points should be viewed by the PD.

For reflection and first order refraction there are four displacement vectors R_{ij} (i=1,2 for beam 1 and 2 and j is used to indicate reflection or refraction) for the two beams. Four vector differences D_{jk} (subscripts j and k indicate independently refraction or reflection for beam 1 and 2 respectively) give the relative displacement of the beams' LoC and describe how they intersect with each other,

$$D_{jk} = R_{1j} - R_{2k} = r_{r1j}n_1 - r_{r2k}n_2 = D_{xjk}\,x + D_{yjk}\,y + D_{zjk}\,z \tag{10}$$

Usually we are interested in intersections of the same type beams' LoC, that is either for reflection or for refraction. Discrepancies though may occur due to the intersection of different type of beams' LoC, as the Gaussian beam effect discussed in the next section.

Differences in y components indicate reduction of the effective control volume as the LoC for intersecting rays are no longer on the same plane. If the y component of the difference is larger than the beam diameter no interference is possible. Differences in the x and z components reveal changes of the point of

intersection. In figure 6 the beam intersection centre **O** corresponds to points **Q₁**
and **Q₂** on the LoC of beam 1 and 2 respectively. Assuming that $D_y = 0$, the
intersection of the LoC of the central rays is at **S**, which corresponds to point **S₁**
and **S₂** on the central rays of beam 1 and 2 respectively. This means that
interference signals will be produced when a particle centre is at **S** and the PD is
viewing points **S₁** and **S₂**. Geometrical considerations show that:

$$S_1 = -\frac{D_x + D_z \tan(\theta/2)}{2}\left(x + \frac{1}{\tan(\theta/2)}z\right) \tag{11}$$

$$S_2 = -\frac{D_x - D_z \tan(\theta/2)}{2}\left(x + \frac{1}{\tan(\theta/2)}z\right) \tag{12}$$

3.2 Locus of "Fringes"

The understanding of the LDA method is significantly facilitated with the
introduction of the control volume fringes produced due to the interference of the
Laser beams. These fringes are the result of the arrival of the incident rays of the
two beams at the control volume points with phase difference which is varying in
the x direction. Since a point scatterer is assumed this phase difference is sensed
by the PD which acts as an optical integrator. This concept is not suited for larger
particles and the corresponding analysis usually concentrates at the spatial fringe
pattern scattered by a particle to the far field.

In the LoC control volume approach we can define, for a specific particle size
and PD location, Locus of "Fringes" (LoF). When a particle centre is in a LoF the
two reflected (or refracted) rays of the two beams arrive at the PD with a phase
difference of 180°. Following van de Hulst (1981) and Bachalo (1987) the optical
path length of a light ray through a sphere of diameter d relative to a reference ray
deflected at the centre of the sphere can be expressed as

$$\eta = \frac{2\pi n_1 d}{\lambda}(\sin\tau - pn_{rel}\sin\tau') \tag{13}$$

where λ is the laser wavelength in vacuum, $p=0$ for reflection and $p=1$ for
refraction, and τ and τ' are the angles between the surface tangent and the incident
and refracted rays respectively.

The rays from each beam are incident upon the sphere at different angles and
therefore reach the PD by different optical paths. Neglecting the phase shifts at
reflection points and focal lines, the phase difference of the two rays is,

$$\delta = \frac{2\pi n_1 d}{\lambda}\left((\sin\tau_1 - \sin\tau_2) - pn_{rel}(\sin\tau'_1 - \sin\tau'_2)\right) + \delta_0 \tag{14}$$

where the subscripts (except in n_1) represent beam 1 and 2 and δ_0 is the phase
difference of the reference rays at the centre of the sphere. Since the angles τ are
fixed by the geometry of the setup the first term in the right hand side of the above

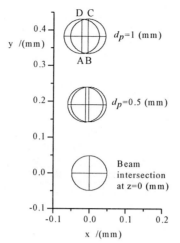

Figure 7. Displacement of the x-y cross section of the LoC control volume for reflecting spheres of d_p=0.5 and 1.0 mm

equation will be constant for a given diameter d. The second term, δ_0, is the phase difference of the incident beams responsible for the fringe pattern in the classical LDA analysis. This means that the LoC fringes will have the same spacing as the normal fringes, that is,

$$d_f = \frac{\lambda}{2\,n_1\,\sin(\theta/2)} \tag{15}$$

and they will be displaced by

$$d_x = \frac{d}{2\sin(\theta/2)}\left(\left(\sin\tau_1 - \sin\tau_2\right) - p n_{rel}\left(\sin\tau_1' - \sin\tau_2'\right)\right) \tag{16}$$

When a particle centre is crossing the LoC intersection, that is the LoC control volume, normal to the fringes, the photodetector is seeing the familiar Doppler burst.

3.3 Effective Control Volume

The effective control volume for a certain particle size is the LoC control volume corresponding to the beam sections viewed by the photodetector. For an optical setup focused on the beam intersection centre the effective control volume is the intersection of the two beam' LoC corresponding to the part of the control volume viewed by the PD. In figure 7 the displacement of the LoC of the x-y cross section of a control volume is shown for reflecting particles of diameter 0.5 and 1.0 mm (θ=1.77°,φ=80°, ψ=2° and beam waist radius w_o=49 µm). The effective control volume for a photodetector focused to a slice of the control volume at z=0, would be significantly reduced for the larger particles. The control volume is decreasing by the volume corresponding to the "rectangle" ABCD in figure 7. An estimate of

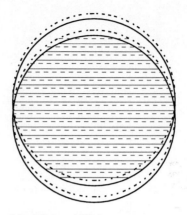

Figure 8. Locus of "fringes" for LDA and PDA

this volume is the product of the absolute value of the corresponding vector $\mathbf{D}_{jk}$ multiplied with the cross sectional area of the control volume, cut by a plane normal to $\mathbf{D}_{jk}$ at midpoint. As it will be further discussed in the last part of this work, the original size of the control volume, for this size class, could be restored if the photodetector was focused about 1mm from the beam intersection centre at the expense of reduced effective control volume for smaller size classes.

3.4 PDA

The spatial displacement of the LoC fringes is used for the evaluation of particle sizes in PDA. A second photodetector at a slightly different position than the first is viewing the LoC fringes displaced by a different distance d_x'. A cross section of the two LoC control volumes and the corresponding LoF are shown in figure 8. The spatial distance of the LoF results in a phase difference between the two photodetectors which is:

$$\Delta\phi = 2\pi\frac{d_x - d_x'}{d_f} \tag{17}$$

Equation (17) for two photodetectors placed symmetrically about the y-z plane at elevation angles $\pm\psi$, leads to the well-known phase difference equations for reflecting and refracting particles (Bauckhage 1988).

$$\Delta\phi = \frac{2\pi d\,n_1}{\lambda}\beta \tag{18}$$

where
$$\beta = \sqrt{2}\left[\left(1+\sin(\theta/2)\sin\psi-\cos(\theta/2)\cos\psi\cos\varphi\right)^{1/2} \\ -\left(1-\sin(\theta/2)\sin\psi-\cos(\theta/2)\cos\psi\cos\varphi\right)^{1/2}\right] \tag{19}$$

for reflection and

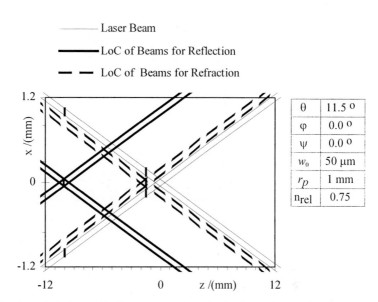

Figure 9. Projections of the beams' LoC on the z-x plane for the forward scattering case

$$\beta = 2\left[\left(1 + n_{rel}^2 - \sqrt{2}\,n_{rel}\left(1 + \sin\frac{\theta}{2}\sin\psi + \cos\frac{\theta}{2}\cos\psi\cos\varphi\right)^{1/2}\right)^{1/2}\right.$$

$$\left. - \left(1 + n_{rel}^2 - \sqrt{2}\,n_{rel}\left(1 - \sin\frac{\theta}{2}\sin\psi + \cos\frac{\theta}{2}\cos\psi\cos\varphi\right)^{1/2}\right)^{1/2}\right] \tag{20}$$

for refraction.

The LoC control volume for PDA measurements is the intersection of the two LDA LoC control volumes viewed by the two photodetectors.

4. Applications of the LoC Method

In this part of the work some optical setups are briefly discussed using the LoC methodology. The diagrams used are projections of the LoC of the beams on the z-x and z-y planes. Sometimes two diagrams are used for the z-x plane to allow for the appropriate resolution. Three types of line are on these diagrams, as indicated in the legend of figure 9. The light continuous line is defining the boundaries of the laser beam. The heavy continuous or dashed lines correspond to the boundaries of the beams' LoC for reflection or refraction respectively.

4.1 Forward scatter LDA in water-air bubble flow.

In figures 9 and 10, the projections of the beams' LoC on the z-x and z-y planes

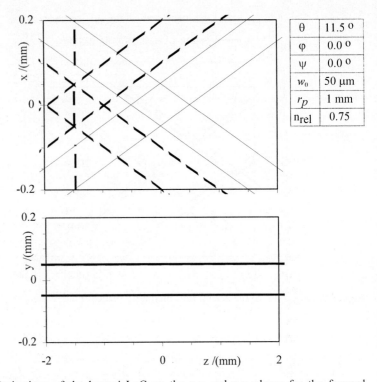

Figure 10. Projections of the beams' LoC on the z-x and z-y planes for the forward scattering case (projection lines as in figure 9)

are shown for a forward scatter LDA measuring in a water-air bubble flow, for bubble size r_p=1 mm. The optical setup is described by the parameters θ=11.5°, φ=ψ=0°, w_o= 50 μm. The projections on the z-y plane indicate that the LoC of the beams remain on the beam plane. The intersection points though change significantly as the z-x projections indicate. For reflection the LoC of the beams intersect at z=-10 mm and the interference of the transmitted rays is impossible due to the separation of the reflecting cross sections (indicated by vertical line segments on the beams). Refraction gives an effective LoC control volume with maximum cross section at z=1.5 mm, not overlapping with the beam intersection control volume (fig. 10). Careful adjustment of the depth of field with the receiving aperture and focusing on the ordinary control volume can be used to eliminate signals from bubbles with radii larger than 1mm. In this way discrimination of phases is achieved and the continuous phase velocity field can be measured using small particles for seeding.

4.2 Increase of the effective volume

The beams' LoC projections of figure 11 are for 1 mm diameter bubbles

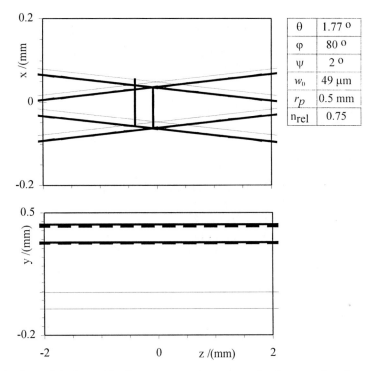

Figure 11. Projections of the beams' LoC on the z-x and z-y planes for the case of section 4.2 (projection lines as in figure 9)

measurements with an optical setup defined by $\theta=1.77°$, $\varphi=80°$, $\psi=2°$ and $w_o=49$ μm. It is the same setup presented in section 3.3 (figure 7). The z-y diagrams show that although the beams' LoC depart from the y=0 plane they remain almost on the same plane. The z-x diagrams indicate that the intersection of the beam' LoC correspond to scattering beams cross sections at z=-0.4 mm. This means that the original size of the maximum cross section area can be restored for the measurement if the PD is focused at the beams at z=-0.4 mm. At this point the beam intersection area is smaller and thus smaller sized particles will be measured with a smaller effective control volume. In fact the effective control volume as a function of particle size will have its larger value for 1 mm particles. The usual procedure of aligning the PD in the final stages for maximum data rate may lead in similar situations the focusing to the cross section most suited to the particles in the flow. Of course the real life procedure does take into account the higher scattering effectiveness of large particles which we have left outside the present analysis.

4.3 Gaussian Beam Effect

The setup described by Grehan et al.(1992) is analysed with the LoC method. The

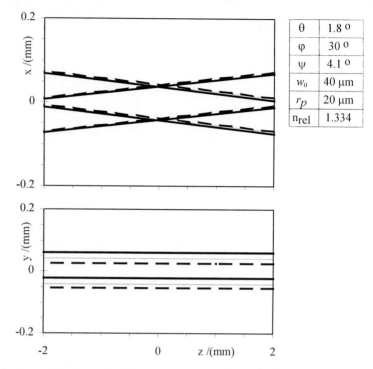

θ	1.8 º
φ	30 º
ψ	4.1 º
w_0	40 μm
r_p	20 μm
n_{rel}	1.334

Figure 12. Projections of the beams' LoC on the z-x and z-y planes for the gaussian beam effect case, droplets (projection lines as in figure 9)

setup used for r_p=20 μm droplet calculations is described by the parameters θ=1.8°, φ=30°, ψ=4.1°, w_0= 40 μm. The LoC of the beams projections diagrams are presented in figure 12. The projections on the z-x plane show very small changes in x and z coordinates of the LoC. Significant changes are shown though in the y direction. The LoC control volume for reflection is displaced towards positive y, contrary to that for refraction which is displaced towards negative y. This behaviour results in a larger effective control volume formed by the union of the refraction and reflection control volumes. In the intersection of the reflection and refraction LoC control volume the location of the LoF is depending on the interference of four rays thus becoming a function of the relative amplitudes as well as of the phase differences. Although the reflection is weaker at this angle, due to the gaussian distribution of the intensity in the beams the high intensity regions of the reflection LoC control volume corresponding to the low intensity regions of the refraction LoC control volume are responsible for significant discrepancies in phase difference in this region.

In bubble flow the gaussian effect is not so pronounced (Grehan et al. 1994). The setup used for r_p=20 μm bubble calculations is described by the parameters θ=1.35°, φ=30°, ψ=4.1°, w_0=40 μm. The projections in figure 13 show that this is

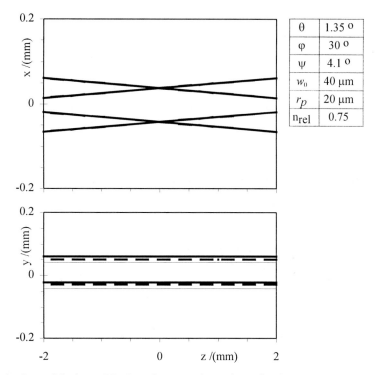

θ	1.35 $^{\circ}$
φ	30 $^{\circ}$
ψ	4.1 $^{\circ}$
w_0	40 μm
r_p	20 μm
n_{rel}	0.75

Figure 13. Projections of the beams' LoC on the z-x and z-y planes for the gaussian beam effect case, bubbles (projection lines as in figure 9)

due to the fact that for $n_{rel}<1$ the LoC of the beams are displaced in the same direction for reflection and refraction. In this way the relative magnitude of the reflection and refraction components remains almost constant throughout the effective control volume and the phase difference is mainly due to the interference of the refracted rays. The gaussian beam effect is expected to be significant though if the two loci depart, that is if the beam waist radius is decreased or the particle size increased as the ratio r_r/r_p for $\gamma=150°$ in figure 3 indicates.

4.4 Planar PDA

The planar PDA was suggested as a solution to the gaussian beam effect by Aizu et al. (1994). The optical setup used is described by the parameters $\theta/2=2.04°$, $\varphi=0°$, $\psi=26.31°$, $w_o=64$ μm, and the particles were droplets with $d_p=95$ μm. As the projections of the beams' LoC in figure 14 show, the two LoC control volumes for refraction and reflection are barely overlapping at the edge. The main LoC control volume is that of refraction, due to the higher intensities of refraction at this angle. Adjacent to it is a reflection LoC control volume responsible for the low amplitude second burst appearing in the experiments.

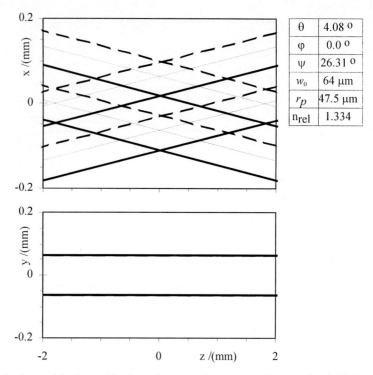

θ	4.08 º
φ	0.0 º
ψ	26.31 º
w_o	64 μm
r_p	47.5 μm
n_{rel}	1.334

Figure 14. Projections of the beams' LoC on the z-x and z-y planes for the planar PDA, d_p=95 μm (projection lines as in figure 9)

For smaller particles the problem of gaussian effect was not alleviated. The projections shown in figure 15 for the same setup and d_p=40 μm droplets suggest that the refraction and reflection LoC control volumes overlap significantly. Due to this fact high intensity regions of the reflection LoC control volume destroy the fringe pattern of the low intensity region of the more important refraction LoC control volume. Following the analysis of the first part of the present work it can be calculated that the distance between LoC control volume centres for γ=150° is 1.8r_p. That is for r_p=20 μm, the distance of the centres is 36 μm and since the waist radius is w_o=64 μm the two LoC control volumes can not be separated.

5. Conclusions

The LoC method presented in this work seems to be promising in the analysis and optimisation of LDA and PDA optical setups for measurements with larger particles. More specifically:
- It gives a good physical overview of the techniques without being too complicated
- It allows a global quantitative understanding of the effects of larger

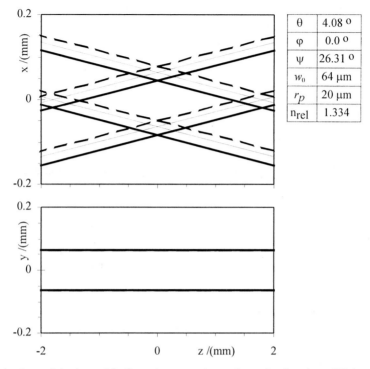

θ	4.08 °
φ	0.0 °
ψ	26.31 °
w_o	64 μm
r_p	20 μm
n_{rel}	1.334

Figure 15. Projections of the beams' LoC on the z-x and z-y planes for the planar PDA, d_p=40 μm (projection lines as in figure 9)

particles on the control volume properties. In this way the choice of the optical setup can be done with the help of soundly based rules.
- It simplifies the investigation of erroneous setups associated with velocity or size measurements due to the distortion of the normal fringe pattern seen by larger particles.
- It sets new standards in the estimation of the active control volume as function of the particle size, which is very important in concentration and particle flux measurements, as for example in bubbly flows.

Acknowledgement

The authors gratefully acknowledge the financial support of the European Commission in the frame of the HCM Network (Contract CT-93-0389).

References

Aizu, Y., Durst, F., Grehan, G., Onofri, F. & Xu, T.-H. 1993, PDA-System without Gaussian Beam Defects, 3rd Intl. Cong. Optical Particle Sizing, Yokohama, Japan, pp.461-470.

Bachalo, W.D. 1987, The Evolution of Particle Size and Velocity Measurement Technology, 2nd Intl. Conf. on Laser Anemometry-Advances and Applications, Strathclyde, UK, pp. 79-99.

Bauckhage, K. 1988, The Phase Doppler Difference Method, a New Laser-Doppler Technique for Simultaneous Size and Velocity Measurements, Part 1: Description of the method, Part. Part. Syst. Charact., Vol. 5, pp. 16-22.

Grehan, G., Gouesbet, G., Naqwi, A. & Durst, F. 1992, On Elimination of the Trajectory Effects in Phase Doppler System, 5th Eur. Symp. Particle Caracterization (PARTEC 92), Nuremberg, Germany, pp. 309-318.

Grehan, G., Onofri, F., Cirasole, T., Gouesbet, G., Durst, F. & Tropea, C. 1994, Measurements of Bubbles by Phase Doppler Technique and Trajectory Ambiguity, 7th Intl. Symp. on Applications of . Laser Techniques to Fluid Mechanics, Lisbon, Portugal, vol. 1, paper 18.2.

Hardalupas, Y. & Taylor, A. M. K. P. 1989., On the Measurement of Particle Concentration Near a Stagnation Point, Experiments in Fluids, Vol. 8, pp. 113-118.

Saffman, M. 1987, Optical Particle Sizing Using the Phase of LDA Signals, Dantec Information, No. 05, pp. 8-13.

Sankar, S. V. & Bachalo, W. D. 1991, Response Characteristics of the Phase-Doppler Particle Analyzer for Sizing Spherical Particles Larger than the Wavelength, Applied Optics, Vol. 30, pp. 1487-1496.

Sommerfeld, M. & Qiu, H.-H. 1995, Particle Concentration Measurements by Phase-Doppler Anemometry in Complex Dispersed Two-Phase Flows, Experiments in Fluids, Vol. 18, pp. 187-198.

Tropea, C., Xu, T.-H., Onofri, F., Grehan, G., Haugen, P. & Stieglmeier, M. 1995, Dual Mode Phase-Doppler Anemometer., 4th Intl. Congress Optical Particle, Pre-prints, pp. 287-296.

Van de Hulst, H. C. 1981, Light Scattering by Small Particles, Dover, New York.

II.6 A Single-Beam Velocimeter Based on Rainbow-Interferometry

J.P.A.J. van Beek and Rieth muller

von Karman Institute for Fluid Dynamics, Chaussée de Waterloo 72, B-1640
Rhoide-Saint-Genèse Belgium Fax: +32 2 3599600 E-mail: vanbeek@vki.ac.be

Abstract. A velocimeter which consists of one laser beam and one photomultiplier is proposed. The instrument is based on the primary rainbow
created by a transparent particle illuminated by a laser beam; this rainbow
moves in front of a photomultiplier when the particle traverses the laser beam.
Frequency analysis of the recorded signal leads to the droplet size and to one
component of the velocity vector. Validation experiments have been carried
out using a spherical glass bead. To improve the measurements with water
droplets, a nonsphericity detection is proposed by adding one photomultiplier.

Keywords. Droplet velocity, droplet size, rainbow technique

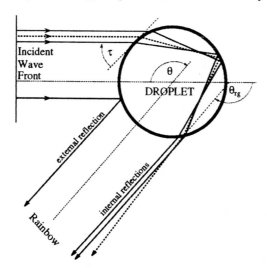

Fig. 1. The dominant geometrical rays that form the primary rainbow.

1 Introduction

In 1988, the rainbow was introduced as a nonintrusive measurement technique to determine the refractive index, i.e. the temperature, of droplets. The initiators N. Roth, K. Anders and A. Frohn (1990, 1991) directly applied the technique to droplet combustion. Shortly after, Sankar et a. (1993, 1994) integrated the rainbow technique with the phase Doppler technique in order to determine simultaneously the droplet size, velocity and temperature.

The rainbow that is used for the technique is created by a single droplet scattering laser light. The technique is based on the primary rainbow which can be explained geometrically by once-internally reflected rays for which the scattering angle θ goes through an extremum with respect to the incidence angle τ; this yields the characteristic high intensity of the rainbow near the extremum which is called the geometrical rainbow angle θ_{rg} (see Fig. 1). From Fig. 1 one can understand that this high intensity pattern consists of fringes because for scattering angles larger than θ_{rg} there exist two parallel rays that interfere at infinity; from the resulting so-called Airy fringes the droplet size can be deduced (J.P.A.J. van Beeck and M.L. Riethmuller (1995)). Moreover, external reflection interferes with internal reflection to form a ripple structure superimposed on the Airy fringes. Frequency analysis of both the ripple structure and the Airy fringes indicates the degree of nonsphericity of the scattering particle (J.P.A.J. van Beeck and M.L. Riethmuller (1996)); this is necessary to know because the droplet temperature and diameter, derived from the rainbow, reveals a strong dependence on the droplet nonsphericity.

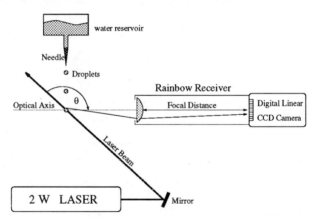

Fig. 2. Sketch of the set-up to detect the rainbow pattern with a linear CCD–camera.

In the present paper an alternative application of the rainbow as a measurement device is proposed, i.e. the measurement of one component of the velocity vector of a moving droplet. This application is also based on frequency analysis of the Airy fringes and the ripple structure. Therefore, the detection of the droplet nonsphericity has to be carried out in a different way.

2 Detection of the monochromatic rainbow by a photo-multiplier

The conventional method to detect a rainbow interference-pattern coming from a single droplet is to use a linear CCD-camera placed at focal length of a positive lens (Fig. 2). Each pixel of the linear CCD-array corresponds to a certain scattering angle. This means that the position of the rainbow pattern on the array is independent of the position of the droplet in the laser beam.

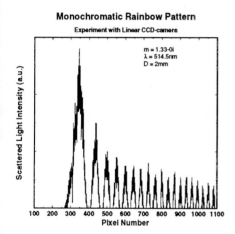
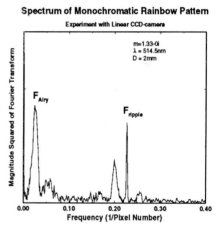

Fig. 3. Typical rainbow signal recorded with a linear CCD-camera. The low-frequency pattern resembles the Airy fringes on top of which a ripple structure is superimposed.

Fig. 4. Spectrum of the rainbow pattern in Fig. 3 revealing the Airy and the ripple frequencies, F_{Airy} and F_{ripple} respectively.

Fig. 3 shows a typical rainbow signal recorded with the CCD-camera. Fig. 4 depicts the spectrum. The indicated peak F_{Airy} originates from the low frequency interference pattern, i.e. the Airy fringes. F_{ripple} corresponds to the ripple structure. Both F_{Airy} and F_{ripple} yield a droplet diameter; if F_{Airy} equals F_{ripple} then the droplet is spherical. Only then a reliable temperature can be determined from the angular position of the main rainbow maximum. The detailed procedure has been described by Van Beeck and Riethmuller (1996).

Because of the limited sensibility of the CCD-camera to the scattered-light intensity, it was decided to study the possibility of using a photomultiplier for the detection of the rainbow signal. When a single falling droplet is concerned, a pin-hole in combination with a photomultiplier is sufficient to constitute a

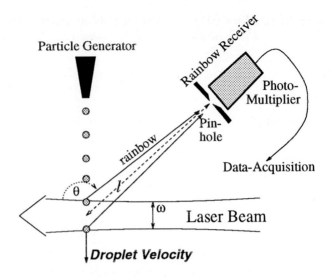

Fig. 5. Outline of the set-up, containing a photomultiplier, to detect the rainbow pattern coming from a single scattering transparent particle. ω is the diameter of the laser beam and l is the distance between the pin hole and the particle in the center of the beam.

rainbow receiver (see Fig. 5). The idea is that the fringes in the rainbow pattern move in front of the pin-hole when the droplet traverses the laser beam. If the pin-hole is more than twice smaller than the smallest fringe spacing, a rainbow pattern in time can be detected. The actual amount of fringes traversing the pin-hole depends on the diameter D of the light-scattering particle, the diameter ω of the laser beam and on the distance l between the pin-hole and the particle.

A simulation of the photomultiplier signal, made with the help of the generalized Lorenz-Mie theory (G. Gouesbet et a. (1988)), shows how the rainbow pattern is modulated with the Gaussian intensity profile of the laser beam (Fig. 6). The simulation was carried out with the following parameters: $D = 1\,mm$, $\omega = 2\,mm$, $l = 0.131\,m$, a wave length of the laser light $\lambda = 514.5\,nm$, a refractive index $m = 1.33 - 0i$ and a velocity $v = 1\,m/s$. Fig. 7 shows the spectrum and the spectrum of the derivative of the rainbow pattern. Similar peaks as in fig. 4 can be identified. However, the difference is that the spectrum of Fig. 7 not only depends on the droplet diameter (and weakly upon the refractive index) but also on the droplet velocity. This implies that the detection of the rainbow pattern with the photomultiplier set-up (Fig. 5) leads to the possibility to determine the droplet velocity from this set-up. This will be explained in the following section.

The main difference between the spectrum and the spectrum of the deriva-

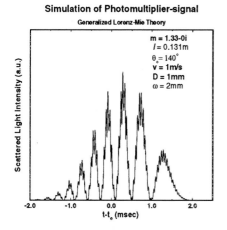

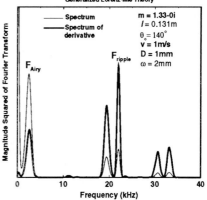

Fig. 6. A simulation of a photomultiplier signal concerning the set-up of Fig. 5. θ_o is the scattering angle that corresponds to the time t_o at which the droplet is positioned at the center of the laser beam.

Fig. 7. Spectrum and spectrum of the derivative of the rainbow pattern in Fig. 6.

tive is that in the latter the peak at $0\,kHz$ has disappeared which makes the Airy frequency F_{Airy} better identifiable. It is important to recognize that the positions of the peaks in both spectra are exactly similar.

3 Determination of the droplet velocity from the rainbow pattern

As suggested in the previous section, the photomultiplier set-up of Fig. 5 can be used to determine the velocity of a moving light-scattering particle. The signal recorded with the photomultiplier reveals clearly the Airy fringes (with the angular Airy frequency F_{Airy}) and the ripple structure (with ripple frequency F_{ripple}) which are so characteristic for the rainbow (Fig. 6). Both interference structures depend on diameter and velocity. Therefore, the droplet diameter has to be known a priori. In order to determine the droplet size, the ratio F_{ripple}/F_{Airy} can be used because it only depends on this parameter. Applying the Airy theory close to the main rainbow maximum, the following expressions for this ratio can be found (van Beeck and Riethmuller (1996)):

$$F_{ripple}/F_{Airy} = 1.089\left(\frac{D}{\lambda}\right)^{\frac{1}{3}} + \frac{1}{2}$$
$$\text{for m=1.333-0i,} \tag{1}$$

$$F_{ripple}/F_{Airy} \;=\; 0.6112\left(\frac{D}{\lambda}\right)^{\frac{1}{3}} + \frac{1}{2}$$
$$\text{for m}=1.517\text{-}0\text{i}. \tag{2}$$

Knowing the diameter, the velocity can be obtained from either the Airy frequency (in Hz):

$$v \;=\; l \cdot F_{Airy} \cdot 1.777\left(\frac{D}{\lambda}\right)^{-\frac{2}{3}}$$
$$\text{for m}=1.333\text{-}0\text{i}, \tag{3}$$

$$v \;=\; l \cdot F_{Airy} \cdot 1.304\left(\frac{D}{\lambda}\right)^{-\frac{2}{3}}$$
$$\text{for m}=1.517\text{-}0\text{i}, \tag{4}$$

or the ripple frequency (also expressed in Hz):

$$v \;=\; l \cdot F_{ripple} \cdot \frac{1}{0.613\left(\frac{D}{\lambda}\right) + 0.281\left(\frac{D}{\lambda}\right)^{\frac{2}{3}}}$$
$$\text{for m}=1.333\text{-}0\text{i}, \tag{5}$$

$$v \;=\; l \cdot F_{ripple} \cdot \frac{1}{0.469\left(\frac{D}{\lambda}\right) + 0.383\left(\frac{D}{\lambda}\right)^{\frac{2}{3}}}$$
$$\text{for m}=1.517\text{-}0\text{i}, \tag{6}$$

where l is the distance between the pin-hole and the location in the laser beam where the scatterer passes. As the Airy theory is used, a systematic error in size and velocity measurements of several percent is introduced. This error is rapidly increasing with decreasing size in the micrometer range (Van Beeck and Riethmuller (1994)); if higher accuracy is requested, the above set of equations has to be based on the Lorenz-Mie theory.

Finally, it is important to realize that the component of the velocity vector, which is measured, lies in the scattering plane and is perpendicular to the optical axis of the rainbow receiver. Therefore, it is suggested to align the optical axis of the rainbow detector perpendicularly to the velocity component that one studies.

4 Experimental results and discussion

The rainbow technique, used as velocimeter, has been compared to the conventional laser Doppler technique. The set-up is depicted in Fig. 8. It contains two crossing He-Ne-laser beams. The Doppler signal is created when a particle passes the cross-over region (i.e. the probe volume); the signal is detected by the photomultiplier "ldv-PM" which is placed after a red filter. Frequency analysis yields the particle velocity, v_{ldv}, which denotes the vertical component of the velocity vector.

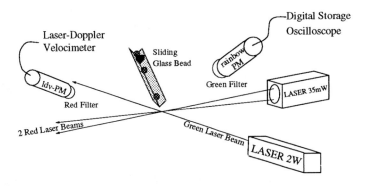

Fig. 8. Schematic of the optical system that served to validate the rainbow technique concerning the velocity measurement.

An expanded green laser beam ($\pm 10\,mm$) crosses the probe volume and is perpendicular to both red laser beams. It generates a rainbow pattern that moves in front of the photomultiplier "rainbow-PM" which is mounted at the geometrical rainbow angle. The pin-hole placed in front of the photomultiplier has a circular aperture of $50\,\mu m$ and is positioned at a distance of $0.83\,m$ from the probe volume. A green optical filter prevents the Doppler-signal from reaching this photomultiplier. The rainbow signal is recorded by a digital storage oscilloscope that is connected to a personal computer. As explained in the previous section, the rainbow pattern in time provides the droplet size D and the velocity component, $v_{rainbow}$, perpendicularly to the optical axis of the rainbow photomultiplier.

Validation of the rainbow-technique has been performed with a spherical glass bead ($D = 5.9\,mm$, $m = 1.517 - 0i$). The bead slides down in an inclined tube having a squared cross-section; the position of the tube is adjustable in the horizontal plane. The inclination is such that the velocity vector of the bead while passing the probe volume is perpendicular to the optical axis of the "rainbow-PM" and lies in the scattering plane of this photomultiplier. Therefore, the velocity measured with the rainbow technique, $v_{rainbow}$, equals the absolute value of the velocity vector. A removable catch inside the tube makes the bead velocity to be repetitive.

Fig. 9 shows an experimental signal detected by the photomultiplier. The Airy fringes are clearly visible and on some of them the ripple structure can be observed. The spectrum (Fig. 10) has been taken from the derivative of the rainbow pattern to better resolve the Airy frequency F_{Airy} (see Sec. 2). The ratio F_{ripple}/F_{Airy} gives a diameter of $6.2 \pm 0.5\,mm$ according to Eq. 2. This compares rather well to the geometrically measured value of $5.9\pm0.1\,mm$. Relationship 4 yields the rainbow velocity based on the Airy frequency: $v_{rainbow} = 0.89 \pm 0.05\,m/s$. In order to obtain the vertical component, $v_{rainbow}$ has to be multiplied by $\cos 22°$, i.e. the angle between the velocity vector and the

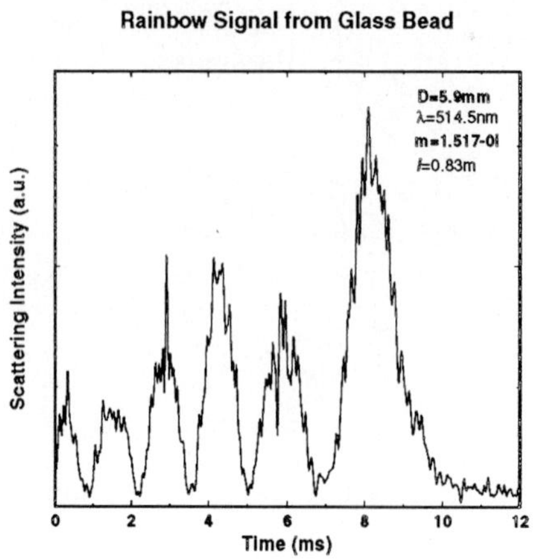

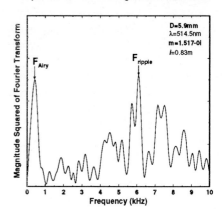

Rainbow Signal from Glass Bead
D=5.9mm
λ=514.5nm
m=1.517-0i
l=0.83m
Scattering Intensity (a.u.)
0
2
4
6
8
10
12
Time (ms)

Fig. 9. Experimental rainbow signal obtained with the rainbow receiver ("rainbow-PM"). The scattering particle was a glass bead.

Spectrum of Rainbow Signal from Glass Bead
D=5.9mm
λ=514.5nm
m=1.517-0i
l=0.83m
F Airy
F ripple
Magnitude Squared of Fourier Transform
0
1
2
3
4
5
6
7
8
9
10
Frequency (kHz)

Fig. 10. Spectrum of the derivative of the rainbow pattern of Fig. 9.

vertical; this results in a velocity of $0.82 \pm 0.05\, m/s$. With the laser-Doppler velocimeter a velocity of $v_{ldv} = 0.85 \pm 0.01\, m/s$ has been measured which agrees with $v_{rainbow}$ within the measurement accuracy.

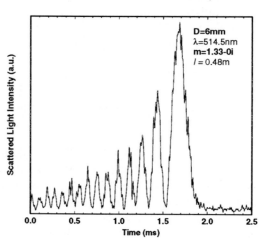

Rainbow Pattern from Water Droplet
D=6mm
λ=514.5nm
m=1.33-0i
l = 0.48m
Scattered Light Intensity (a.u.)
0.0
0.5
1.0
1.5
2.0
2.5
Time (ms)

Fig. 11. Rainbow pattern coming from a single falling water droplet.

While performing measurements with the rainbow technique it is important to recognize that the laser beam has to be as collimated as possible. For the present set-up, a beam divergence of 1 *mrad* would already have resulted in a velocity bias of −6 %. Furthermore, from Fig. 1 one understands that the diameter of the laser beam has to be larger than the droplet size in order to create the entire rainbow structure, that is to say the Airy fringes and the ripple structure. Yet, looking to Figs. 9 and 10, it is clear that it is not easy to detect the ripple structure because of the rather high level of noise. Therefore, a proper detection of the rainbow requires a high signal to noise ratio. This ratio is determined by the photomultiplier-voltage, the laser power, the size of the droplet, the size of the pin-hole and the distance l between scatterer and probe volume.

5 Nonsphericity detection using the photomultiplier set-up

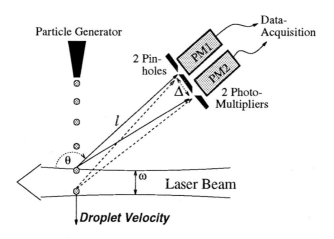

Fig. 12. Modification of the photomultiplier set-up of Fig. 5 in order to detect the nonsphericity of the scattering particle.

Fig. 11 shows a rainbow signal using the set-up depicted in Fig. 8. The scattering particle was a single falling water droplet of about 6 *mm* in diameter and having a velocity of $v_{ldv} = 2.6 \, m/s$ while passing the probe volume. The ratio F_{ripple}/F_{Airy} results in a droplet diameter of only $D = 356 \, \mu m$. If with this erroneous diameter the velocity is calculated, then Eq. 3 would yield $v = 44 \pm 3 \, m/s$. A large free falling droplet at about 37 *cm* from the needle-exit is oscillating, thus only temporarily sufficient spherical to obtain reliable physical quantities from the technique rainbow (Van Beeck & Riethmuller (1995)). The nonsphericity detection method using the CCD-camera compares the diameter from the Airy frequency with the diameter obtained from the ripple frequency (Sec. 2). For the photomultiplier set-up of Fig. 5 this method can not be applied as only one diameter is obtained from the ratio of both frequencies F_{ripple} and F_{Airy}. Therefore a modification of the set-up is presented in Fig. 12. It consists of two photomultipliers instead of one, sepa-

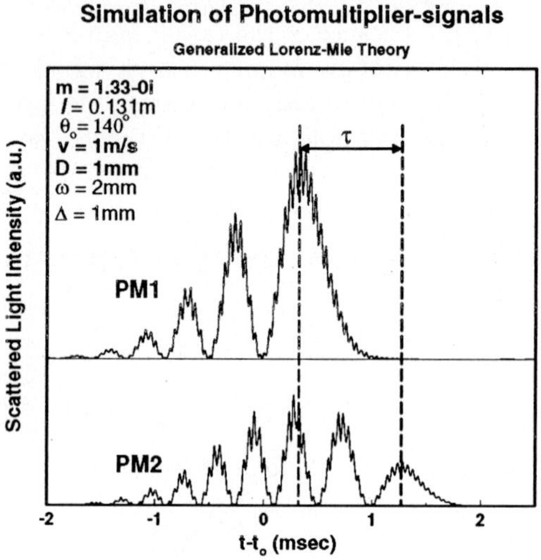

Simulation of Photomultiplier-signals

Generalized Lorenz-Mie Theory

Fig. 13. Simulation of the photomultiplier-signals in the set-up of Fig. 12 using the generalized Lorenz-Mie theory (Gouesbet et a. (1988)).

rated by a distance Δ. Fig. 13 shows a simulation of the signals as recorded by the two photomultipliers PM1 and PM2; the set of parameters is the same as the simulation of Fig. 6. Both photomultipliers are placed such that they observe the main rainbow maximum and several supernumerary Airy fringes. Due to the distance Δ, the rainbows are shifted by a time τ and are modulated in a different way by the Gaussian laser beam intensity profile. The velocity can directly be deduced from τ and Δ:

$$v = \frac{\Delta}{\tau} . \tag{7}$$

Subsequently, one diameter, D_{Airy}, can be found from the Airy and another, D_{ripple}, from the ripple frequency applying respectively Eq. 3 and Eq. 5. When the diameters disagree, the droplet is nonspherical. This method to detect the droplet nonsphericity can be applied to the signal in Fig. 11 using the velocity obtained by the laser Doppler velocimeter, $v_{ldv} = 2.6\,m/s$. It results in $D_{Airy} = 25\,mm$ and $D_{ripple} = 8\,mm$. Subsequently, the droplet was nonspherical. It is very interesting to notice that, unlike the diameter, the droplet velocity obtained from Eq. 7 is probably independent of the droplet nonsphericity as the velocity is not derived from the Airy theory which is only valid for spherical particles. This statement deserves certainly further study because it would mean that, using the photomultiplier set-up of Fig. 12, also reliable velocity measurements can be made in a medium with only nonspherical scatterers.

6 Conclusions

A velocimeter based on rainbow-interferometry has been proposed. The instrument can be used to measure simultaneously the velocity and the size of droplets. The detection of the rainbow is performed by a photomultiplier. The technique has been validated for spherical glass particles by comparing it with the performance of the established laser Doppler velocimeter. The detection of the nonsphericity of water droplets was proved to be necessary in order to obtain reliable quantities from the rainbow technique. A promising method to select spherical droplets has been proposed making use of an additional photomultiplier.

7 Acknowledgments

The authors are grateful to J-Ph. Pages for his contributions to the experiments presented in this paper.

References

Roth, N., Anders, K. & Frohn, A. 1990, Simultaneous Measurement of Temperature and Size of Droplets in the Micrometer Range, Journal of Laser Applications, vol. 2, number 1, pp. 37-42.

Roth, N., Anders, K. & Frohn, A. 1991, Refractive-index Measurements for the Correction of Particle Sizing Methods, Appl. Opt., vol. 30, number 33, pp. 4960-4965.

Sankar, S.V., Ibrahim, K.H., Buermann, D.H., Fidrich, M.J. & Bachalo, W.D. 1993, An Integrated Phase Doppler/Rainbow Refractometer System for Simultaneous Measurement of Droplet Size, Velocity, and Refractive Index, Proc. of The Third International Congress on Particle Sizing, Yokohama, Japan.

Sankar, S.V., Buermann, D.H. & Bachalo, W.D. 1994, Simultaneous Measurements of Droplet Size, Velocity, and Temperature in a Swirl-Stabilized Spray Flame, Proc. of Seventh International Symposium on Applications of Laser Techniques to Fluid Mechanics, Lisbon, Portugal.

van Beeck, J.P.A.J. & Riethmuller, M.L. 1995, Non-intrusive Measurements of Temperature and Size of Raindrops, Appl. Opt., vol. 34, number 10, pp. 1633-1639.

van Beeck, J.P.A.J. & Riethmuller, M.L. 1996, Rainbow Phenomena applied to the Measurement of Droplet Size and Velocity and to the Detection of Nonsphericity, Appl. Opt., vol. 35, number 13, pp. 2259-2266.

Gouesbet, G., Maheu, B. & Gréhan, G. 1988, Light Scattering from a Sphere Arbitrarily Located in a Gaussian Beam using a Bromwich Formulation, J. Opt. Soc. Am. A, vol. 5, number 9, pp. 1427-1443.

van Beeck, J.P.A.J. & Riethmuller, M.L. 1994, Simultaneous Determination of Temperature and Size of Droplets from the Rainbow using Airy Theory, Seventh International Symposium on Applications of Laser Techniques to Fluid Mechanics, Lisbon, Portugal, also published in "Developments in Laser Techniques and Applications to Fluid Mechanics", R.J. Adrian et al. (eds.), Springer, pp. 330-339.

II.7 Accuracy of Particle Flux and Volume Fraction Measurement by Shadow Doppler Velocimetry

H. Morikita[1], Ilias Prassas[2] and A.M.K.P. Taylor[2]

[1] Keio University, Department of Mechanical Engineering
    3-14-1 Hiyoshi, Kohoku-ku, Yokohama 223, Japan.
[2] Imperial College, Department of Mechanical Engineering
    Exhibition Road, London SW7 2BX, UK.

Abstract. This paper reports on the accuracy of measurement of volume fraction and flux of a dispersed phase by Shadow Doppler Velocimetry (SDV). The SDV is an imaging technique, based on a conventional laser Doppler velocimeter and a linear photodiode array, for simultaneous particle- size and velocity measurement. The SDV also provides a measurement of the spatial position of a particle as it travels through the probe volume; as a result, it is not necessary to establish the dependence of the cross-sectional area of the probe volume on particle size for flux and volume fraction measurements. Instead, this area is defined by the operator of the instrument and is independent of the irradiance distribution of the incident laser beams, unlike - for example - phase Doppler anemometry (PDA). The accuracy of the flux measurement has been assessed from integration of flux and volume fraction measurements across a fully developed 20×20 mm turbulent water channel flow flowing at 0.16 m/s, laden with a known fraction of quasi-neutrally buoyant polyethylene spheres of nominal mean diameter of 100 μm, and the random uncertainty in the derived particle volume fraction was no more than 20% for particle volume fractions up to 0.005% and typically 10%. The integration procedure contributed a random uncertainty of less than 12%, the determination of particle volume a systematic uncertainty of 6% and the determination of the area of the sampling-space a further 8% systematic uncertainty, whereas the random uncertainty in determining the area of the sampling-space was estimated to be approximately 10%. For particle volume fraction larger than 0.005% and up to 0.05%, signal identification was affected by turbidity and multiple occupancy effects and thus the SDV measurement was compensated by the signal error rate, resulting in a particle volume fraction accurate to better than 40%.

Keywords.   Shadow Doppler Velocimetry, Flux and volume fraction measurement, Irregular particles

1. Introduction

The measurement of flux and volume fraction of a dispersed phase represents an outstanding difficulty for optical single particle counters. For both cases of spherical and irregular particles, which are typically measured by phase Doppler anemometry (PDA) and diffracted amplitude [e.g. Hirleman *et al.*, 1982; Morikita *et al.*, 1993; Orfanoudakis and Taylor, 1992 & 1995; Yeoman *et al.*, 1982] respectively, the determination of the local particle flux or concentration is associated with higher uncertainties than is the case for the size and the velocity. This is because flux and concentration are not directly measured, but calculated from the particle velocity and size and a corresponding cross-sectional area, or volume, of the optical probe volume. It is the uncertainty in the latter two dimensions, and particularly their dependence on particle size due to the Gaussian irradiance of the incident laser beams, which are the major contributors to the uncertainties (see Taylor [1995], for example, for a review of the definitions).

In the case of amplitude-based techniques which use diffracted light, the sizing error due to the shape of the particles is typically 20% and can be as high as 70% for 20 µm ellipsoids of aspect ratio of 2.0 [Hardalupas *et al.*, 1995]. There is little recent published work on flux measurements by such a technique with the exception of Orfanoudakis [1994], who used the work by Hardalupas and Taylor [1989], as developed for PDA. Nevertheless, the accuracy of flux measurement is expected to be similar to that for PDA, discussed below. In PDA, the size of the optical probe volume is preferably determined *in-situ* from each measurement [e.g. Saffman, 1987; Hardalupas and Taylor, 1989; Schöne *et al.*, 1994]. The extensive experience in the use of PDA has proven that the uncertainties are usually of the order of several tens of percent [e.g. Hardalupas and Taylor, 1994; Maeda *et al.*, 1996], although these may be as low as 5% under ideal conditions [Qiu & Sommerfeld, 1992; Sommerfeld & Qiu, 1995]. Theoretical investigations show, in addition, that the so-called 'trajectory ambiguity effect' [Gréhan *et al.*, 1992], also due to the Gaussian irradiance, and the so-called 'slit effect' [Xu and Tropea, 1994] can result in an effective probe volume different from that estimated by the formulae of, for example, Saffman [1987], and this error source is a large potential contributor to the observed discrepancies. In addition, determination of the probe volume dimensions can be erroneous in the case of measurement of a two-dimensional flow using a single-channel PDA, because in the latter case the formula applied for the size of the probe volume assumes that the flow is normal to the fringes [Saffman, 1987].

Due to the fact that amplitude-based size measurement of irregular particles is less straightforward and accurate than for spherical ones, imaging techniques like shadow Doppler velocimetry [Hardalupas *et al.*, 1994] have been recently developed. The sizing accuracy of the shadow Doppler velocimeter (SDV) instrument has been evaluated in comparison with microscope measurements and it was found that the sizing accuracy was better than 10% and a further inaccuracy ±5% was caused by so-called "defocusing" of the particle from the focal plane of the receiving optics by up to ±500 μm [Hardalupas et al., 1994]. The instrument was later improved [Morikita, 1995] by reducing the response time of the photodiode array and by increasing the number of active segments of the detector, thereby increasing the maximum measurable particle velocity and the particle size dynamic range, respectively. These features of the technique were exploited to measure the location of the particle during its trajectory through the probe volume and to measure the angles of particle trajectories relative to the axis of the photodiode array [Morikita et al., 1995]. Additionally, the accuracy of size measurement made by this improved version was assessed in the case where the beams had to pass through optical windows and through the variable refractive index fields produced by flames, situations where amplitude-based instruments can yield unacceptably large sizing errors due to the uncontrollable thermal movement of the windows. Hishida et al. [1995] found that the sizing uncertainties do not exceed -12% in the former case and 15% in the latter. Maeda et al. [1995] demonstrated the ability of the instrument to measure the two-dimensional motion of *burning* coal particles near the exit of an asymmetrical laboratory-scale confined coal combustor, although no particle flux or concentration was reported.

For particle mass flux and concentration measurement using SDV, the size of the effective probe volume (hereafter termed as 'sampling-space' to indicate that it is explicitly user-delimited, rather than set by a combination the Gaussian irradiance profile and electronic trigger levels with the resulting sampling-space size dependence estimated from on-line measurement, as in the case of PDA) must be accurately defined. However, the procedure of determining the sampling-space suggested by Morikita et al. [1995] is complicated for non-spherical particles in omni-directional flows due to the dependence of the sampling-space size on the particle size, shape, orientation and trajectory, which could cause a systematic bias in the validation rate, especially for trajectories near the edge of the sampling-space. In addition, the design of the signal processor they used implies larger probability for larger particles to trigger the instrument than smaller particles because they used a single diode as a trigger-channel. Although correction for this bias *was* included in the post-processing of their data, it is not a practically convenient design for the measurement of polydisperse particles. Therefore simplifications in the procedure for the calculation of particle mass flux and concentration are required and should be tested experimentally in order to enable accurate flux and concentration measurements in complex flows.

The purpose of the present contribution is to improve the procedure for the calculation of the size of the sampling-space of the SDV and, using this procedure, to assess the accuracy of the flux and volume fraction measurement in a unidirectional turbulent channel water flow. Monodisperse spheres with density similar to that of water were used in the experiment, and the measurement accuracy was tested as a function of particle volume fraction.

The procedure to determine the size of the sampling-space was simplified relative to Morikita *et al.* [1995] by providing a *user-defined* spatial limitation to eliminate the effect of the particle size, shape, orientation and particularly trajectory angle.   The signal processor was redesigned to achieve the necessary higher data rates, relative to the previous versions, for the measurement of flux and volume fraction and was also improved in respect of the data rates associated with smaller particles, by providing multiple trigger channels.   In addition, the accuracy of the measurement of the particle two-dimensional location in the probe volume, upon which the delimited size of the sampling-space depends, was assessed before measuring particle flux and volume fraction.

The paper is structured as follows: first, a brief description of the principle, the optics and the electronics of the current version of the instrument is presented.   The paper continues with the theoretical background for the measurement of flux and volume fraction using SDV and the evaluation of the method by comparison with experiments.   Suggestions are also offered for the extension and improvement of the volume fraction measurement.   The text concludes with the most important findings from the present investigation.

2. Shadow Doppler Velocimetry

2.1 Principle of the Shadow Doppler Velocimetry

Figure 1 briefly illustrates the principle of the technique along with the Cartesian xyz co-ordinate system which will assist us in the description.   An ellipsoidal probe volume is formed from the intersection of the two laser beams of a conventional LDV system and its $1/e^2$ dimensions are denoted D_x, D_y and D_z.   The SDV receiving optics, described in detail in Hardalupas *et al.* [1994], serve to magnify and project the image of the probe volume, along with the moving shadows of any passing particles, onto a linear photodiode array.   The array, once the instrument is triggered, samples one-dimensional slices of the moving two-dimensional shadow at a pre-defined sampling frequency.   The time series of the one-dimensional slices and the simultaneously measured particle velocity are used for the reconstruction of the two-dimensional particle shadow, from the area of which the particle size is determined.   Due to the one-dimensional sampling, the shape of the recorded image, *before correction* for the "aspect ratio" [Morikita *et al.*, 1995], depends on the particle location in the measurement volume and the angle of the particle trajectory relative to the x-y plane, as illustrated for spherical particles in figure 1.

Table 1 shows the optical parameters of the present configuration. Particle shadows in the probe volume were magnified 82 times after projection on the photodiode array.   Each of the 32, 1 mm wide, segments of the photodiode array received light over a distance $y= 12.2$ μm in the probe volume for the present magnification.

Table 1. Optical parameters

Laser Wavelength [nm]	514.5 (Green)
Beam intersection angle 2γ [deg]	5.72
Probe volume size at e^{-2}, D_x, D_y, D_z [μm]	$200 \times 600 \times 5600$
Fringe spacing δ [μm]	5.16
Magnification of receiving optics [-]	82
Pitch of elements of diode array [mm]	1

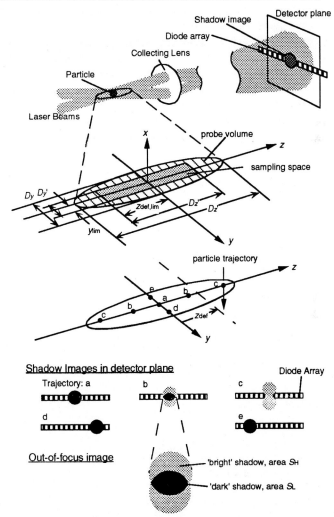

Figure 1. Principle of Shadow Doppler Velocimetry. The figure shows the geometry of the probe volume, of the SDV sampling-space, the Cartesian coordinate system attached to it and examples of shadow images on the detector plane from particles passing through the sampling-space at various locations. Trajectories b and c give "out-of-focus" images as shown at the bottom of the figure.

When a particle is at $z=0$, which corresponds to the focal plane of the collecting optics, the shadow projected onto the detector has exactly the same shape as the outline of the particle (case a) produced by projecting it onto a plane normal to the z axis. As the particle distance from the focal plane, which is referred to as *defocus*, z_{def}, increases, two superimposed shadows produced by each of the two beams gradually separate (case b) and, for sufficiently large defocus distances, the two shadows do not overlap at all (case c). In case b, the shadow consists of a darker part (subsequently referred to as the 'dark' shadow) which occupies the overlapping region of the two shadows from each beam (figure 1) and a brighter part ('bright' shadow) at the non-overlapping region. Based on previous investigations [Hardalupas *et al.*, 1994; Morikita *et al.*, 1995], which showed that the sizing error can be smaller than 10% when only shadows which *contain* a darker overlapping area are recorded, the signal processor has been designed to be triggered only by particles with trajectories such as a and b of figure 1. Thus, a limitation on acceptable defocus distance, $z_{def,lim}$, was applied and here was 300 μm.

The projected area-equivalent diameter, d_a, is defined from the average size of the two types of shadow, S_p, as follows

$$d_a = \sqrt{\frac{4}{\pi} S_p}, \quad S_p = \frac{2S_L + S_U}{2}$$

(1)

where S_L and S_U represent the area of the 'dark' and the 'bright' shadows in figure 1 respectively. Note that the magnitude of S_P is not affected by small defocus distances z_{def} [Hardalupas *et al.*, 1994]. Because the shadow Doppler velocimeter measures a projected area, other diameters could be used, such as a Feret diameter.

If the particle has a y component velocity, as well as an x component, the centroid of the projected shadow travels parallel to the detector axis during sampling. In this case, although the reconstructed shadow image *before correction* for this effect is distorted, the areas of the shadows S_U and S_L and accordingly the equivalent diameter d_a are not affected. Moreover, if the defocus of the particle z_{def} is large enough ($z_{def} > 100$ μm in the present configuration), *i.e.* case b in figure 1, the displacement between the two shadows can be used to obtain the trajectory angle of the particle in the x-y plane [Morikita *et al.*, 1995]. This feature is ultimately used to calculate the velocity component in the y direction, from the trajectory angle measured by the SDV and the velocity component along the x direction, measured by LDV. It is important, in the context of one of the potential error sources examined below, to recall that the defocus distance was measured using a cross-correlation technique on the measured images [Morikita *et al.*, 1995] and that the resolution limitation of the array associated with the pitch between diodes implies a corresponding resolution limitation in the measurement of the defocus location in the z direction. If ΔP represents the diode pitch spacing of the *image* of the array in the sampling-space, 12.2 μm in this case, then the *z*-resolution limitation, Δz, is

$$\Delta z = \frac{\Delta P}{2 \tan \gamma} \cong 120 \, [\mu m] \qquad (2)$$

The nominal resolution of the cross-correlation function thus also corresponds to 120 μm which is too coarse to be useful and hence an interpolation technique, based on the "adjusted Gaussian fitting" [Kobashi *et al.*, 1990], was used to provide better accuracy for the measurement of the defocus distance, z_{def}.

2.2 Signal Processor

In order to make flux and volume fraction measurements feasible, the speed of the processor was increased from the configuration used by Maeda *et al.* [1995] and the amount of data transferred to the host computer was reduced. A different method of processing was adopted in this work: a new data recorder was developed, consisting of 2-bit analogue window comparators and digital transient recorder, shown in figure 2, with a single channel A/D converter for Doppler signals (not shown) to replace the 8 bit A/D converters used in all previous work reviewed in §1. The window comparators were aligned to distinguish three different irradiance levels, two of which corresponded to the 'dark' and 'bright' shadows, and the third to the absence of shadow on the detector. The threshold levels of the comparators were set empirically at 75 and 40% of the maximum amplitude [Hardalupas *et al.*, 1994]: the maximum amplitude corresponded to irradiance levels on the detector in the absence of particles in the probe volume, and the two threshold levels were used to discriminate between dark shadows (irradiance levels smaller than 40% of maximum) and bright shadows (between 40 and 75%). This arrangement allowed faster, hardware processing of the images, performed by the comparators rather than software processing of the A/D converted signals and also the amount of the data transferred to the host computer was reduced by 75% compared to the 8 bit A/D converted signals. In addition, the data transfer rate from the processor to the host computer was increased by a factor of 10 compared to all previous work and, hence, the maximum measurable particle number density was further increased.

The processor had two independent sets of memory with 4096 samples/channel for 32 channels for storage of the digital signals from the comparators. The detailed specifications of the processor are listed in Table 2. The maximum data rate of the present hardware was about 300 particles per second and was limited by the data transfer to the host computer rather than by the sampling rate.

A further improvement to the processor was to increase the number of diode segments available for triggering. An arbitrary number of trigger channels could be used and this increased the data rate associated with small particles: here we used 8 contiguous trigger channels. In addition, the analysis of §2.4 will show that this simplified the calculation of the area of the sampling-space and minimised the potential particle shape- and trajectory-biasing in the measurement of the flux and the volume fraction.

Table 2.   Specifications of the SDV signal processor

Input Channels	32 (can be extended up to 128)
Sampling Frequency for Shadow [MHz]	10
Sampling Frequency for LDA [MHz]	10
Data transfer rate [Mbyte/s]	2.0
Interface	IBM-PC/AT (ISA)

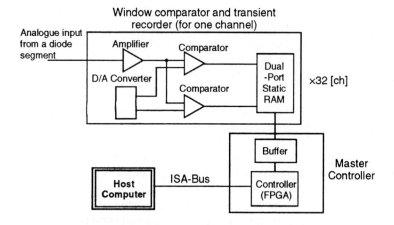

Figure 2.   Schematic of the signal processor.

2.3 Method of Flux and Volume Fraction Measurement by SDV

The *mean* mass flux of particulate phase, $G_\perp$ [kgm^{-2}s^{-1}], a vector defined in the direction normal to the interference fringes in the LDV probe volume, is defined by the amount of particle mass crossing unit area per unit time [Hardalupas et al., 1994]:

$$G_\perp = \frac{\rho_p}{T_s} \sum_i \left[\frac{V(i)}{A_\perp(i)} \sum_{j=1}^{n(i)} \frac{u_\perp(i,j)}{|u_\perp(i,j)|} \right] \tag{3}$$

where T_s is the total sampling time of the measurement, ρ_p is particle density, $V(i)$ is volume of the particle, $u_\perp(i,j)$ is velocity of particle j in size class i, $n(i)$ is the total number of particles in size class i and $A_\perp(i)$ is the area of sampling-space through which particles flow with unit normal perpendicular to the fringes. The cross-sectional area $A_\perp(i)$ is a function of the particle size and the method for its calculation will be described in §2.4. The particle volume $V(i)$ in the case of irregular particles can be estimated only from the projected area equivalent diameter, d_a; however, spherical particles have been used in the present study to avoid this additional uncertainty in the calculation of particulate volume fraction.

There are two different methods to determine the mean particle volume fraction with LDV-based techniques. The first, using the LDV measured particle velocity perpendicular to the plane of the fringes, $U_\perp$, assumes predominantly unidirectional flow normal to the plane of the fringes from which the particle volume fraction C_v is obtained as follows

$$C_v = \frac{1}{T_s}\sum_i \frac{V(i)}{A_\perp(i)|U_\perp(i)|} \tag{4}$$

where $U_\perp(i)$ is the ensemble-averaged velocity for size class i,

$$|U_\perp(i)| = \frac{n(i)}{\sum_j^{n(i)} |u_\perp(i,j)|^{-1}} \tag{5}$$

where $u_\perp(i,j)$ is velocity of particle j in size class i and $n(i)$ is the total number of particles in size class i, measured over T_s.

This is an approximation which can yield indefinitely large errors in recirculating flows near a stagnation point, because the denominator in eqn. (4) approaches zero. Hardalupas and Taylor [1989], based on the suggestion of Capp [1983], proposed an alternative method to overcome this problem. This method uses the residence time of a particle in the sampling-space:

$$C_v = \sum_i \left\{ \frac{\sum_j^{n(i)} \tau(i,j)}{T_s \cdot V_s(i)} \right\} \tag{6}$$

where $V_s(i)$ is the volume of the sampling-space for particles in size class i and $\tau(i,j)$ is the particle residence time in the probe volume.

The definition in eqn. (6) is, in principle, a better estimation of the volume fraction than eqn. (4), because it represents the duration of particle occupancy in the probe volume, which is equivalent to volume fraction independent of the velocity direction. However, if the flow is predominantly one-directional, the former *is* equivalent to the latter. Since the present measurements were made in a uni-directional flow, the former method was preferred. An extension of the residence time method suitable for SDV measurements is proposed in §3.3.

2.4 Size of the Sampling-space of SDV

To measure the correct size distribution, Morikita *et al.* [1995] proposed a procedure to correct the biasing caused by the dependence of the size of the *sampling-space* inside the probe volume on particle size and defocus. The sampling-space is defined as that region of space over which a particle generated two superimposed shadows, the overlapping region of which over some part of its trajectory passes over the trigger diode and which is not vignetted by the finite width of the linear array. These statements are quantified below but Morikita *et al.* [1995] showed that the sampling-space has, in principle, elliptical shape in the *y-z* plane (figure 1). Their definition was, however, appropriate only for *spherical* particles travelling *normally* to the axis of the diode array when *only one* trigger channel of the diode array was used.

The equations proposed by Morikita *et al.* [1995] for the dependence of the width of the elliptical sampling-space on particle size, assuming spherical particles, are here extended to include the effect of multiple (contiguous) trigger channels, the width of the sampling-space as a function of the defocus distance z_{def}, the particle diameter d_p, the laser beam half-angle of intersection γ, and d_t, the width of the *image* of the triggering area in the LDV probe volume (given by the width of the eight trigger elements of the diode array divided by magnification of the collection optics, $d_t = 12.2 \times 8 \cong 98 \mu m$). Assume, for the moment, that the linear array and that D_y (figure 1) are indefinitely large. Then the width of the cross-sectional area of the sampling-space $d_w{}'$ is:

$$d_w'\left(d_p, z_{def}\right) = \sqrt{d_p^2 - 4 z_{def}^2 \tan\gamma} + d_t$$

(7)

Figure 3 shows the increase of $d_w{}'$ with particle size, calculated for defocus distance z_{def} of 0 and 300 µm using eqn. (7) for the present optical parameters. Owing to the small beam intersection angle 2γ, $d_w{}'$ is almost independent of particle size, for sizes over about 50 µm, and of the defocus distance, z_{def}.

We now take into account that the diode array *has* finite length and thus can image only a finite portion of the optical probe volume. Let W be the width of the *image* of the detector, corresponding to the 30 segments of the diode array, in the sampling-space of the velocimeter and which depends on the magnification of the collection optics. Morikita *et al.* [1995] showed that the width of the cross-sectional area of the sampling-space, $d_w{}''$, *decreases* linearly with particle size to avoid vignetting:

$$d_w''(d_p) = W - d_p \tag{8}$$

Note that d_t does not appear in this equation. The calculated values for d_w'' from eqn. (8) are also plotted, with open circles, in figure 3. The net value of the width of the sampling-space, d_w, for any particle size is given by

$$d_w(d_p, z_{def}) = min\left(d_w'(d_p, z_{def}), d_w''(d_p)\right) \tag{9}$$

An implication of eqns. (7) and (8) is that the area $A_\perp(d_p)$ of the sampling-space for particle size d_p with defocus limitation at $z_{def,lim}$,

$$A_\perp(d_p) = \int_{-z_{def,lim}}^{z_{def,lim}} d_w(d_p, z_{def}) dz_{def} \tag{10}$$

is a non-monotonic function of particle diameter. The variation of the area of the sampling-space was estimated to be about 20% between 80 and 120 µm particle sizes in the experiment described in section 3.2, with a triggering width, d_t, of 98 µm and a defocus limit, $z_{def,lim}$, of 300 µm, for reasons given in §3.2 below.

In general, the validatable sampling-space will be *smaller* than that predicted by eqn. (10) if there are particles of random shapes and with random orientations (which is *not* the case in these experiments) and particularly if the

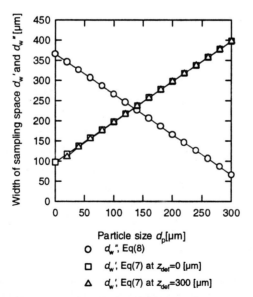

Figure 3. Calculated variation of the width of sampling-space d_w' and d_w'' with particle diameter d_p.

trajectories relative to the axis of the photodiode array are random (which *is* true in any turbulent flow). The latter will be vignetted by the detector, and consequently rejected by software, which is not accounted for in eqns. (8) and (10) [Hardalupas *et al.*, 1994]. Figure 4 shows a pseudo-image, *i.e.* the output of the linear detector array as a function of time, which illustrates these cases: in figures 4 (a) and (b) a spherical particle passes through the sampling-space at trajectory angles relative to the axis of the array at 90°, case (a) and other than 90°, case (b). The trajectory in figure 4(a) will be accepted but that of figure 4 (b) will be rejected by software because during the whole trajectory part of the shadow image is vignetted and not recorded, although the particle, *at the time of triggering*, lies within the area defined by eqn. (10). Figures 4 (c) and (d) illustrate the first of two possible effects of particle shape and orientation for the particular case of identical ellipsoids with an aspect ratio of about two and trajectory normal to the diode array axis, passing through the centre of the probe volume. The choice of ellipsoids is for the sake of example only and does not limit the generality of application of the conclusions drawn here. The width of the validatable cross-sectional area of the sampling-space in figure 4 (c) is larger than in figure 4 (d).

Figure 4 (e) demonstrates the second effect, which is to limit the maximum defocus distance up to the point at which the two shadows just overlap. Particles with the orientation of figure 4(e) can have larger maximum defocus distances than with the orientation of figure 4(f) before they fail to trigger the instrument, and the length of the validatable area of the sampling-space will be therefore bigger. The magnitude of these two effects

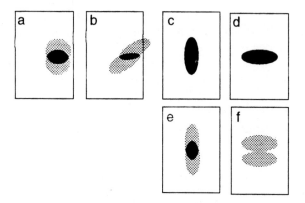

Figure 4. SDV pseudo-image as a function of time for different cases: spherical particle at (a) 90° and (b) other than 90°; ellipsoid in-focus(c) and (d) and defocused (e) and (f). The width of the frame surrounding each image represents the width of the active elements of the linear diode array. The height represents samples of the array as a function of time

depend on the details of particle shape and orientation. The effect of trajectory, figure 4(b), can also affect that of orientation and shape of figures 4(c) to (f) and must either be accounted for, or avoided.

In order to avoid biasing of the flux and volume fraction measurements due to the effects illustrated in the preceding paragraph, an additional spatial limitation was introduced in *post-processing software* to restrict the sampling-space along the y axis, which also simplified the shape of the sampling-space to a rectangle (figure 1). Thus, only images which lay within $y<d_w$ were accepted in the *post-processing* software, where d_w was calculated from:

$$d_w = 2y_{lim}$$
(11)

and y_{lim} is a limiting distance along the y axis, measured from the centre of the array. This dimension was set at 100 µm, empirically, by finding the area of the sampling-space over which the measured particle arrival rate (number of measurements made over a given time) was uniform. By introducing the spatial limit, y_{lim}, the area of the sampling-space varied by only 15% over the measured size range in the experiment described in section 3.2.

3. Experimental Results

3.1 Accuracy of Estimation of the size of the Sampling-space

In order to apply the method described in §2.4 for the calculation of the sampling-space, it is first necessary to establish the error of measurement of the location of the particle in the sampling-space. The particle position in the y direction of figure 1 can be measured from the position of the recorded two dimensional image, by calculating the position of the 'weighted-centre' of the shadow image. The uncertainty was due to the discretisation error, corresponding to the half-pitch of the detector, in determining shadow displacements in the y direction: for the magnification used in this experiment, this corresponded to an uncertainty of ±6 µm in the present configuration, or ±3% of the width of sampling-space, from eqn. (11), used in the present experiments.

The position in the z direction (defocus, z_{def}) was deduced from the measurement of the separation between the two shadow images described in §2.1. However, the uncertainty of the measurement of the defocus was expected to be worse than that in the y direction due to the resolution Δz (§2.1), as a result of the small intersection angle. In order to quantify the uncertainties, an experiment was carried out using a 100 µm pinhole. The size of the pinhole was arbitrarily chosen because the error on the defocus measurement is dependent only on the separation between the two shadow images. The pinhole image, which extended over approximately 8 segments

246

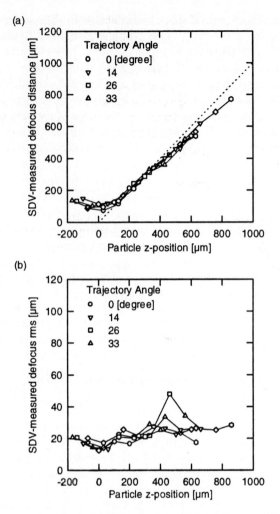

Figure 5. Mean (a) and rms (b) SDV-measured defocus distance of a pinhole shadow image for four trajectory angles relative to the axis of the array: 0° (circles), 14° (down-triangle), 26° (square) and 33° (up-triangle)

of the diode array, was fixed on an optical flat on a rotating disk driven by a servo-motor mounted on a three-dimensional traverse to provide a stable and reliable signal source. The pinhole was rotated with a velocity about 1.5 m/s and the rotational drifting was smaller than 2%. This technique was proposed by Hovenac & Hirleman [1991].

Figures 5 a and b compare the mean and rms of the SDV-measured defocus with the true defocus for four trajectory angles, which were varied by traversing the rotating disk. The measured mean pinhole defocus distance, over 100 measurements, collapsed on a single curve for defocus distances between 100 and 500 μm, with systematic error smaller than 8% except in the region with defocus smaller than 100 μm. This systematic error was probably due to error in measurement of the intersection angle which, for practical reasons, had to be measured from triangulation limited to the transmitting optics side. The large departures from the actual defocus distance in the region of 0 to 100 μm were due to the response time of the detector array, which was not fast enough to detect the sharp signal changes from the shadow passage. The random error was always smaller than 50 μm and typically 20 μm, which corresponded to 7% if the sampling-space dimension in z direction was larger than 600 μm. In any case, a random error does not affect the measurement of $G_\perp$ (less than 1% if the sample size was more than 100) and, only the systematic error of 8% was considered in the further discussion. As a consequence, the systematic error in the determination of $A_\perp$ was 8%.

3.2. Flux and Volume Fraction Measurement

The accuracy of flux measurement was estimated by comparison with measurements in a flow with known particulate mass flux and volume fraction. In the present study, a water channel flow laden with polyethylene spheres was used, as shown in Figure 6. A closed water circuit was arranged by using a vertical rectangular duct 550 mm long with 20×20 mm^2 square cross-section connected to an electronically controlled pump. The bulk water velocity was 0.16 m/s and the corresponding Reynolds number 3.2×10^3. Pre-weighed

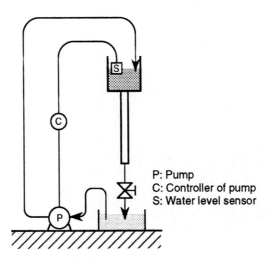

P: Pump
C: Controller of pump
S: Water level sensor

Figure 6.   Schematic of the flow channel.

batches of monodisperse polyethylene particles (SB-100S, Asahi Kasei, Japan) of 100 μm nominal, company-quoted diameter were mixed in a known water volume and thoroughly stirred to form a uniform mixture. The uncertainty of the precision balance was smaller than 0.1% of the total particulate mass suspended in the water. The drift velocity between the continuous and the dispersed phase was expected to be negligible, because the density of the particles was 1055 kg/m^3, very close to that of water and it is therefore a reasonable approximation to assume that the velocity of the continuous phase was that of the particles.

Particle mass flow rate was evaluated by integration of the local flux at 20 points over the cross sectional area at 450 mm below the inlet of the duct, as illustrated in figure 7, for the case of bulk volume fraction, or loading, of 0.005%. The error due to numerical integration, which was estimated from comparison between best and worst results using numerical integration schemes of first and second order of accuracy, was estimated to be smaller than 12% and was considered to be random. In addition, from the integration procedure it was estimated that the measured particle loading on the centreline was larger than the bulk value by a factor of 1.3 as might be expected from the ratio of centreline to bulk velocity in a turbulent pipe flow. The bulk volume fraction was then derived by dividing the mass flow rate by the bulk velocity, based on the bulk particle velocity established by integration and the assumption that the particle and continuous phase velocity were identical, and the known cross-sectional area of the duct. The bulk particle volume fraction was *estimated* from a single flux measurement in the centre of the duct at all other loadings, by correction with the empirical factor estimated for the case of 0.005% volume loading. This was a reasonable approximation, since the volume fraction

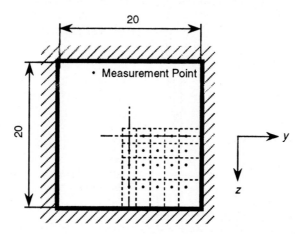

Figure 7. Location of measurement points in a cross section of the flow channel of figure 6 used for the integration of the local flux.

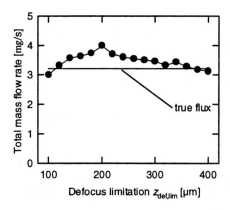

Figure 8. Variation of particulate mass flow rate (circles) as a function of defocus limitation, $z_{def,lim}$, in comparison with the true flow rate (solid line).

profiles were expected to be similar for the range of volume loadings considered here because particle-particle interaction is negligible. For the determination of the local flux, eqn. (3) was applied. The size of the area of the sampling-space was calculated from eqn. (11) but, following the analysis of §2.4, y_{lim} in eqn. (11) was chosen to be 100 µm and that resulted in a rectangular cross-sectional area for the sampling-space which satisfied the requirements for minimum width of the sampling-space from eqns. (7), (8) and (11).

In principle, the measured mass flux should be independent of the size of the sampling-space, because the particle density is presumably uniform within the probe volume. However, because of the finite response time of the detector and the so-called cockscomb appearance of the leakage error associated with the use of the "adjusted Gaussian fitting" functions for interpolation of the correlation-based measurement of defocus distance [Kobashi et al., 1990; Ibrahim et al., 1990], the measured flux varied with defocus limitation. Thus, for a uniform distribution of particles in the z direction, the probability density function of *measured* defocus distance would *not* be uniform but could have a quasi-sinusoidal cockscomb variation about the mean, with wavelength related to Δz. To minimise the error associated with this variation, it is necessary to make the value of $z_{def,lim}$ as large as possible. Figure 8 presents the measured particulate mass flow rate (circles), over 1000 measurements for each point, as a function of defocus limitation, compared to the mass flow rate (solid line, calculated from the particle mass and water volume) of 3.21 mg/s for the case of 0.005% volume fraction and 0.119 mm^{-3} calculated number concentration. The raw data, stored in the computer, for this figure was *the same for each point* and the defocus limitation was applied in post-processing software: the calculated particulate flux was within -7% and +25% of the SDV-measured value and the variation is due to the cockscomb leakage effect. It is observed that the flow rate increases linearly with defocus for distances smaller than 200 μm, due to leakage effects and the use of $\Delta z \approx 100$ μm, and then decreases towards the true value. For defocus distances larger than 500 μm, the sizing error becomes larger than the 10% random error in the estimation of the area of the sampling-space, which again results in high uncertainties in the calculation of the flux. As a rule of thumb, the minimum defocus distance, $z_{def,lim}$, should not be smaller than that corresponding to separation between the two shadows equivalent to 2 diode segments (about 250 μm in the present configuration) and not larger than 500 μm, because the sizing error becomes comparable to the error in calculating the sampling-space. A value of $z_{def,lim}$ of 300 μm was therefore adopted in this work.

The uncertainties of the quantities related to the mass flux measurement are summarised in table 3. The *systematic* uncertainty of the integrated particle volume was estimated from comparison of the SDV-measured mean diameter and the maker-specified diameter. The *random* uncertainties in the measurement of a particle diameter, the position along the detector and the defocus distance which were (0.5/8) or 6%, 3% and 7% respectively, do *not* contribute to the total uncertainty in the flux measurement, because volume flux was estimated from average quantities over a large sample and according to the central limit theorem, the variance of the mean (*i.e.* the standard error of the mean) decreases with increasing sample size.

Table 3. Estimated systematic uncertainties of the mass flux measurement at a point.

Particle volume $V(i)$	6%
Area of sampling-space $A_\perp (i)$	8%
Mass flux $G_\perp(i)$ Particle volume fraction C_v	$\cong 14\%$

Figure 9 shows comparison between the SDV-measured (blocked circles) against the actual (solid line) particulate integrated volume fraction on log-log axes. The vertical uncertainty bands denote the expected systematic uncertainty according to the table 3. We chose to base the comparison on integrated particle volume fraction, rather than on mass flow rate, because the reference values were a known mass of particles, of known material density, and a known volume of water in the flow circuit, so that the bulk volume fraction is more accurately known than is the mass flow rate of particles in the duct.

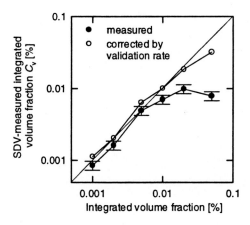

Figure 9. Comparison between the measured (blocked circles) against the actual (solid line) integrated particulate volume fraction. Open circles correspond to the fraction corrected by the signal error rate of the SDV given in figure 10.

For dilute flows with volume fraction up to 0.005%, the discrepancy did not exceed 20% and the observed discrepancy in figure 9 is, taking into account the systematic errors listed in Table 3 together with 12% of the systematic error due to flux profile integration, no worse than is to be expected. However, the measured flux was consistently smaller than the actual value for volume fractions larger than 0.005%. This difference was caused by the turbidity of the flow and most of the rejected signals were interpreted by the software as multiple occupancy of the sampling-space; it was even observable by eye that the incident laser beams were partly extinguished by their passage through the flow. The dense-flow limitation of the instrument on the basis of the present measurements was estimated in the order of 0.005% in volume fraction: to provide a comparison, this would be equivalent to a mass loading of 5% for water droplets in an air flow in the duct.

The larger discrepancy in figure 9 for integrated volume fraction higher than 0.005% is likely to be a result of beam extinction due to multiple particle occupancy along the beam path. As a consequence, the discrepancy should correlate with the number of rejected measurements. Indeed, figure 10 shows that the signal error rate (rejected signals / validated signals, where signals were rejected *either* because the particle location was outside the user-set defocus limitation, $z_{def,lim}$, *or* because the cross-correlation routine suggested multiple-occupancy of the probe volume). In figure 10, it should be noted that, for volume fractions below about 0.01%, the signal error rate of about 20-30% was overwhelmingly due to particles being beyond $z_{def,lim}$ and hence the signal error rate does not, in principle, contribute to an error in mass flux measurement. In contrast, the signal error rate increased rapidly with volume

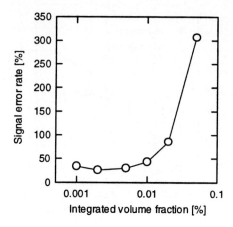

Figure 10.   Signal error rate as a function of volume fraction.

fraction above a volume fraction of 0.01% and this was overwhelmingly due to multiple occupancy.   This result suggested that correction of the flux measurement by the signal error rate can provide a reasonable estimate of the flux particularly for fractions above 0.01% v/v; the open circles in figure 9 correspond to the corrected values of the measured fraction.   As expected, better agreement for the dense limit of the present measurements was achieved from the correction procedure and the maximum discrepancy did not exceed 40% for the worst case, compared with the almost 100% for the uncorrected values.

3.3   Extension to Residence-Time Based Volume Fraction Measurement

The features of the method described in earlier sections imply that it is possible to calculate residence time-based volume fraction using eqn. (6).   In that case, the particle residence time $\tau(i,j)$ through the sampling-space must be determined from the analogue signal, *i.e.* the signal before passing through the window comparators, which in turn implies that this time will be a function of the threshold levels used for the discrimination of the shadows, figure 11.   The uncertainty in the determination of the residence time arises from the shape of the analogue signal which rises and falls more sharply than the raw signals from LDV.   Thus, according to the results of Hardalupas *et al.* [1994], the error in the estimation of a transit time will be exactly that involved in the determination of diameter and hence will not exceed 10% for irregular, powder-like particles.   However, the method of §2.4 must now be extended to calculate the volume $V_s(i)$ of the sampling-space as a function of the trajectory and the particle shape and size.

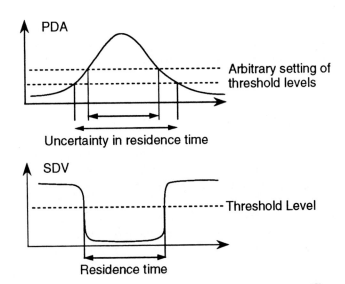

Figure 11. Schematic of residence time calculation as a function threshold levels, for PDA and SDV respectively.

The volume of the effective sampling-space of the SDV can be determined from the area of the sampling-space, following the analysis of §2.4, together with the relevant dimension in the x direction of figure 1. This dimension will be the size of a sampled particle, because the instrument is triggered as long as there is particle passing through the sampling-space. The volume thus is given by

$$V_s(i) = D_y' \cdot D_z' \cdot d_p(i) \tag{12}$$

where $D_y'=2y_{lim}$ and $D_z'=2z_{def.lim}$ are the sizes of the sampling-space in the y and z direction, respectively, as shown in figure 1 and $d_p(i)$ is the nominal diameter of particle size class i. As for single-component LDV and PDA systems, this analysis holds if the x velocity component is larger that the y and z components, or, in other words, if the flow is almost normal to the axis of the photodiode array.

When the particle is non-spherical, asphericity of the particle could cause measurement error on flux, similar to the effect illustrated in figures 4(c) and (d). To avoid this effect, the area-equivalent diameter, d_a, in eqn. (1) can be used instead of nominal diameter, d_p, in eqn. (7), assuming that the particle passes the probe volume with random orientations. Although the volume fraction defined by eqn. (6) is equivalent to the result from eqn. (4) *in the present condition*, since the particle used in the experiment is spherical and the flow was uni-directional, the technique can be used to obtain residence-time based volume fraction measurement, as well as the velocity-based value.

4. Conclusions

A method for the measurement of local mass flux and volume fraction by shadow Doppler velocimetry was proposed and evaluated by using a turbulent water channel flow with known particle volume fraction.   We summarise the most important conclusions from the present investigation:

1.   The accuracy of the measurement of particle location, z_{def}, in the probe volume of SDV was tested with a mounted 100 μm pinhole which corresponded to about 8 diode segments.   The random uncertainty corresponded to the equivalent of ±0.5 pitch along the laser beams which is less than 7% of the typical size of the sampling-space of SDV.

2.   Mass flux was measured by defining a defocus limitation, $z_{def,lim}$, of the sampling-space.   The measured volume flux was almost independent of the size of the sampling-space, $A_\perp(i)$, when the defocus limit was set larger than 300 μm to minimise the leakage error of the cross-correlation technique used to measure defocus.

3.   The accuracy of the integrated particle volume fraction, C_v, from the SDV was better than 20% and typically 10% at particle volume fractions of 0.005% or less.   Of this, random uncertainties of 12% were attributable to the integration procedure, systematic of 8% to the angle measurement for determining $A_\perp(i)$ and a further systematic 6% contribution was due to inaccuracy in the measurement of the particle volume $V(i)$.   In the denser flow, the accuracy decreased and hence the measured mass flux was compensated by the signal error rate.   The majority of invalid measurements were due to multiple occupancy, *i.e.* two or more particles in the sampling-space.   The corrected values of C_v were within 40% of the true values, for particulate volume fractions up to 0.05 %.

4.   The volume of the sampling-space used for residence time-based volume fraction measurements, eqn. (6), can be estimated from the area of the sampling-space and the size of the measured particle.   The uncertainty, for flows which are almost normal to the axis of the array, is expected not to exceed 10%.   Unlike the PDA, the volume of the sampling-space *is not affected* by particle size-dependent scattering phenomena due to the Gaussian irradiance of the incident laser beams.

5. Acknowledgements

The authors are grateful to Mr. K Suzuki for his help with the measurements. The groups at Imperial and Keio are in receipt of a British Council (Tokyo Office) travel grant for collaborative research which facilitated the production of this publication. One of the authors (H.M.) is supported by the Japan Society for the Promotion of Science (grant No. 4927).

6. References

Capp, S. P.: Experimental investigation of the turbulent axisymmetric jet, PhD thesis, State University of New York at Buffalo, 1983.

Gréhan, G., Gouesbet, G., Naqwi, A. and Durst, F., 1992, Trajectory ambiguities in phase Doppler systems: use of polarizers and additional detectors to suppress the effect, Proceedings of 6th international symposium on Application of Laser Techniques to Fluid Mechanics, Lisbon, Portugal, paper 12.1.

Hardalupas, Y. and Taylor, A. M. K. P., 1989, On the measurement of particle concentration near a stagnation point, Exp. Fluids , Vol 8, pp. 113-118.

Hardalupas, Y., Hishida K., Maeda M., Morikita, H., Taylor A. M. K. P. and Whitelaw, J. H., 1994, Shadow Doppler Technique for Sizing Particles of Arbitrary Shape, Appl. Optics, Vol. 33, No. 36, pp. 8417-8426.

Hardalupas, Y., Taylor A. M. K. P. and Whitelaw, J. H., 1994, Mass flux, mass fraction and concentration measurement of liquid fuel in a swirl-stabilised flame, Int. J. Multiphase flow, Vol. 20, pp. 233-259.

Hirleman, E. D. and Moon, H. K., 1982, Response Characteristics of the Multiple-Ratio Single-Particle Counter, J. of Colloid and Interface Sci, Vol. 87, No. 1 pp.124-139.

Hishida, K., Maeda, M., Morikita, H., Taylor, A. M. K. P. and Whitelaw, J. H., 1995, Particle sizing in a confined reacting flow using Shadow Doppler Anemometry, 4th Int. Congr. on Optical Particle Sizing, Nürnberg, Germany.

Hovenac, E. A. and Hirleman, E. D., 1991, Use of Pinholes and Reticles for Calibration of Cloud Droplet Instrumentation, J. of Oceanic Tech., Vol. 8, No. 1, pp.166-171.

Ibrahim, K. M., Werthimer, G. D. and Bachalo, W. D., 1990, Signal processing considerations for laser Doppler and phase Doppler applications, Applications of Laser Techniques to Fluid Mechanics, Springer-Verlag, pp. 291, 316.

Kobashi, K., Hishida, H. and Maeda, M., 1990, Measurement of fuel injector spray flow of I.C. engine by FFT based phase Doppler anemometer- an approach to the time series measurement of size and velocity, Applications of Laser Techniques to Fluid Mechanics, Springer-Verlag, pp. 268, 287.

Maeda, M., Morikita, H., Prassas, I., Taylor, A. M. K. P. and Whitelaw, J. H., 1995, Size and Velocity Measurement by Shadow Doppler Velocimetry within a Pulverised Coal-Fired Furnace, HTD-Vol. 321/FED-Vol. 233, pp.351-360.

Maeda, T., Morikita, H., Hishida, K. and Maeda, M.:, 1996, Determination of effective measurement area in a conventional phase-Doppler anemometry, 8th Int. Symp. on applications of laser techniques to fluid mechanics, Lisbon.

Morikita, H., 1995, personal communication.

Morikita, H., Hishida, K. and Maeda, M., 1995, Measurement of Size and Velocity of Arbitrarily Shaped Particles by LDA Based Shadow Image Technique, Developments in Laser Techniques and Applications to Fluid Mechanics, Springer-Verlag, pp. 354-375.

Morikita, H., Hishida, K. and Maeda, M., 1994, Simultaneous Measurement of Velocity and Equivalent Diameter of Non-spherical Particles, Part. Part. Syst. Charact., Vol. 11, No. 3, pp. 227-234.

Orfanoudakis, N. G. and Taylor, A. M. K. P., 1992, The effect of particle shape on the amplitude of scattered light for a sizing instrument, Part. Part. Syst. Charact., Vol. 9, pp. 223-230.

Orfanoudakis, N. G., 1994, Measurements of size and velocity of burning coal, PhD Thesis, Univ. of London.

Orfanoudakis, N. G. and Taylor, A. M. K. P., 1995, Evaluation of an amplitude sizing anemometer and application to a pulverised coal burner. Comb. Sci. and Tech., Vol. 108 , pp. 255-277.

Qiu, H. H. & Sommerfeld, M., 1992, A reliable method for determine the measurement volume size and particle mass fluxes using phase-Doppler anemometry, Exp. Fluids, 13, pp. 393-404.

Saffman, M., 1987, Automatic calibration of LDA measurement volume size, Appl. Optics, Vol. 26, No. 13, pp. 2592.

Sankar, S. V. and Bachalo, W. D., 1991, Response characteristics of the phase Doppler particle analyser for sizing spherical particles larger than the light wavelength, Appl. Optics, Vol. 30, No. 12, pp.1487-1496.

Sommerfeld, M. and Qiu, H.-H., 1995, Particle concentration measurement by phase Doppler anemometry in complex dispersed two-phase flows, Exp. Fluids, 18, pp. 187-198.

Schöne, F., Bauckhage, K. and Wriedt, T., 1994, "Size of the detection area of a phase-Doppler anemometer for Reflecting and Refracting particles", Part. Part. Syst. Charact., 11, 1994, pp. 327-338.

Taylor, A. M. K. P., 1995, Experimental techniques for dispersed two phase flows, Annual ASME/JSME Fluid Eng. Meeting, Hilton Head, SC, USA.

Yeoman, M. L., Azzopardi, B. J., White, H. J., Bates, C. J. and Roberts, P. J., 1982, Optical development and application of a two colour LDA system for the simultaneous measurement of particle size and particle velocity, Engineering Applications of Laser Velocimetry, Winter Annual Meeting ASME, Phoenix, Arizona.

Xu, T-H and Tropea, C., 1994, Improving the performance of two-component phase Doppler anemometers, Meas. Sci. Technol., Vol. 5, pp. 969-975.

II.8 Structure of a Turbulent Boundary Layer Using a Stereoscopic, Large Format Video-PIV

Z.C. Liu[1], R.J. Adrian[1], C.D. Meinhard[2] and W. Lai[3]

[1] Department of Theoretical and Applied Mechanics
University of Illinois, Urbana, Illinois 61801
[2] Department of Mechanical and Environmental Engineering
University of California, Santa Barbara, California 93106
[3] TSI, Inc., St. Paul, Minnesota 55164

Abstract. Development of a stereoscopic particle image velocimeter for the measurement of three-dimensional vectors on a planar domain is described. The camera is based on two large format (2k x 2k) video cameras. Experiments in a turbulent boundary layer at $Re_\theta = 2525$ demonstrate its ability to measure three-dimensional turbulent flow. In addition to the quantitative value of the out-of-plane component, it is found that having the complete three-dimensional vector also significantly improves the qualitative visualization of the flow.

Keywords. particle image velocimetry, boundary layer, three-dimensional, stereoscopic, videographic, turbulence

1 Introduction

The extension of particle image velocimetry to measurement of three-dimensional velocity vector fields is both desirable and achievable. Hinsch (1995) reviews many of the techniques used for three-dimensional measurements. In general, there are two classes of measurement techniques: those which measure three-dimensional velocity vectors on full three-dimensional domains , i.e. volumes, and those which measure three-dimensional vectors on planar domains, i.e. light sheets. Three-dimensional volumetric measurements can be performed by using holographic techniques (Barnhart et al. 1994, Meng and Hussain 1994), by using photogrammetric particle tracking techniques (Maas, Gruen and Papantonoiu 1993, Brodkey 1977, Nishino and Kasagi 1991), or by scanning a laser light sheet rapidly (Brucker 1995). Photogrammetric methods use three or four cameras to view particles from several different directions. Volumetric measurements are in general difficult and require equipment that is rather different than the standard planar PIV equipment which uses a laser light sheet and a single camera. Holographic systems require a unique set of techniques and apparatus, and three-dimensional particle tracking systems involve multiple cameras and specialized software, plus they place demands on the optical access required to view the flow field.

However, the most demanding aspect of volumetric measurements is the sheer number of velocity vectors that are obtained from the measurements. A modest 100 x 100 x 100 velocity grid measurement yields one million vectors, and if these measurements are to be repeated hundreds or even thousands of times for the purpose of statistical averaging, the amount of data is overwhelming. Thus, while volumetric measurements can be extraordinarily valuable for studying instantan-eous flow field structures, they are often too rich in data to be used on a routine basis with present equipment.

An intermediate approach to three-dimensional volumetric measurements with PIV is to perform measurements of a three-dimensional velocity vector filed on a planar domain. In this way, the number of vectors is the same as in ordinary two-dimensional planar PIV, but the third component contributes significantly to the experimenter's capability to visualize the flow and it is valuable for the purposes of quantitative analysis of the flow. Further, full measurement of the three-dimensional vectors eliminates the perspective error that is inherent in monocular PIV systems (Adrian 1991, Prasad and Adrian 1993). This error can be quite significant if the out-of-plane velocity component has non-negligible magnitude relative to the in-plane components and/or the angular field-of-view is not small.

Systems operating with this capability can be designed using various aspects of image correlation (Robinson and Rockwell 1993), or using the change in image magnification (Willert and Gharib 1992). However, the most common technique is the well-known stereoscopic method whereby one obtains a pair of image planes, each image plane viewing the particles in the illuminated light sheet plane from a different direction. Displacements of the particle images in the two different views differ because of the different viewing angles, and measurements of these displacements can be used to solve for the full three-dimensional displacement of particles. This technique can be applied to individual particles, as in particle tracking systems, or it can be used in combination with correlation techniques that measure the displacements of groups of particles, or even continuous grey level patterns. The correlation approach is desirable because the three-dimensional velocity vectors can be obtained on uniform grids, and the problem of matching individual pairs is solved automatically. The limitation of the stereo technique is that to assure good measurements of the out-of-plane component of velocity, the viewing angles between the stereo lenses must be substantial, of the order of thirty degrees, and the time between exposures of the images must be small enough that the particles remain within the thickness of the light sheet. These requirements have been explored amply and shown to be achievable and not inconvenient in practice (Prasad and Adrian 1993, Troy and Adrian 1996)).

Stereographic systems of conventional form have been used by Arroyo and Greated (1991), Prasad and Adrian (1993), and Troy and Adrian (1996). These experiments each used two photographic recordings, interrogated the recordings to obtain displacements and then solved the stereo equations to obtain three-dimensional velocity vectors. The difficulty with using photographic film is the problem of registering the two pieces of film with respect to each other so that measurements of displacements locations in the first image can be identified precisely with locations in the second. This problem must be faced for each pair of photographic exposures that are taken in a stereo-photographic system. The registration process is more laborious and time consuming than developing the film in itself, and it places a fundamental constraint on the utility of this method.

To make feasible extensive quantitative analysis of three-dimensional vector fields, it is necessary to make the image acquisition analysis process easy enough and fast enough to permit thousands of images to be taken for averaging purposes. To this end, the photographic recording must be eliminated, and replaced by videographic recording. Video cameras in a stereo system can be registered during the construction of the stereo camera and by this means they can be aligned once and for all, thereby eliminating the need for any future efforts to register the two images. The registration can be accomplished by means of mechanical alignment during construction, or by electronic image processing alignment during the experimental stage. The electronic alignment is necessary in situations where an aberrating medium is placed between the object plane and the cameras, so that the registration that was originally achieved in the construction of the camera is distorted by the medium. In this case, a calibration procedure can be used to correct for aberrations, provided they are not so severe as to completely destroy the images of the individual particles.

In this paper, the design and construction of a stereographic camera based on video recording is described, and some of the procedures used in the aberration calibration and correction process are explained. The system is applied to the visualization of three-dimensional vector field structure in a low Reynolds number turbulent boundary layer.

2 Stereoscopic PIV System

The stereoscopic camera is shown schematically in Fig. 1. The camera lenses L_1 and L_2 lie in a plane parallel to the object plane defined by the laser light sheet. They are offset laterally by d_{L1} and d_{L2}, respectively. Since the camera arrays also lie in a plane parallel to the object plane, the magnification of each lens is constant, equal to

$$M_O = d_i / d_o \tag{1}$$

where d_i and d_o are the image and object distances, respectively. Within the angular field indicated by the solid lines each lens images point objects with resolution and distortion that are within the specification of the lens. In the object plane the size of the region that can be imaged with good resolution by both lenses, the 'joint field', is determined by the intersection of the respective angular fields. The lateral offsets d_{L1} and d_{L2} are chosen to be the largest values that still permit the angular fields to overlap over the desired region in the object plane, i.e. to make the joint field as large as possible while still keeping the angle between

262

the lenses large enough to achieve small measurement error for the z-component of velocity. Nominally, d_{L1} and d_{L2} are equal, and for the present system the angle between the axes of the two lenses is 24.4 degrees. According to Prasad and Adrian (1993) the error in the z-component of velocity is about twice the error in the in-plane components of velocity when the angle between the axes of the lenses exceeds 30 degrees.

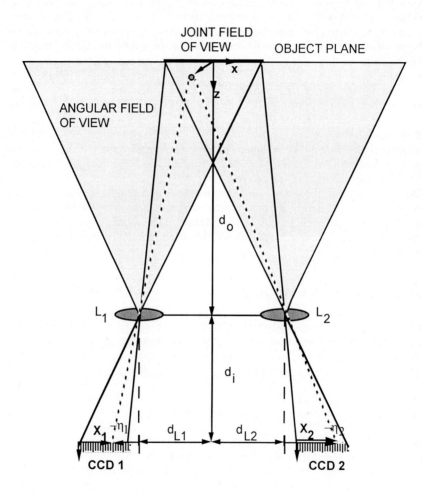

Figure 1. Schematic of the stereoscopic camera.

The field of view of each CCD array is simply the area of the array mapped onto the object plane. As in Prasad and Adrian (1993) and Troy and Adrian (1996), offsetting the cameras laterally in the image plane overlaps their respective fields of view. Each camera sees nearly the same region in the flow, thereby maximizing the size of the joint field of view. This design is relatively simple to lay out, but care must be taken to ensure that tolerances in the focal lengths of the lenses and in the orientations of the individual elements of the system are taken into account, either mechanically or by careful calibration.

The joint field of view increases by more than a factor of two if the lenses are tilted towards each other so as to completely overlay their angular fields. This method, called 'angular displacement', has been discussed by Gauthier and Riethmuller (1988) and more recently by Prasad and Jensen (1996). Its principle advantage is the larger joint field of view, and its principal disadvantage is the more complicated alignment that is needed, which also involves using magnification that varies across the field of view. Only the simpler lateral offset method will be considered here, although angular offset can be easily accommodated by the present camera.

The camera uses two Kodak *Megaplus 4.2* CCD video cameras, each with 2029 x 2044 pixels resolution. Since the camera is intended to image a 100-300 mm field of view in the flow, and the CCD chip dimension is only 18 mm, the magnification of the system is substantially less than unity. M_o= 0.23 for the present system. Two Nikon *EL-Nikkor* 302.5 mm F/5.6 enlarging lenses that are optimized for normal operation at large magnification are mounted in the reverse direction to provide the appropriate small magnification with little aberration, minimal distortion and large aperture (53.6 mm). The angular field of view of this lens is 57 degrees. Images are normally acquired at F/8. The video cameras and two lenses are mounted on a 3-D translation stage. The stage provides the six degrees of freedom needed to position and focus the cameras and to accurately register them with respect to each other. For the wind tunnel experiments, $d_o = 1617$ mm and $d_i = 372$mm.

The rest of the stereoscopic system is shown in Fig. 2. Illumination of the flow field is provided by two Continuum Nd:YAG with a wavelength of 532 nm at pulse frequency of 10 Hz. Each laser delivers up to 200 mJ/pulse of energy with a pulse duration of 8 nsec. The laser beam is formed into a sheet of about 1 mm thickness in the test section. To resolve directional ambiguity, image shifting is normally needed. However image shifting was not used for the present measurements because of the strong mean flow in comparison with the turbulence intensity in the boundary layer. If directional ambiguity were a problem the cameras could be replaced by 1k x 1k cross-correlating cameras. Timing for the sequence of laser pulses and the image capture with the cameras is provided by a TSI, Inc. *Laser Pulse* synchronizer box.

3 Image Analysis

The image analysis procedure is described in Figure 3. The capture and analysis of images is controlled by *INSIGHT,* a Windows-based software package supplied by

TSI, Inc. The software controls the simultaneous image capture by the video cameras. Image analysis to determine the velocity vectors on each image plane is performed by the autocorrelation technique. The outcome of the analysis is a pair of displacement vector fields, $\Delta \mathbf{X}_1$ and $\Delta \mathbf{X}_2$ that are functions of the camera coordinates, $\mathbf{X}_1$ and $\mathbf{X}_2$. The vectors in these fields are tested for validity using procedures described by Meinhart, *et al.* (1994), and invalid vectors are either replaced by second or third choice vectors from the autocorrelation interrogation, replaced by interpolated values, or simply removed. Typically the fraction of invalid vectors is less than 2%. Corrections for the mis-registration of the images, and the variation of the magnification factors are all performed by post-processing software. Effectively, this step maps the displacement vectors from a grid in each of the CCD array image planes onto a common grid in the object plane. This step is necessary in general if there is any distorting medium between the object plane and the camera, such as a water-air interface (Prasad and Adrian 1993). In the present experiment the 6.35 mm thick glass windows of the wind tunnel cause the images to be displaced laterally by 0.55 mm. Since this shift is nearly constant, it is most easily accounted for by mechanical alignment in the initial layout of the cameras. In more general cases of nonlinear distortion it is better to calibrate the distortion and correct for it in software. The procedures for doing this will be discussed in another paper.

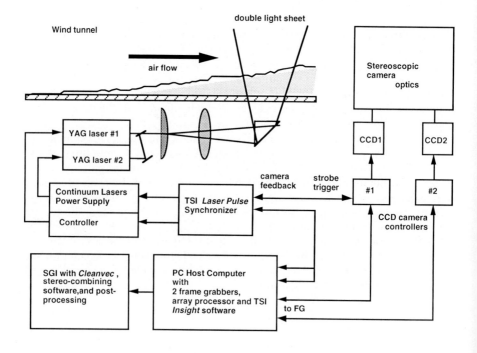

Figure 2. Stereoscopic system.

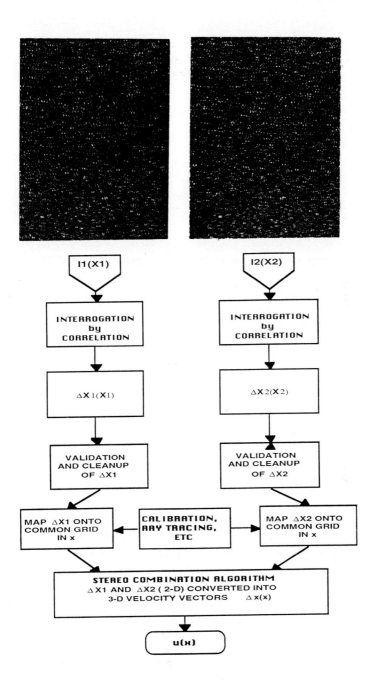

Figure 3. Image analysis procedure.

Lastly, the stereoscopic equations are applied to compute the third-component of the velocity vectors. For the ideal camera system shown in Fig. 1 the image of a point particle at $\mathbf{x}$ is mapped by lens L_ℓ onto a location η_ℓ in the $\mathbf{X}_\ell$-plane on camera ℓ given by

$$-\eta_1 = \frac{-d_i}{d_o - z}[(x + d_{L1})\hat{\mathbf{x}} + y\hat{\mathbf{y}}], \tag{2a}$$

$$-\eta_2 = \frac{-d_i}{d_o - z}[(x - d_{L2})\hat{\mathbf{x}} + y\hat{\mathbf{y}}] \tag{2b}$$

where carats denote unit vectors. Three-dimensional displacement of an image by an amount $\Delta\mathbf{x}$ results in two-dimensional displacements $\Delta\eta_1$ and $\Delta\eta_2$ that are found by taking the increment of (2). The results depend on the z-location of the particle at the time of the first exposure. However, PIV correlation analysis gives an estimate of the *volume average* of the displacements of the particles that lie in the measurement volume defined by the thickness of the light sheet and the area of the interrogation spot. By integrating the equation for the image displacement over such a volume and by making use of the normally large ratio of the object distance to the light sheet thickness, Prasad and Adrian (1993) have shown that the volume averaged displacements (denoted by overbars) are given by

$$\Delta\overline{\eta}_1 = M_o\left[\overline{\Delta x} + \frac{(x + d_{L1})}{d_o}\overline{\Delta z}\right]\hat{\mathbf{x}} + M_o\left[\overline{\Delta y} + \frac{y}{d_o}\overline{\Delta z}\right]\hat{\mathbf{y}} \tag{3a}$$

$$\Delta\overline{\eta}_2 = M_o\left[\overline{\Delta x} + \frac{(x - d_{L2})}{d_o}\overline{\Delta z}\right]\hat{\mathbf{x}} + M_o\left[\overline{\Delta y} + \frac{y}{d_o}\overline{\Delta z}\right]\hat{\mathbf{y}} \tag{3b}$$

Given the two two-dimensional vector displacements $\overline{\Delta\eta_1}, \overline{\Delta\eta_2}$ three of the equations in (3) can be solved for $\overline{\Delta x}, \overline{\Delta y}, \overline{\Delta z}$, and the fourth can be used to obtain a second estimate of $\overline{\Delta y}$. Alternatively, the two $\overline{\Delta y}$-equations can be added together to obtain a slightly more robust equation. The three-dimensional velocity vector is, of course, found by dividing three dimensional displacement by the time between exposures.

Figures 4a and 4b show the displacement fields that result on cameras 1 and 2, respectively when the object field undergoes a pure, uniform displacement in the z-direction. Solving equation (3) for the three-dimensional particle displacement yields the field shown in Fig. 4c.

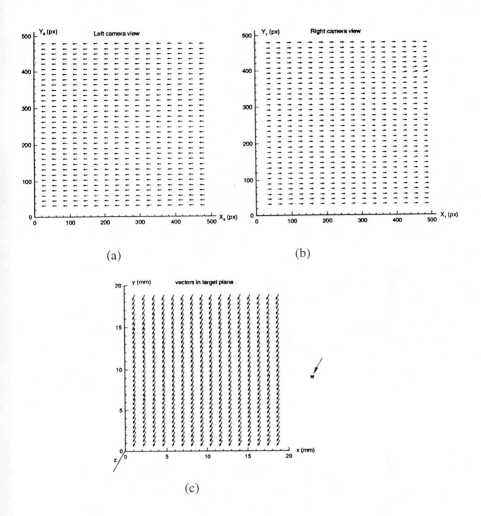

Figure 4. Image displacement fields due to a pure out-of-plane translation. (a) Camera 1, $\Delta\mathbf{X}_1(\mathbf{X}_1)$; (b) $\Delta\mathbf{X}_2(\mathbf{X}_2)$; (c) Three-dimensional displacement field $\Delta\mathbf{x}(\mathbf{X})$.

4. Turbulent boundary layer measurements

The turbulent boundary layer measurements were performed in an eiffel-type low-turbulence boundary layer wind tunnel with a working section 914 mm wide x 457 mm high x 6.096 m long. The free stream turbulence intensity at the test section inlet is less than 0.2% for free stream velocities less than 10 ms⁻¹. The turbulent boundary layer was produced on a flat plate placed 100 mm above the floor of the test section. The boundary layer was tripped by the a 4.7 mm diameter wire which spanned the boundary layer plate just downstream of the leading edge. Optical access to the boundary layer was provided from the side by float glass windows, and from below by 610 mm wide x 2.748 m long float-glass windows embedded in the boundary layer plate. The stereo PIV measurements presented here were performed at a free stream velocity of 3.4 ms⁻¹, which produced a Reynolds number based on the momentum thickness θ of $Re_\theta = 2575$ at the location of the measurements. All measurements were made with the light sheet in the x-y plane of the flow, where x is streamwise and y is normal to the wall.

Figure 5 shows the (u,v) components of one realization of the flow field. A constant convection velocity of 3000 mm s⁻¹ has been subtracted to make the

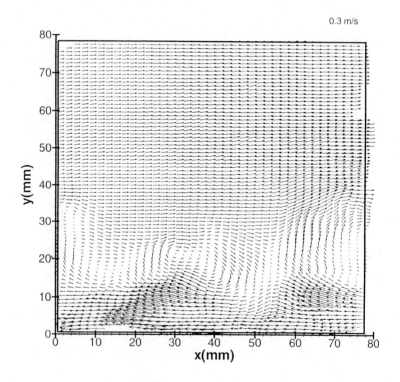

Figure 5. Sample realization of the streamwise-wall normal $(x$-$y)$ vector field of the turbulent boundary layer. $Re_\theta = 2525$. The free stream velocity is 3.4 ms⁻¹, and a constant convection velocity of 3.0 ms⁻¹ has been subtracted to reveal the eddy structure more clearly. The vectors are the projection of the three-dimensional vectors onto the x-y plane.

fluctuations more visible. The edge of the turbulent boundary layer lies about 50 mm above the wall in this realization. As observed in earlier work, there is an internal layer character close to the wall that grows in a manner similar to that of an ordinary boundary layer. This internal layer, seen as a region of uniformly low momentum that extends from the wall up to about 20 mm, is capped by a collection of several intense spanwise vortices.

The w-component of velocity is shown in Figure 6 by plotting its contours in gray-level form. It is immediately clear that the values of the w-component are frequently as large or larger than the other components. There is a tendency for the w-contours to align along the same ~30 degree inclination that is observed in the u-v field. There is also a tendency for the sign of the w-component to be positive on one side of a 30 degree line and negative on the other, indicating a large-scale rotation about an inclined axis. The strength of the w-component and its coherent organization shows that one must be very cautious in interpreting two-dimensional projections of the three-dimensional vectors such as the field in Figure 5.

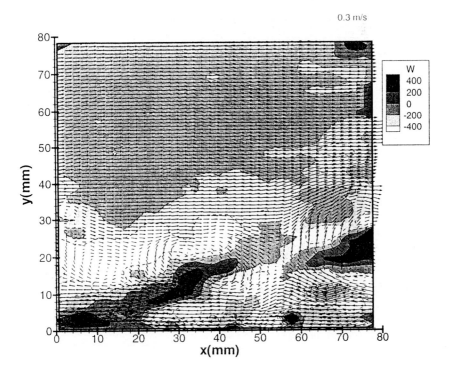

Figure 6. Sample realization of the streamwise-wall normal $(x$-$y)$ vector field from Fig. 5 with the field of the spanwise turbulent velocity superimposed on it in the form of grey-level contours. The heavy line denotes the zero value of the w-component, and dashed lines denote negative values.

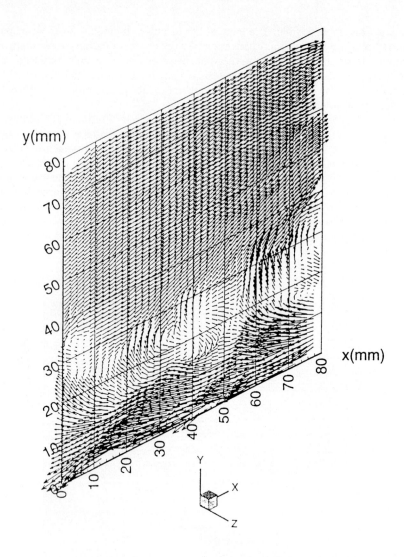

Figure 7. Three-dimensional vectors on the *x-y* plane in oblique view. All parameters as in Fig. 5.

The entire plane of three-dimensional vectors is shown in oblique view in Figure 7. The growth of an internal low-momentum layer is very evident. The pattern is attributed to the alignment of inclined hairpin vortices (c.f. Meinhart and Adrian 1997, Zhou, Adrian and Balachandar 1997) In Figure 8 profiles of the *v-w* components of the velocity vector are plotted for several different *x*-locations.

These vectors demonstrate very clearly the oscillation in the sign of the *w*-component with increasing distance above the wall, and the inclination of the pattern at about thirty degrees with respect to the wall. The oscillation is consistent with the presence of inclined, hairpin-like vortices, as observed by Meinhart (1994) and Meinhart and Adrian (1997).

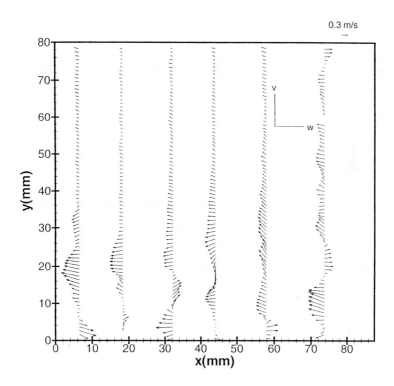

Figure 8. Profiles of *(v(x,y), w(x,y))* at several *x*-locations from the vector field in Fig. 7.

5 Summary

Stereoscopic PIV corrects for perspective error as well as providing the out-of-plane component. It requires a special camera, but otherwise it uses essentially the same hardware as a monocular PIV. Videographic recording increases the ease of use substantially by eliminating the necessity of registering photographic images. The present system has been used to measure thousands of frames in experiments that required only a few days to set up. Although the video-based PIV system has less resolution than a photographic system, it is encouraging that it is still able to resolve many of the important features of the turbulent boundary layer that have only recently been discovered by photographic PIV. The results are very similar to those obtained by photographic PIV. Thus, while the video system clearly has a role to play in the acquisition of data for analysis by statistical averaging, it also provides enough resolution to be of considerable value in visualizing the flow.

Acknowledgments

This work was supported by the United States Air Force Office of Scientific Research, the United States Office of Naval Research, and TSI, Inc.

References

Adrian, R. J. 1991, Particle-imaging techniques for experimental fluid mechanics, *Ann. Rev. Fluid Mech.,* vol. 23, pp. 261-304.

Arroyo, M.P. and Greater, C.A., 1991, Stereoscopic particle image velocimetry, *Meas. Sci. Tech.* Vol. 2, pp. 1181-1186.

Barnhart, D.H., Adrian, R.J. and Papen, G.C., 1994, Phase-conjugate holographic system for high resolution particle image velocimetry, *Appl. Opt.,* Vol. 33, pp. 7159-7170.

Brodkey, R.S., 1977, Stereoscopic visual studies of complex turbulence shear flows, *Flow Visualization,* (Ed. Asanuma, T.), New York: McGraw-Hill, pp. 117-122.

Brucker, C., 1995, 3D-PIV using stereoscopy and a scanning light-sheet: application to the 3D unsteady sphere wake flow. *Flow Visualization VII,* (Ed. Crowder, J.P.), New York: Begell House, pp. 715-720.

Gauthier, V. and Riethmuller, M.R., 1988, Application of PIDV to complex flows: measurements of the third component. *Von Karman Institute Lecture Series on Particle Image Velocimetry.* Rhodes-Saint-Genese: von Karman Institute for Fluid Dymanics.

Hinsch, K.D., 1995, Three-dimensional particle velocimetry, *Meas, Sci. Tech.* Vol. 6, pp. 741-753.

Mass, H. G., Gruen, A, and Papantonoiu, D., 1993, Particle Tracking velocimetry in three-dimensional flows. Part 1. Photogrammetric determination of particle coordinates. *Exp. fluids,* Vol. 15, pp. 133-146.

Meinhart, C.D., 1994, Investigation of turbulent boundary-layer using particle image velocimetry, Ph. D. thesis, University of Illinois.

Meinhart, C.D. and Adrian, R.J., 1997, The coherent structure of the overlap region in a turbulent boundary layer, (manuscript in preparation).

Meng, H. and Hussain, F., 1995, In-line recording and off-axis viewing (IROV) technique for holographic particle velocimetry, *Appl. Opt.* Vol. 34. pp. 1827-1840.

Nishino, Y., Kasagi, S. and Hirata, M., 1989, Three-dimensional particle tracking velocimetry based on automated digital image processing. *ASME J.* Vol. 111, pp. 384-391.

Prasad, A.K., and Adrian, R.J., 1993, Stereoscopic particle image velocimetry applied to liquid flows, *Exp. Fluids,* Vol. 15, pp. 49-60.

Prasad, A.K., Jensen, K., 1995, Scheimpflug stereocamera for particle image velocimetry in liquid flows, *Appl. Opt.,* Vol. 34, pp. 7092-7099.

Robinson, O., and Rockwell, D., 1993, Construction of three-dimensional images of flow structure via particle tracking techniques, *Exp. Fluids,* Vol. 14, pp. 257-270.

Troy, V. and Adrian R. J., 1996, Investigation of turbulent penetrative convection, 1996, *Eighth International Symposium on Applications of Laser Techniques to Fluid Mechanics*, Lisbon, Portugal, pp.23.1.1-23.1.8.15

Willert, C.E. and Gharib, M., 1991, Digital particle image velocimetry, *Exp. Fluids,* Vol. 10, pp. 181-193.

Zhou J., Adrian R. J., Balachandar S. and Kendall T., 1997, Hairpin vortices in near-wall turbulence and their regeneration mechanisms, (manuscript in preparation).

II.9 Three Dimensional Velocity Measurements Using Hybrid HPIV

Jingyi Zhang, Bo Tao and Joseph Katz

Mechanical Engineering Department
The Johns Hopkins University
Baltimore, MD 21218, USA

Abstract. A hybrid holographic particle image velocimetry system has been developed and implemented while studying fully developed turbulent flow in a rectangular channel. The system combines the advantages of both in-line and off-axis holography without having their drawbacks. It improves the signal to noise ratio of the reconstructed images, allows use of 3 - 15 μm particles in water at high density, and achieves large dynamic ranges both in velocity and space. An automated image acquisition system and correlation based software are used for analyzing data. In a sample volume of 47 x 47 x 42 mm^3 the data consist of 97 x 97 x 87 vectors. The success rate exceeds 80% of the total sample points. Sample results are provided.

Keywords. Holography, Particle Image Velocimetry, whole field measurement

1. INTRODUCTION

Holographic particle image velocimetry (HPIV) is a natural extension of the two dimensional particle image velocimetry (PIV) and it is based on the same principles. When double exposure images of a flow field seeded with small particles are recorded, the local velocities are determined by measuring the displacements of the particles. The advantages of HPIV over PIV is that it records data in a three dimensional (3-D) space and thus renders 3-D velocity distributions. Barnhart & *et al.* (1994), Hussain & *et al.* (1994) and other authors had reported successful 3-D flow measurements using HPIV systems developed in their labs.

The two common optical arrangements in HPIV are in-line and off-axis holography. An in-line scheme uses a single beam for both subject illumination (forward scattering) and a reference, which makes this approach simple to implement. However the reconstructed images are subject to speckle noise, mostly due to the overlap of the reference beam, real and virtual images (Collier & et al.,1971). In addition, the particle population in in-line holography is relatively low because the reference beam, which must maintain its coherence, passes through the interrogation volume. With increasing particle density the reference beam quality deteriorates. An off-axis holography system employs a separate reference beam. With proper design it eliminates the overlap of the reference and the two images and hence reduces the speckle noise substantially. The separate reference also allows higher population of seeding particles which yields higher spatial resolution. However, a typical off-axis scheme utilizes side-scattering of light from particles which is 2 to 3 orders of magnitude lower than forward scattering (depending on particle size and scattering angle). Low scattering efficiency requires use of considerably larger particles, a setback as far as velocity measurements are concerned.

In a previous paper, Zhang & Katz (1994) proposed a hybrid optical system that combined the advantages of both in-line and off-axis holography without having their drawbacks. In this system, Fig.1, the subject beam is arranged for forward scattering. Two relay lenses are inserted in the path between the interrogation volume and the film drive. A spatial high pass filter (a small pin) is installed at the focus of the first relay lens, which blocks the undisturbed portion of the subject beam (DC components in the Fourier space - Goodman, 1968). However, light scattered from particles, AC components, can reach the film with minimum obstruction. A separate off-axis beam serves as the reference. The hybrid system eliminates the speckle noise associated with in-line holography, yet keeps its high efficiency of forward scattering (the zero-degree forward scattering is cut by the high pass filter). It allows an increase in the energy of subject beam without over exposing the film, since most of the beam is cut by the filter. This feature increases the intensity of light scattered from the particles - an advantage while using small particles. As a result, the signal to noise ratio (SNR) of the reconstructed images is greatly improved. Using this approach, images of flows seeded with dense population of particles in the range of 3 - 15 μm can be reconstructed successfully.

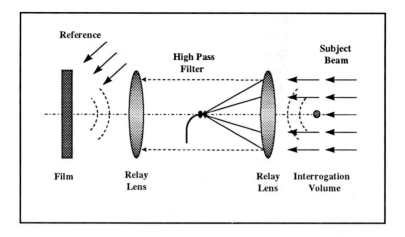

Fig. 1. The principles of the hybrid HPIV system. Solid lines indicate undisturbed subject beam. Dashed lines indicate scattered light from particle.

In the present study, the hybrid technique is further improved with the implementation of phase conjugate image reconstruction. A phase conjugate wave is used to illuminate the hologram from the opposite direction of the recording reference beam. The reconstructed wave front (real image of particles) propagates back through hologram that are located at the exact same positions as in the recording setup. This modification eliminates the aberrations introduced by the relay lenses that cause severe distortions or even disappearance of particles located far from the optical axis in the reconstructed images. With conjugate reconstruction the images are aberration free and the sample volume with high image quality extends to the very edge of the subject beam. Thus, a large dynamic range in space is achieved.

Using the approach described above it is possible to determine the two velocity components perpendicular to the optical axis at a high accuracy (1~2%). However an attempt to determine the third, axial, velocity component by means of stereoscopy, i.e. by examining the reconstructed field with two cameras arranged at an angle (Meng and Hussain, 1995), has been unsuccessful. The main causes of the poor results are the relay lenses and the water-glass-air interfaces at the walls of the channel. The relay lenses allow only near forward light to pass and as a result, the cameras of the stereo system can only be arranged at a small angle, which adversely affects the accuracy of the axial velocity. The water-glass-air interfaces cause severe aberration during observations at angles exceeding a few degrees. Although in principle it can be corrected by using prisms, there are still many prohibitive practical difficulties. Thus, in order to obtain all three velocity components at the same level of accuracy it is necessary to use two hybrid systems that intersect each other at a large angle (90° in the present study). Each system measures two velocity components and 3-D data are obtained by combining them together. As the next section shows, this dual system has been implemented successfully to measure fully developed, turbulent, channel flows.

2. EXPERIMENTAL SETUP

2.1 Test Facility

A schematic description of the test facility is shown in Fig. 2. The measurements are performed in a 57 × 57 mm^2, square vertical channel and the test section is located 2032 mm downstream of the entrance reducer. The wall of the test volume are made of flush mounted glass windows, that create an unobstructed interrogation volume of 57 × 57 × 50 mm^3. The entire loop is kept under low positive pressure to prevent bubbles formation.

The water is filtered and seeded with 15 μm diameter, polystyrene spheres that have a specific gravity of 1.06. Based on the experimental results, the seeding density in the test section varies between 1 - 8 particles/mm^3. The present measurements are performed at a mean velocity of 2.1 m/s. The corresponding Reynolds number is 1.04 × 10^5.

2.2 Optical Setup

A schematic description of the optical setup for recording the holograms is shown in Fig. 3. The light source is a ruby laser that has a wavelength of 694 nm, coherence length of about 1 m, pulse width of about 20 ns and energy level of 25 mJ per pulse. Only about 10% of this energy is needed for recording a hologram. This laser can generate multiple pulses during a single flash and the interval between pulses can be adjusted from about 20 to 500 μs. The delay between pulses while recording the present data is 40 μs.

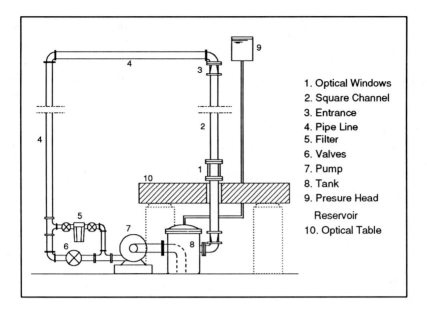

Fig. 2. Schematic description of the test facility.

The laser beam is pre-expanded by a lens pair and is split into two beams with the same intensity by the beam splitter B1. Each beam is further separated by splitters B2 or B3 that have transmission to reflection ratios of 90% to 10%. The two 90% transmitted subject beams are expanded to 70 mm diameter, collimated and directed to the test section at 90° to each other. The 10% reflected beams are spatially filtered, expanded, collimated and directed to the film drives at an angle of 15°. The holograms are recorded on AGFA, HOLOTEST 10E75, 70 mm film.

In each subject beam path, two identical achromatic doublet lenses, separated by twice their focal lengths, are inserted to form a relay. A 2 mm diameter, L-shaped pin, installed at the focus of the first relay lens, serves as a spatial, high-pass filter. Each set of relay lenses, high pass filter and film drive are mounted on a single rail. The entire unit is transported to another optical table during reconstruction.

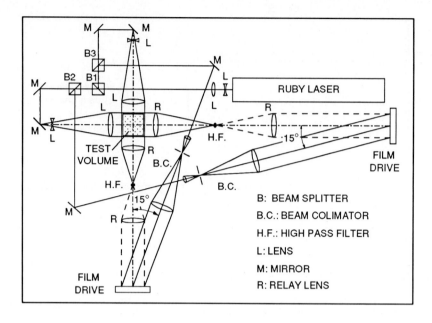

Fig. 3. Optical setup for recording holograms. The neutral density filters used to adjust the beam intensity are not shown.

2.3 Image Reconstruction

The image reconstruction system is shown schematically in Fig. 4. The light source is a 633 nm, 5 mW He-Ne laser, whose beam is spatially filtered, expanded and collimated.

To implement the phase conjugate image reconstruction, the relay lenses along with the film drive mounted on the rail are transferred to the reconstruction table. The film drive is removed and replaced by an adapter that ensures that the developed hologram is remounted at its exact location during exposures. The hologram is illuminated from a direction which is opposite to that of the recording reference (conjugate wave). The reconstructed wave-front propagates through the relay lenses and form a 3-D image in front of a video camera (Panasonic WV-BD400), that scans the image. The camera is equipped with an objective zoom lens, and is mounted on X-Y-Z translation stages driven by computer   controlled   stepping motors. A   PC based image grabbing system, CORECO OC-F/64 DSP, digitizes the images and saves them on a hard disk. The files are then sent via

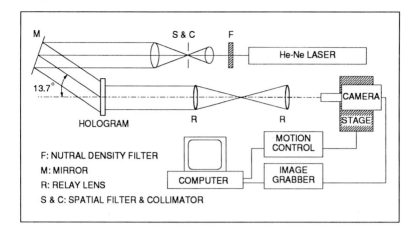

Fig. 4. Image reconstruction and data acquisition system.

local network to a SGI Power Challenger for data analysis. The entire process of camera translation and image acquisition is automated.

2.4 Data Acquisition and Processing

In the present paper we present data on the turbulent flow outside of the immediate vicinity (5 mm) of the wall. The sample volume is 46.60 × 46.60 × 42.25 mm^3. Measurements very close to the wall require longer delays between exposures, denser seeding (allowing smaller windows), filtering of noise associated with imperfections or dirt on the glass windows and careful alignment of the beam relative to the walls (to within 0.2°). These requirements can be satisfied with careful preparations, but they are not within the scope of the present paper that deals with the recording technique.

Each 640 pixel × 480 pixel image frame corresponds to a physical dimension of 3.11 × 2.35 mm^2. The camera scans an entire plane normal to the optical axis (that will be identified as a slice), changes its axial location and acquires the next slice. Each slice consists of 270 image frames, that are patched together. For convenience, the analysis is performed using files that contain quarter slices. The images are enhanced and the velocity is computed using auto-correlation analysis adopted from two-dimensional PIV (Roth et al., 1995). The window size is 192 pixel × 192 pixel (0.93 × 0.93 mm^2) and the spacing between windows is 96 pixels (0.466 mm), i.e., a 50% overlap between windows. The entire interrogation volume contains 97 x 97 x 87 sample points, providing a total of 818,583 vectors. Using one processor of the SGI Power Challenger, it takes about 80 minutes to complete the processing of a single slice (8439 vectors), including frame patching, enhancement, auto-correlation (using look-up- table method instead of FFT), vector extraction and vector map patching. Each hologram generates 97 such slices. The final 3-D vector field is obtained by combining the two sets of data generated from two holograms. The redundant vectors in the main flow direction are used for precision matching of the two holograms and for evaluating the data quality.

3. RESULTS AND DISCUSSIONS

A sample reconstructed image obtained using the hybrid system is shown in Fig. 5. For comparison, Fig. 6 gives an image of a similar area recorded using in-line holography. It is evident that the background speckle noise that dominates Fig. 6 is essentially non-existent in Fig. 5. As a result, the characteristic SNR of Figure 5 is 12 dB compared to a mere 3 dB in Fig. 6. As noted before, the higher SNR of the hybrid scheme is achieved by eliminating the overlap of the reference, real and virtual images and by the introduction of high pass filter, which allows forward scattering with much higher intensity of the subject beam. The entire imaging system has a lateral resolution of about 4.9 μm per pixel. The average measured particle diameter is around 3 pixels, which is in good agreement with nominal size of the present particles, 15 μm. When a hologram is reconstructed particles show elongated shapes in the depth direction, a phenomenon known as the "depth of focus" (Barnhart *et al.*, 1994, Hussain *et al.*, 1994). The ratio of the elongated axial extent of particle image to its diameter ranges from several hundreds in off-axis holography to over one thousand for in-line holography (Barnhart, *et al.*, 1994).

In the present setup, observations show that when the window size (0.93 mm) is used as the interval between adjacent slices, the majority of the particles appear only in one slice. This trend is illustrated in Fig. 7. There is only a very small fraction of large particles that appear on two, three or even more adjacent slices. Thus, for most of the data, successive slices contain different particles. This observation suggests that for most of the particles the elongation ratio is less than 62, which is smaller than results

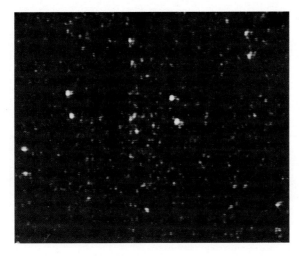

Fig. 5. Reconstructed image of 15 µm particles obtained with the hybrid system. Sharp particle pairs are in focus. The area covered is about 1.2 x 1.0 mm².

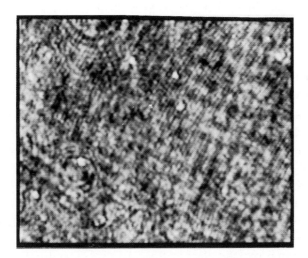

Fig. 6. Reconstructed image of particles obtained using in-line holography. The speckle noise is evident.

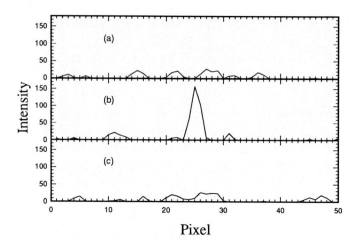

Fig. 7. Sample intensity distributions in adjacent slices separated by 0.93 mm. The plots are along lines with the same coordinates. In slice b the line crosses a particle. This particle does not leave any trace in slice a and c.

reported by Barnhart *et al.* (1994). A plausible (but speculative) explanation for the smaller ratio is the removal of the zero-degree forward scattered light from the particles by the inserted high pass filter. In in-line holography this zero-degree, forward scattered light is dominant, which may account for the associated high elongation ratio. Hussain *et al.* (1994), however, reported an elongation ratio of 15 - 20 using in-line holography. We could not achieve such a ratio using in-line holography in water.

Fig. 8 shows a sample 2-D velocity distribution of a single slice located at the center plane of the channel. The mean velocity in the channel (not in this plane in particular) is subtracted from each vector. As noted before, the data includes only the central 47×42 mm^2 of the sample volume (about 5 mm from the walls). Within this volume the characteristic displacement between exposures varies from 13 pixels at the edge to 21 pixels at the center (before subtracting the mean). The vector production rate, namely the fraction of the sampling nodes providing reliable data, exceeds 80%. Most of the windows with bad or no data do not contain any particles. At this stage we make no attempt to interpolate or filter the data. The decision on whether data is acceptable depends on the magnitude of the auto-correlation peak and limitations on the allowed difference in velocity between adjacent windows.

A 3-D vector field of the sample volume is presented in Fig. 9. The data are neither interpolated nor filtered. Clearly, the hybrid HPIV method allows us to map the three dimensional velocity distribution within water at a high Reynolds number. This data will be used mostly for turbulence modeling.

4. CONCLUSIONS

The hybrid HPIV system described in this paper is capable of measuring the three dimensional velocity distribution in liquid at high Reynolds numbers. It is characterized by a fairly simple optical setup. Its features include an off axis reference beam, the use of relay lenses together with high pass filters in the paths of the subject beams. This arrangement still allows forward scattering of light from the particles, more intense illumination of the sample volume (by an order of magnitude higher than in-line holography) and separations of the reference, real and virtual images, which substantially reduces the speckle noise. Consequently, the SNR of the hybrid system is considerably higher than in-line holography. It allows to clearly resolve in water particles with diameters ranging from 3 to 15 μm. The depth of focus is reduced substantially. Wide dynamic ranges in both space and velocity are achieved. Combined with an automated image acquisition system and an existing correlation based data analysis software, this system has been used for measuring a three dimensional vector array containing almost one million vectors.

5. ACKNOWLEDGMENT

This work was supported in part by the Office of Naval Research under Contract Numbers N-00014-92-J-1109 and N00014-93-1-1359, and in part by National Science Foundation under contract No. OCE-9107564 and CTS 9506077. The authors would also like to thank Professor Charles Meneveau for his close involvement and valuable discussions.

REFERENCES

Barnhart, D.H., Adrian, R.J. & Papen, G.C. 1994, Phase-conjugate holographic system for high-resolution particle-image velocimetry, Appl. Opt. vol. 33, pp. 7159-7170.
Collier, R., Burckhardt, C. and Lin, L. 1971, Optical Holography, pp. 74-78, Academic Press,

286

Goodman, J. W. 1968, Introduction to Fourier Optics, pp. 83-90, McGraw-Hill, New York.

Hussain, F., Meng, H., Liu, D., Zimin, V., Simmons, S. & Zhou, C. 1994, Recent Innovations in Holographic Particle Velocimetry, Proc. 7th ONR Propulsion Meeting, Eds. G. Roy & P. Givi, pp. 233-249.

Meng, H. and Hussain, F. 1995, Instantaneous flow field in an unstable vortex ring measured by holographic particle velocimetry, Phys. Fluids vol. 7, pp. 9-11.

Roth, G., Hart, D., Katz, J. 1995, Feasibility of Using The L64720 Video Motion Estimation Processor (MEP) To Increase Efficiency of Velocity Map Generation for PIV, Proc. of the ASME/EALA Sixth International Symposium on Laser Anemometry, FED vol. 229, pp. 387-396.

Zhang, J. & Katz, J. 1994, Off-axis HPIV with forward light scattering from particles, Proc. ASME Fluids Engineering Division Summer Meeting, Lake Tahoe, vol. 191 pp. 173-177.

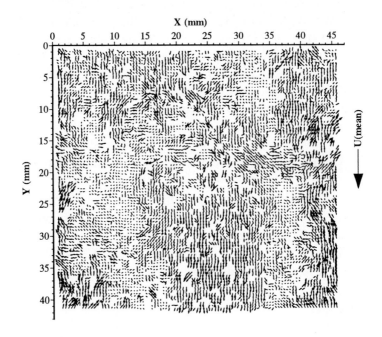

Fig. 8. The 2-D velocity distribution in the center slice after subtracting the mean velocity from each vector. The large arrow on right indicates the direction and magnitude of the mean velocity.

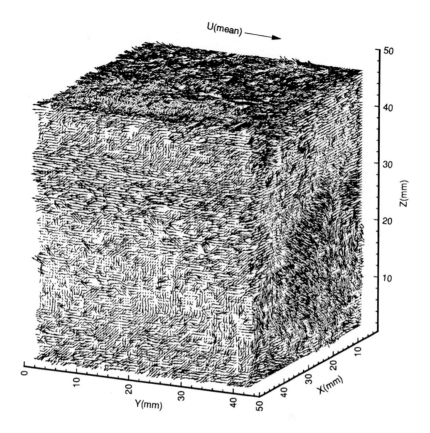

Fig. 9. A 3-D vector map of the velocity distribution within the channel. For clarity, only the data on the three surfaces of the sample volume are shown. The large arrow on top shows the direction and magnitude of the mean velocity which has been subtracted from each vector.

II.10 Improvements of Holographic Particle Image Velocimetry (HPIV) by Light-in-Flight Holography

M. Böhmer, H. Hinrichs, K.D. Hinsch, J. Kickstein

Carl von Ossietzky Universität Oldenburg
FB 8 -.Physik, D-26111 Oldenburg

Abstract. Holography allows to extend particle image velocimetry (PIV) to deep flow volumes. Interrogation of the reconstructed image volume by focusing on a specific depth layer, however, is limited severely by noise from out-of-focus regions. The problem is overcome by light-in-flight holography (LIF), utilizing the limited coherence of the light source. The scheme allows to reconstruct thin layers in depth without disturbance by the rest of the field. Their location is selected by the position of the reconstructing aperture on the hologram.

Keywords. Holographic Particle Image Velocimetry (HPIV), Light-in-Flight (LiF), 3-D velocity fields

1. Introduction

Many fluid dynamic situations, in unstable boundary layers or turbulent structures, for example, show three-dimensional behavior and therefore require not only instantaneous recording but also recording of a whole 3-D volume. The great interest in utilizing PIV for these problems is expressed by the publication of an increasing number of articles about 3-D registration methods [Adrian, Hinsch].

A straightforward approach is to expand a light sheet in depth and take a holographic record of the particles illuminated by a double pulse from a coherent source. The whole volume is then reconstructed and interrogated plane-wise. When focusing on a certain plane in depth, however, it can not be avoided to encounter noise from out-of-focus particles. Due to the coherent superposition the noise should be similar to speckle noise. A modified holographic recording technique will allow to overcome the problems investigated in more detail with numerical and experimental simulations.

2. Numerical and experimental simulation

To obtain a basic performance analysis of HPIV we used a computer-based simulation of the optical registration of particles in a deep volume. The volume is divided into equally spaced sheets and the images of particles from all sheets are added up coherently. This allows to concentrate on the basic noise problem while eliminating any additional noise sources due to imperfect experimental realizations of the holographic recording process. All other sheets are shifted by random amount. Focusing on the sheet with the known displacement results in a PIV-record, which is subjected to a correlation evaluation. This is repeated a hundred times for a given configuration. The number of erroneous displacement values is used to estimate a validation rate. The depth of the volume, aperture of the imaging system and type of correlation technique (auto or cross) are varied.

To verify the simulations we have made experiments in a model environment. A batch of transparent plastic sheets (thickness 1.3 mm) with embedded particles are arranged in depth in an index matching liquid. This model is illuminated with laser light parallel to the sheet orientation, and a CCD-Camera is focused on the central sheet. For a second exposure the central sheet is shifted by a known amount while all others are shifted randomly. The data are used for double-exposure autocorrelation or crosscorrelation interrogation as in the computer simulation. Since straightforward video recording is involved, there is still a difference to autocorrelation evaluation of the reconstruction of a double-exposure hologram, where both records add up coherently. Yet, this is taken care of in the calculation. We find good agreement of the experimental data with the results of the computer simulations and thus have good reason to believe in the applicability of the computer simulation for further system analysis (fig. 1).

The simulations as well as the model experiments demonstrate how the success of an interrogation just by focusing on the plane of interest is limited by the depth of the illuminated flow. In autocorrelation processing the depth limit is a few centimeters, in crosscorrelation the valid detection is improved by a factor of about two. It is important to note that all simulations imply an uncorrelated change of the flow structure in depth. This means, that neighboring planes are spaced according to the sampling theorem, i.e., the separation of planes is just matched to size of the flow structures in depth. There is no redundancy in the information, and data from neighboring planes are not suited for data validation. This is contrary to often applied interrogation schemes, where velocity values on neighboring grid points are used to validate the data, made possible only by oversampling. The success in the holographic flow analysis by Adrian and Barnhart can be understood on this ground [Adrian et al.]. However, when data reduction becomes important as in large three-dimensional data sets, oversampling may become a luxury. A compromise would be to retain the familiar oversampling within each plane to improve validation rates, but to sample depth at the limit required for spatial resolution.

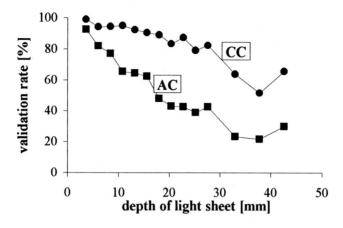

Fig. 1: Validation, i.e. percentage of correct velocity evaluations vs. depth of the illuminated flow field for autocorrelation (AC) and crosscorrelation (CC) PIV processing.

To avoid oversampling and to interroagte deeper volumes successfully a technique allowing reconstruction of a single plane without adding any noise from out of focus particles is recommended. We already introduced such a holographic method [Hinrichs & Hinsch], but the main disadvantage is its complexity. For each light sheet in depth one has to choose an extra reference beam and carefully match coherence length of the light source and optical paths of the various beams. Thus, the number of lightsheets is limited by holographic considerations. Furthermore, the sampling positions in depth are fixed by the optical setup and can be adapted to changing details of the flow structure only with much more effort.

3. Basic Concept

We now present a new concept to overcome these limitations. Abramson [Abramson] has introduced the technique of light-in-flight holography where the finite coherence length of a light wave is employed to freeze wavefronts in space. When a hologram is recorded with an obliquely incident reference beam, each area on the hologram registers only that region in object space that conforms to the requirement that the optical path of reference and object wave be equal within the coherence length. With a properly small coherence length successive parts of the hologram show the wavefront as it proceeds in space. The method has been demonstrated quite illustratively on light waves passing through optical components. The exact relation between the position in the hologram and the corresponding region in the object depends on the geometry of the setup. In particle velocimetry this scheme can be applied as sketched in the following figure.

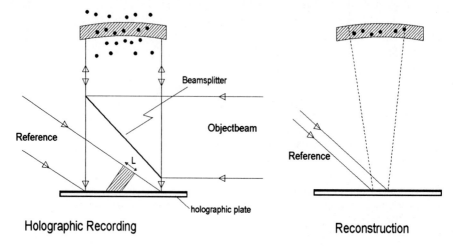

Holographic Recording

Reconstruction

Fig.2: Principal setup for recording and reconstructing particle images by light-in-flight holography (top view)

In the holographic recording the particle field is illuminated from the direction of observation through a beam splitter. This is useful for the light-in-flight geometry and also takes advantage of the higher backscattering light intensity of particles as compared to the 90 degree viewing in PIV. The effective recording aperture determined by the angular characteristics of the scattering should be somewhat better than in forward scattering as often used in in-line holography. The reference beam is obliquely incident on the recording plate. Now, the coherence length L of the laser in use is limited to a few millimeter. For a given position on the holographic plate, this length determines the thickness of a sheet in the object volume that can be recorded holographically at this location. Correspondingly in the reconstruction, when viewing through a narrow vertical strip on the hologram, only a limited band in the volume is seen. By moving the viewing aperture, the observed band moves in depth. To obtain thin bands a light source of small coherence length is needed. This can be a Nd:YAG-laser or an argonion-laser operated without any etalons and with a broad band mirror. Both laser types provide a coherence of about a centimeter.

4. Light-in-flight holography

The following description shows how to measure the coherence length of the ruby laser in use. The object volume - later seeded with particles - is illuminated with a plane wave from a direction as close as possible to viewing. The obliquely incident reference wave reaches the lower part of the holographic plate in Fig. 3 first and arrives latest at the top of the plate. The object light reaching the plate first originates from region O3 and is made equal in travel time to the first reference light.. Light from O1 on the other hand arrives just in time with the reference light at the top part of the plate. This can is indicated by the three photos taken of the reconstructed ruler bar taken through the different slits H1-H3 as the numbers indicate cm values.

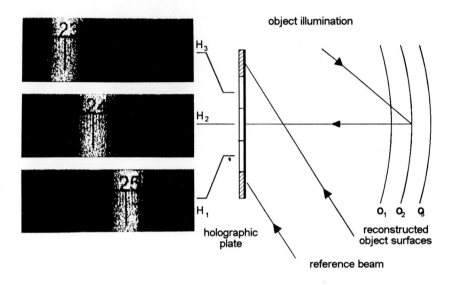

Fig. 3: Measurement of coherence depth in a setup of light-in-flight holography. Photos show the ruler bar as seen through the indicated apertures on the hologram.

The coherence length of the laser we have is shorter than from the 1 cm seen on the ruler bar in each photo. We have to take into account the projecting effect because of the 45^0 orientation in the object field and the convolution with the slit aperture of 8 mm width on the holographic plate. This leads to an estimate of a coherence length of about 3 mm which is in good agreement with a theoretical prediction assuming about 200 modes in a resonator of optical path length of about 80 cm. A closer look at the ruler images even reveals low intensity secondary maxima at both sides.

5. Recording of particle field in a fluid

To demonstrate the performance of the technique and the improvement in the recorded particle image fields the ruler bar in Fig. 3 is replaced by a water tank containing tracer particles of a mean size of 5 μm. Such particles are even smaller than most particles used in liquid flows and are typical in size for slow flows in air.

Fig. 4: Video recording of 8 x 8 mm^2 section from a particle field as seen from an observer through a 10 mm wide aperture in the hologram. The image quality is comparable to that in ordinary PIV.

The depth of the illuminated object region was some 200 mm. The reconstructed virtual images of the particle fields were recorded with a video camera for further inspection. Figure 4 shows a video record of a 8 x 8 mm^2 area in object space reconstructed with a with a 10 mm slit aperture. Comparison with the ordinary PIV images of 1mm deep light sheet shows no detectable difference in the images. The particles are resolved within the diffraction limit.

To prove that we reconstruct different layers in depth from different positions of the slit on the hologram a correlation was performed that were gained between images by successively moving camera and 10mm slit mask in steps of 2.5 mm. A 3 x 3 mm^2 region of the field was investigated. With the moving aperture the layer in the particle field that is reconstructed also moves in depth and gradually a different particle ensemble is seen.

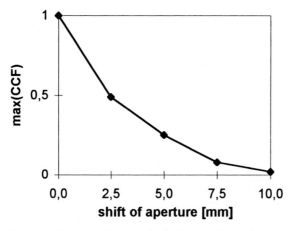

Fig. 5: Correlation coefficient indicating similarity of video recordings of the same 3 x 3 mm² large area in the particle field when the reconstructing aperture slit of 10 mm width is displaced.

The similarity between the images should vanish accordingly as more and more particles seen in the first image do not show up seen in a later image while particles from the progressing corresponding depth are added. The effect is seen in Fig. 5 where the correlation coefficient is plotted versus the shift of the aperture. It is obvious that adjacent 10 mm wide apertures reconstruct completely different particle fields. This method can serve as a precise way to measure the coherence depth of the laser employed.

6. Conclusions

The application of light-in-flight holography in PIV offers a new approach in quantitative three-dimensional flow analysis. The particle field in object space can be reconstructed layer after layer without any disturbances by neighboring out-of-focus particle images. This allows the study of deeper flow fields. The novel setup is greatly simplified compared to earlier proposal of sampling a volume by a set of fixed light sheets that needed separate reference beams and obey complicated path length requirements. In the new method the sampling separation in depth is no longer fixed in the recording, but can be adjusted during reconstruction accordingly to the individual flow structures. Thus, the amount of measurement data is kept at the allowable and minimum oversampling for validation purposes can be avoided due to a better signal to noise ratio.

First experiments with light-in-flight holographic particle recording have substantiated the suitability of the method for velocimetry (LIF-PIV). In continuing investigations quantitative flow analysis is under way. Furthermore, the suitability of Nd:YAG lasers that are often available in fluid dynamically labs is studied in more detail. First experiences indicate that the coherence depth involved will be slightly larger than in the ruby laser. Finally, more basic studies concern the limits set by the finite slit aperture or size of the hologram as well as refined optical setups to extend these limits.

References

Abramson N: "Light-in-flight recording by holography." Optics Letters 3.4 (1978), 121-123.

Adrian R.J., Meinhart C.D., Barnhart D.H., Papen G.C.: "An HPIV system for turbulence research." Holographic Particle Image Velocimetry (ASME FED 148) ed. E.P. Rood, American Society of Mechanical Engineers, (New York / 1993), pp 17-21.

Hinrichs H., Hinsch K.D.: "Multiple light sheet particle holography for 3-D flow velocimetry." Proc. Seventh Int. Symp. On Applications of Laser Techniques to Fluid Mechanics (Lisbon / 1994), paper 26.2.

Hinsch K.D., Hinrichs H., Roshop A, Dreesen F.: "Holographic and Stereoscopic Advances in 3-D PIV." (ASME FED 148) ed. E.P. Rood, American Society of Mechanical Engineers, (New York / 1993), pp 33-36.

Hinsch K.D.: "Three-dimensional particle velocimetry." Meas. Sci. Technol. 6 (1995), 742-753.

Keane R.D., Adrian R.J.: "Optimization of particle image velocimetry. Part I: double pulsed systems." Meas. Sci. Technol. 1 (1990), 1202-1215.

Chapter III

Wall-Bounded Flows

III.1 Effects of Centrebody Rotation on Laminar Flow Through an Eccentric Annulus

M P Escudier and I W Gouldson

University of Liverpool, Department of Mechanical Engineering,
Brownlow Street, Liverpool, L69 3GH, England

Abstract   The work discussed here is part of a programme of research, motivated by drilling hydraulics, in which the annular flow of both Newtonian and non-Newtonian liquids is being investigated.   The present results are limited to fully developed laminar flow of a Newtonian liquid in an eccentric annulus with a) an imposed bulk flow without centrebody rotation (axial velocity profiles), b) centrebody rotation with zero bulk flow (tangential velocity profiles) and c) combined bulk flow and centrebody rotation (axial and tangential velocity profiles).   In the absence of an imposed bulk flow, the critical Taylor number at which Taylor vortices appeared in the eccentric annulus geometry was determined by monitoring the onset of an axial velocity in the widest sector of the annulus.   Detailed measurements are restricted to sub-critical Taylor numbers.   The measured velocity profiles are in excellent agreement with available theory.

1.   Introduction

During wellbore drilling operations a fluid (drilling mud) is pumped down the rotating drillstring, through nozzles in the drillbit, and then back to the surface through the annulus between the drillstring and the wellbore wall.   The drilling mud has to satisfy several different requirements, including cooling and lubricating the drillbit, cleaning the workface, carrying drilled cuttings to the surface, preventing ingress of formation fluids into the wellbore, and preventing wellbore collapse.   Operational constraints in wells of great depth or extended reach, particularly the need to limit mud pressure to avoid fracturing the rock surrounding the borehole, often restrict the flow of a drilling mud in the annulus (usually eccentric) to the laminar regime (Ooms and Kampman-Reinhartz, 1996).   Such flows thus represent a relatively rare example of laminar flows with major industrial relevance.   As a consequence of the multiplicity of operational requirements, the rheology of a drilling mud is generally non-Newtonian in character: almost invariably shear thinning, and often exhibiting viscoelastic and thixotropic properties as well as a yield stress.   The combination of a fluid exhibiting all of these characteristics and an eccentric annulus with centrebody rotation represents a flow of considerable complexity, the more so if the rotation rate exceeds the critical condition at which Taylor vortices are generated.

Various elements of the flow of liquids in annular geometries are being tackled in an extensive programme of research at the University of Liverpool, motivated by wellbore hydraulics. Escudier et al (1995 a, b) have presented detailed measurements of the characteristics of Taylor vortices produced in both Newtonian and non-Newtonian fluids by centrebody rotation with no throughflow in a concentric annulus. Other work has emphasised the turbulent flow aspects of both Newtonian and shear-thinning fluids in a concentric annulus without centrebody rotation (Escudier et al, 1995c) and also with rotation (Escudier and Gouldson, 1994, 1995).

The present contribution is limited to fully developed laminar flow of a Newtonian liquid in an annulus of radius ratio 0.506 with an offset centrebody rotated at subcritical speeds. The experimental results compare well with the classical analytical expressions for flow in a concentric annulus without rotation, for circular Couette flow in a concentric annulus in the absence of an imposed axial flow, and with the recent numerical calculations of Manglik and Fang (1995) for axial flow in an eccentric annulus without rotation. There is also good qualitative agreement with the work of Bakhtiyarov and Siginer (1995) on the flow of a linear fluidity fluid in an eccentric annulus with centrebody rotation but zero bulk flow.

2. Experimental Conditions

Measurements were performed for flow through a 5.8 m long pipe consisting of six modules of precision-bore glass tubing (100.4 mm ID) with a stainless steel centrebody (50.8 mm OD) providing an annulus with a radius ratio $\kappa = 0.506$. The glass tubing was supported on linear bearings which permitted the geometry to be eccentered between $\varepsilon = 0$ and $\varepsilon = 1$. A DC motor and gearbox allowed the centrebody to be rotated at speeds of up to 126 rpm, monitored by an optical encoder with a resolution of 0.1 rpm. Bulk flow through the annulus was provided by a progressive cavity Mono pump, with the flowrate monitored by a Fischer and Porter electromagnetic flowmeter. A Dantec fibreflow LDA system, with a probe volume of length 0.19 mm in water, was used to measure the radial variation of axial and tangential velocities. Signal processing was carried out using a Dantec BSA unit. At a location 90 hydraulic diameters (4.5 m) downstream of the inlet to the annulus, measurements were made in the four orthogonal planes shown in **Figure 1** (sectors A-D) of axial velocities with and without rotation, and of tangential velocities with and without axial flow. It may be noted that the measurement planes are diametral planes with respect to the outer pipe so that as the LDA probe is moved radially inwards it senses a progressively decreasing fraction of the tangential velocity with respect to the centrebody as the centrebody offset is increased. The test fluid used was a 2.5:1 w/w glucose-water mixture with density $\rho = 1290$ kg/m^3, viscosity $\mu = 0.125$ Pa.s at 20°C and essentially the same refractive index as water. The bulk axial velocity was nominally 0.22 m/s and the rotation speed 78 rpm which corresponds to $U/\omega R_i$ of 1.06. The corresponding bulk flow Reynolds number Re was 113, the Taylor number Ta = 2900, and the rotational Reynolds number Ro = 54. Table 1 gives the exact values for Re, Ro, Ta and $U/\omega R_i$ for each individual experiment.

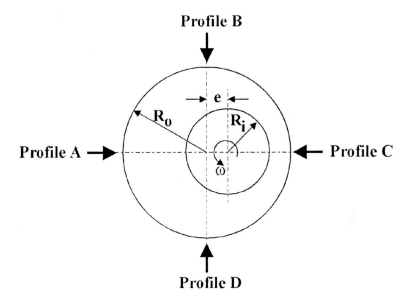

Figure 1. Definition of eccentric geometry and circumferential locations at which velocity profiles were measured.

Figure No.	ε	Re	Ro	Ta	$U/\omega R_i$
2(a)	0	105	-	-	-
2(b)	0.2	105	-	-	-
2(c)	0.5	105	-	-	-
2(d)	0.8	110	-	-	-
6(a)	0	-	50	2300	-
6(b)	0.2	-	51	2430	-
6(c)	0.5	-	53	2600	-
6(d)	0.8	-	53	2600	
7(a) & 8(a)	0	105	55	2800	0.99
7(b) & 8(b)	0.2	105	55	2700	1.00
7(c) & 8(c)	0.5	115	57	3000	1.03
7(d) & 8(d)	0.8	120	56	2900	1.08

Table 1. Reynolds and Taylor numbers for individual experiments.

3. Experimental results

3.1 Bulk axial flow without rotation

The integrity of the measurements is evident from the agreement of the measured velocity profile for $\varepsilon = 0$ (**Figure 2a**) with the theoretical profile for fully developed laminar flow in a concentric annulus:

$$\frac{u}{U} = \frac{2[1-(r/R_o)^2-(1-\kappa^2)ln(r/R_o)/ln\kappa]}{(1+\kappa^2)+(1-\kappa^2)/ln\kappa} \tag{1}$$

That even slight eccentricity has a major influence on the azimuthal variation of the axial velocity is apparent from **Figure 2b** ($\varepsilon = 0.2$). The highest velocities occur in the widest part of the annulus (A) with a peak velocity twice the bulk average and 30% higher than for the concentric case. The corresponding reduction in the velocities in the narrowest part of the annulus (C) is a major source of problems in the cement/mud - displacement process which follows the drilling of a well (Tehrani et al, 1993). In fact, for a shear-thinning non-Newtonian fluid, the situation is even worse because the fluid viscosity increases in regions of low shear rate and the fluid may even gel. The results for $\varepsilon = 0.2$ are in good agreement with the numerical calculations of Manglik and Fang (1995) for $\kappa = 0.5$ who find $u_{max}/U = 2$ for sector A and $u_{max}/U = 0.92$ for sector C. Slight deviations from symmetry are evident in planes B and D, probably attributable to minor imperfections in the geometry, but the profiles are little changed from those for $\varepsilon = 0$. The trends of **Figure 2b** are further exaggerated for $\varepsilon = 0.5$ (**Figure 2c**), whilst for $\varepsilon = 0.8$ (**Figure 2d**) the fluid in the narrow part of the annulus is practically at rest. As for $\varepsilon = 0.2$, the agreement with Manglik and Fang's calculations is very good. The tendency for the measurements to fall slightly below the theoretical curves is probably attributable to an overestimate of the bulk velocity, a consequence of the need to operate close to the lower limit of resolution (about 5% of full scale) of the flowmeter.

Frictional pressure drop data represented as $f.Re$ versus ε (**Figure 3**) are in reasonable agreement with the numerical calculations of Manglik and Fang for $0 < \varepsilon < 0.6$ and also with the earlier work of Shah and London (1978) which extends to higher values of ε.

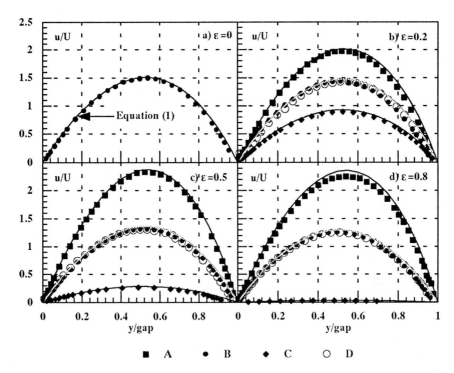

Figure 2. Influence of eccentricity on axial velocity profiles for Ta = 0, Re ≈ 106. Theoretical curves for ε > 0 from Manglik and Fang (1995).

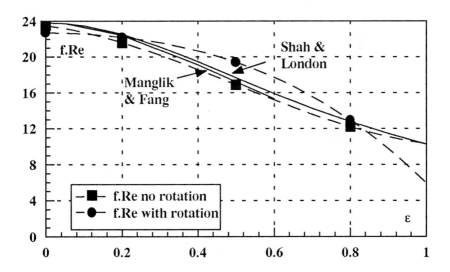

Figure 3. Variation of f.Re with eccentricity: ■ Ta = 0, ● Ta ≈ 2500.

304

3.2 Rotation without bulk axial flow

The variation of critical Taylor number Ta_c with eccentricity (**Figure 5**) was determined by monitoring the onset of an axial velocity in sector A at a point 2/3 g from the outer wall (**Figure 4**). This onset is very well defined and so a sensitive indicator of Taylor vortex motion. The results show a progressive increase in stability with increasing offset, in fair agreement with the numerical calculations of Lockett (1992) which he showed are well represented by

$$\frac{Ta_c}{Ta_o} = 1 + 2.6185\,\varepsilon^2 + 2.13\,\varepsilon^4 + 12.1\,\varepsilon^6, \tag{2}$$

independent of κ. All other measurements were made for Taylor numbers in the range 2430-3000, well below the critical value for any eccentricity since $Ta_o \approx 4500$.

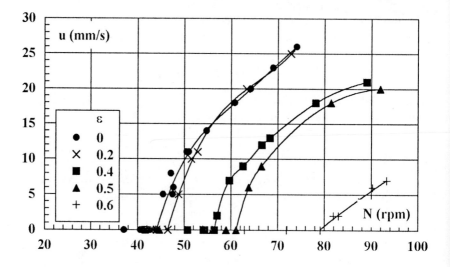

Figure 4. Influence of eccentricity on Taylor vortex onset.

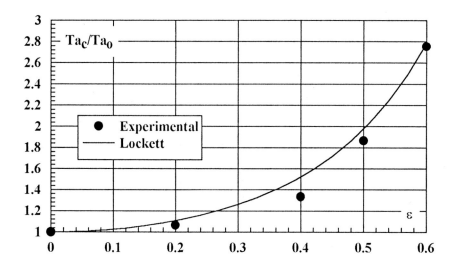

Figure 5. Dependence of critical Taylor number on eccentricity

For ε = 0 (**Figure 6a**) agreement with the theoretical profile for circular Couette flow

$$\text{i. e.} \quad \frac{w}{\omega R_i} = \frac{R_i / r - \kappa^2 r / R_i}{1 - \kappa^2} \tag{3}$$

is excellent. The effect of slight eccentricity (ε = 0.2, **Figure 6b**) is again evident with increased tangential velocities in the narrow side (sector C) of the annulus and decreases in the other three sectors compared with the zero offset case. The situation is reminiscent of plane Couette flow with an imposed longitudinal pressure gradient from which we infer that rotation generates a significant azimuthal pressure variation in the annulus, much as in a journal bearing. The influence of eccentricity is considerably enhanced for ε = 0.5 (**Figure 6c**) with evidence of a counter-rotating vortex (w<0) centred 40% from the outer wall in the wide side (sector A) of the annulus. Since the orientation of the LDA system was not sensitive to the radial (with respect of the outer wall) component of ωR_i, the reduced velocity on the surface of the inner cylinder in sectors B and D is a consequence of the eccentric geometry. The trends of **Figure 6c** continue with increasing eccentricity: for ε = 0.8 (**Figure 6d**) a counter-rotating kidney-shaped vortex has penetrated sectors B and D and a clear asymmetry has developed. Although detailed comparisons are difficult to make, the observed influence of eccentricity is in complete qualitative agreement with the theoretical calculations of San Andreas and Szeri (1984) and also with the observations and calculations of Bakhtiyarov and Siginer (1995).

3.3. Combined bulk axial flow and rotation

For all eccentricities, the distributions of tangential velocity (**Figure 7**) for the situation of combined bulk axial flow and centrebody rotation reveal only slight quantitative differences compared with the situation for zero bulk axial flow. This result is as it should be since, for fully developed laminar flow of a Newtonian fluid, the azimuthal flow is decoupled from the axial flow. This decoupling does not apply to the axial flow (or to a non-Newtonian fluid) and, in contrast to the tangential velocities, with the exception of the concentric geometry, the axial velocity distributions (**Figure 8**) are markedly different from their zero-rotation counterparts. The differences are particularly apparent in the case for the lowest eccentricity ($\varepsilon = 0.2$, **Figure 8b**) which has a very significant influence on the axial flow. The symmetry about AC is replaced by an apparent AB/DC symmetry, i.e. symmetries in the profile shape and magnitude of the velocities in sectors A and D, and B and C. The overall flow distribution is closer to axisymmetric, with a reduction of about 15% in the highest velocities and an increase of 30% in the lowest compared with the case for zero rotation. Since no calculations including the effects of rotation are available for direct comparison, curves representing the calculations of Manglik and Fang (1995) are included as a reference. For $\varepsilon = 0.5$ (**Figure 8c**) and, to a lesser degree, for $\varepsilon = 0.8$ (**Figure 8d**) the influence of rotation is significant but less dramatic: the highest velocities are again in the widest part of the annulus, the lowest in the narrowest, and the peak locations are shifted radially. It is apparent that the essential influence of rotation is to move fluid of high axial momentum into regions which, in its absence, were momentum deficient.

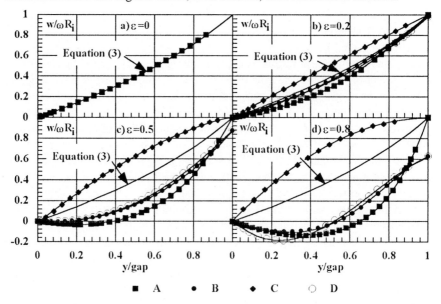

Figure 6. Influence of eccentricity on tangential velocity profiles for Ta ≈ 2500, Ro ≈ 52, Re = 0

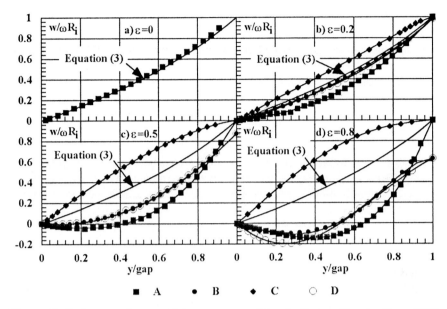

Figure 7. Influence of eccentricity on tangential velocity profiles for Ta ≈ 2850, Ro ≈ 56, Re ≈ 110, U/ωRᵢ ≈ 1

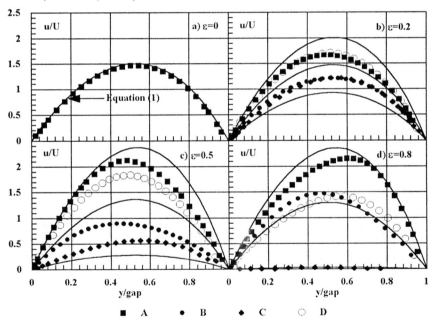

Figure 8. Influence of eccentricity on axial velocity profiles for Ta ≈ 2850, Ro ≈ 56, Re ≈ 110, U/ωRᵢ ≈ 1. Theoretical curves for ε > 0 from Manglik and Fang (1995).

The influence of rotation on $f.Re$ (**Figure 3**) is greatest for $\varepsilon = 0.5$ where there is a 15% increase compared with the zero rotation case. Significant changes in $f.Re$ were to be anticipated given the strong effect of rotation on the axial velocity profiles. The laminar flow situation may be contrasted with turbulent flow where the effect of rotation on both the axial flow and $f(Re)$ is negligibly small (Escudier and Gouldson, 1995).

4. Conclusions

In the absence of centrebody rotation, increasing eccentricity results in progressive departures from axisymmetry for axial flow through an annulus with $\kappa = 0.506$. For $\varepsilon = 0.8$ the peak velocity in the widest part of the annulus is 50% higher than for a concentric geometry whilst the velocity is practically zero in the narrowest section. These results are in excellent agreement with the numerical studies of Manglik and Fang (1995).

Centrebody rotation at sub-critical Taylor numbers, in the absence of a bulk throughflow, generates a radial/tangential velocity field with a counter-rotating kidney shaped vortex in the widest part of the annulus for eccentricities of about 0.5 or higher.

As expected from theory, which indicates that the two are decoupled, bulk axial flow at the low Reynolds numbers investigated here ($\sim 10^2$) has no significant influence on the radial/tangential velocity field generated by centrebody rotation.

At all eccentricities, and particularly the lowest value investigated (0.2), centrebody rotation produces a significant reduction in the azimuthal variation in the axial velocity distributions.

In agreement with the numerical work of Lockett (1992), the onset of Taylor vortices is delayed to higher Taylor numbers as the eccentricity is increased, i.e. eccentricity has a stabilising influence on the flow.

Acknowledgements

The authors gratefully acknowledge the financial support of EPSRC/MTD (research grant GR/F87813), Shell Research BV and British Gas plc. We are also indebted to Professor R Manglik, University of Cincinnati for extending and providing full details of the computations reported in Manglik and Fang (1995).

Nomenclature and Definitions

d   mean annular gap R_o-R_i (m)
e   displacement of centrebody axis from outer pipe axis (m)
g   gap width (A, B, C or D)

R_i	outer radius of centrebody (m)
R_o	inner radius of outer tube (m)
Re	bulk axial Reynolds number $2\rho Ud/\mu$
Ro	rotational Reynolds number $\rho\omega R_i^2/\mu$
Ta	Taylor number $(\rho\omega/\mu)^2 R_i d^3$
Ta_c	critical Taylor number
Ta_o	value of Ta_c for $\epsilon = 0$
u	axial component of velocity (m/s)
u_{max}	peak value of u in any sector of the annulus
U	bulk axial velocity (m/s)
w	tangential component of velocity (m/s)
y	distance from inner wall of outer tube (m)
ϵ	eccentricity e/d
κ	radius ratio R_i/R_o
μ	fluid dynamic viscosity (N.s/m^2)
ξ	non-dimensional distance from outer wall y/g
ρ	fluid density (kg/m^3)
ω	angular velocity of centrebody (rad/s)

7. References

Bakhtiyarov S and Siginer D A, 1995, Flow of linear fluidity fluids in eccentric annuli, Proc ASME Symposium Developments and Applications of non-Newtonian Flows, FED-Vol 231/MD-Vol 68.

Escudier M P and Gouldson I W, 1994, Annular Flow of Shear-Thinning Fluids with Centrebody Rotation, Paper 4.6, Seventh International Symposium on Application of Laser Techniques to Fluid Mechanics. Lisbon.

Escudier M P, Gouldson I W and Jones D M, 1995 a, Circular Couette Flow and Taylor Vortices in Shear-Thinning Liquids. Developments in Laser Techniques and Applications to Fluid Mechanics. Proc. 7th Int. Symp. Lisbon 1994. Springer. Eds. R J Adrian, D F G Durao, F Durst, M V Heitor, M Maeda and J H Whitelaw.

Escudier M P, Gouldson I W and Jones D M, 1995 b, Taylor Vortices in Newtonian and Shear-Thinning Liquids, Proc Roy Soc Lond. Series A 449, (1935), pp 155-176.

Escudier M P, Gouldson I W and Jones D M, 1995 c Flow of Shear-Thinning Fluids in a Concentric Annulus, Experiments in Fluids, 18, (4), pp 225-238.

Escudier M P and Gouldson I W, 1995, Concentric annular flow with centrebody rotation of a Newtonian and a shear-thinning liquid, Int. J. Ht and Fluid Flow, 16, no 3, pp 156-162.

310

Lockett T J, 1992, Numerical simulation of inelastic non-Newtonian fluid flows in annuli, PhD Thesis Imperial College of Science, Technology and Medicine.

Manglik R M and Fang P P, 1995, Effects of eccentricity and thermal boundary conditions on laminar and fully developed flow in annular ducts, Int. J. Ht. and Fluid Flow, **16**, No 4, pp 298-306.

Ooms G and Kampman-Reinhartz B E, 1996, Influence of drillpipe rotation and eccentricity on pressure drop over borehole during drilling, European J. Mechanics. B./Fluids **15** No 5, pp 695-711.

San Andreas A and Szeri A Z, 1984, Flow between eccentric rotating cylinders, ASME J. Appl. Mech., **51**, pp 859-878.

Shah R K and London A L, 1978, Laminar flow forced convection in ducts, Advances in Heat Transfer, Academic, New York.

Tehrani M A, Bittleston S H and Long P J G, 1993, Flow instabilities during annular displacement of one non-Newtonian fluid by another, Experiments in Fluids **14**, No 4, pp 246-256.

III.2 Measurement of Temporal and Spatial Evolution of Transitional Pipe Flow with PIV

J. Westerweel and A.A. Draad

Laboratory for Aero and Hydrodynamics, Delft University of Technology, The Netherlands

Abstract. Turbulent 'slugs' were generated by injection of a small amount of fluid into a fully-developed laminar pipe flow at a Reynolds number of 5,800. Digital *particle image velocimetry* (PIV) was applied to measure the instantaneous flow field at different downstream locations with respect to the point of injection. At each location the passage of the turbulent slug could be recorded in a sequence of images. The results show the development in space and time of the turbulent slug as it progresses inside the pipe. For one location the consecutive PIV results were compiled into a single data set that represents a (quasi-) instantaneous cross section of the turbulent slug. From this data set it is possible to identify the flow structures that occur in the leading, central and trailing parts of the turbulent slug.

Keywords. particle image velocimetry, pipe flow, transition, turbulence

1 Introduction

The transition to turbulence in laminar pipe flows is still far from being fully understood. In the past measurements of transitional flow structures were carried out mainly with single-point measurement techniques in combination with conditional averaging or phase averaging. It is anticipated that the measurement of the temporal and spatial development of the transitional flow field may shed more light on understanding the mechanisms involved in the transition to turbulence and in the re-laminarization (Darbyshire & Mullin 1995).

We used particle image velocimetry (PIV) to study the laminar-turbulent transition process in a pipe flow. For these experiments we used a digital image acquisition system that was designed especially for the purpose of recording time sequences for PIV measurements in (turbulent) flows; in a previous measurement of a fully-developed turbulent pipe flow it was demonstrated that the accuracy and spatial resolution of this system are comparable to those of a photographic-based system (Westerweel *et al.* 1996). The experiments presented in this paper are a continuation of earlier (preliminary) digital PIV measurements in transitional pipe flow (Westerweel *et al.* 1994; Draad *et al.* 1995).

In Section 2 we describe the experimental facility that was used for these measurements, and in Section 3 we give a detailed description of the mea-

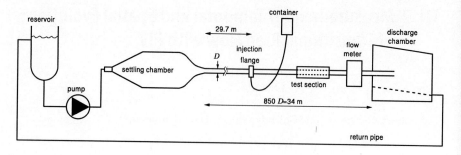

Figure 1: Schematic of the pipe flow facility (not to scale).

surements. The results are presented in Section 4, and Section 5 summarizes the main conclusions from this study.

2 Flow facility

The pipe flow facility that was used for this experiment was designed especially for the purpose of studying transitional flows. In this section we give a brief description of this flow facility; further details can be found in the thesis of Draad (1996).

The main flow line of the facility consists of a smooth pipe with an inner diameter of 40 mm, and a total length of 34 m. A sketch of the pipe flow facility is given in Figure 1. The working fluid is water, and the flow in the facility remains laminar over the full length of the pipe for Reynolds numbers up to 60,000. During the experiments the flow rate was kept at a constant value, which is monitored directly with a magnetic-inductive flow meter. The flow rate during the measurements was 609±3 liters per hour (l/h), which corresponds to a Reynolds number of about 5,800 (based on the pipe diameter and the bulk velocity).

A transition to a turbulent flow state was induced by injection of fluid at a distance of 29 m from the pipe inlet. At this distance, which is 725 pipe diameters from the pipe inlet, the laminar flow can be considered to be completely developed, so that the transition to turbulence is induced by the injection of the fluid alone.

The fluid was injected through a 1.0 mm hole in the pipe wall and the duration of the injection was 0.5 s. The jet was fed by a small container placed above the pipe and the injection speed could be manipulated using a regulator. The injection velocity is 5.2 times the centerline velocity in the pipe; nonetheless, the mean volume flow rate of the injected fluid was only 0.6% of the bulk flow rate. The injection velocity is slightly larger than the critical injection velocity for a the development of a turbulent slug ((Draad et al. 1995)).

Table 1: An overview of relevant experimental conditions.

Pipe:	diameter	40.0	mm
	length	34	m
	wall thickness	0.19	mm
Flow:	kinematic viscosity	0.9327	mm^2/s
	bulk flow rate	609	l/h
	bulk velocity	135	mm/s
	Reynolds number	5,790	
Seeding:	nominal diameter	30	μm
Light sheet:	power	0.5	W
	thickness	0.5	mm
	exposure time delay	1.725	ms
	number of exposures	8	
Recording:	resolution	1000×1016	pixels
	lens focal length	50	mm
	numerical aperture	2.8	
	image magnification	0.305	
	viewing area	39.4×40.0	mm^2
Interrogation:	resolution	32×32	px
	area	1.26×1.26	mm^2
Data set:	vectors/image	3,481	
	images/data set	131	
	number of data sets	45	

The disturbed flow advects along the downstream section of the pipe, and its development into a turbulent 'slug' was be observed at four measurement locations.

3   Measurements

In this section we describe the measurements and experimental conditions; an overview of the relevant experimental parameters is presented in Table 1.

To facilitate measurements with PIV, the flow was seeded with small tracer particles (Optimage) which have a nominal diameter of 30 μm. The flow was illuminated with a scanning-beam light sheet using an argon-ion cw-laser and a rotating polygon mirror. The width of the light sheet is approximately 0.5 mm, and the plane of light sheet coincided with the centerline of the pipe. A schematic of the optical configuration is shown in Figure 2. The light-sheet scan time is 1.725 ms, and each image is exposed 8 times. We could thus achieve a high image density, while keeping the volume fraction and number density of the tracer particles at a minimum to avoid two-phase effects of the seeding on the flow ((Westerweel et al. 1996)).

The measurements were carried out at four downstream locations in the pipe, i.e.: 365, 655, 870 and 1660 mm downstream from the point of injection. At the measurement location the pipe was enclosed in a rectangular box filled with water. The pipe wall inside this box was replaced by a thin sheet with a thickness 0.19 mm and with a refractive index that is very close to that of water. We could thus record images with minimal distortion due to the curvature of the pipe wall and differences in refractive index of water and air.

The PIV images were recorded with a CCD camera that has a spatial resolution of 1000×1016 pixels. The field-of-view of the camera is exactly one pipe diameter (in the radial direction). The camera was connected to an acquisition system which could capture sequences of 131 images at a rate of 10 images per second. The recording of the first image of each sequence coincided with the end of the injection period. Further details regarding this acquisition system are given by Westerweel *et al.* (1996).

We interrogated the digital images in 32×32-pixel sub-images with a spacing of 16 pixels (i.e., a 50% overlap). The pattern of interrogation locations was chosen such that it is symmetric with respect to the centerline of the pipe. Each image yielded a total of (59×59=) 3,481 velocity vectors, and each measurement — consisting of 131 images — yielded a total of 456,011 velocity measurements.

All measured velocity fields were tested for spurious velocity measurements using a local median test (Westerweel 1994). The average fraction of spurious vectors was less then 8%, and most of the spurious vectors occurred near the pipe wall, where the velocity gradient is very large and where there is a high probability that subsequent particle images overlap.

The number of spurious vectors at the upper wall was higher in comparison to that at the lower wall, which could be explained by the fact that the flow in the pipe is mostly laminar. In a turbulent flow the tracer particles are mixed continuously by the flow, so that a homogeneous distribution of the particles is maintained. This mixing process is absent in a laminar flow, so particles that do not exactly match the density of the ambient fluid will

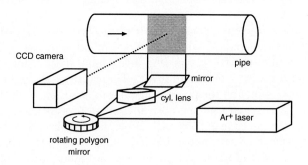

Figure 2: Optical configuration for the PIV measurements. (The optical box that encloses the pipe is not drawn.)

settle. In this particular experiment the flow is laminar between the inlet of the pipe and the injection location (i.e., 29 m), and it takes (on average) about 200 seconds for a tracer particle carried by the flow to travel this distance. As a result the concentration of tracer particles near the upper wall is lower in comparison with that near the lower wall, and explains a higher occurrence of spurious vectors near the upper wall.

The exposure time delay was chosen in accordance with the 'one-quarter rule' (Keane & Adrian 1991) for the particle-image displacement at the pipe centerline for the *turbulent* flow state. Consequently, the displacement in the central section of the pipe in the *laminar* flow state is considerably larger than 1/4th of the size of the interrogation window. On the other hand, the out-of-plane displacement in the laminar flow state is practically zero, whereas it makes a significant contribution to the loss of correlation in the turbulent flow state. So, although we allowed for a relatively large displacement in the laminar flow state, this did not result in a higher fraction of spurious velocity measurements.

4 Results

In total 45 image sequences (with a total volume of 6 Gigabyte) were recorded (which include measurements at different downstream locations and at each location for four different circumferential injection positions). Here we will only present the results that were obtained from a set of four image sequences that were recorded at the four measurement locations under equal injection conditions. The axial coordinate in the direction of the flow is indicated by x, and the radial coordinate is indicated by r; a negative radial coordinate refers to the lower section of the pipe.

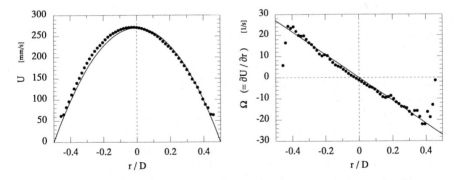

Figure 3: Profiles of the axial velocity (U) and the vorticity (Ω) as a function of the distance (r) from the centerline with respect to the pipe diameter (D), for the laminar flow state. The solid lines represent the Poiseuille profiles for a water flow rate of 609 l/h.

4.1 Poiseuille flow

In the fully-developed laminar flow state the velocity field only depends on the radial coordinate. We could therefore determine the laminar velocity profile by averaging the velocity data over each row, and then ensemble averaging the data over all images for which the flow was in the laminar flow state.

Figure 3 shows the results for the radial profiles of the axial velocity and the shear rate (viz., vorticity); the solid lines represent the corresponding profiles for a Poiseuille flow with a flow rate of 609 liters per hour. The differences between the experimental results and the Poiseuille profiles are small. However, the experimental result for the laminar velocity profile is not exactly symmetric. This small skew appears to be caused by the Earth's rotation, and it has not been reported in experiments by others. This may be partly due to the fact that most other facilities contain air flows in which the skew effect is much smaller due to the larger kinematic viscosity of air with respect to that of water. Also, the development of the corresponding secondary flow pattern takes a long time, so that the effect only occurs in very long pipes. These aspects are discussed in further detail by Draad (1996).

The measured shear rate coincides with the expected shear rate over almost the full pipe diameter ($|r| < 0.4D$). The deviations of the data points close to the pipe wall is directly related to the presence of spurious measurements that remained undetected.

4.2 Transitional flow

To visualize both the spatial and temporal development of the disturbance, we determined for each PIV measurement the axial velocity profile over each individual image, i.e.:

$$U_i(r) = \frac{1}{L} \sum_j \tilde{u}_i(r, x_j) \Delta x_j \tag{1}$$

where the index i denotes the number of the frame, $\tilde{u}$ is the instantaneous velocity, L the width of the image, and where r and x are the radial and axial coordinates respectively. In this particular situation we have: $L \approx D$. In Figure 4 are plotted the combined results for $U_i(r)$ for all frames at the four measurement locations. Note that the disturbances near the centerline of the pipe arrive earlier at the first measurement station than the disturbances close to the wall. When the disturbance arrives at the second measurement location, it fills the complete cross-section of the pipe, and it remains turbulent also at the third and forth measurement stations. It should be noted the results shown in Fig. 4 are measurements of four individually generated disturbances. Occasionally, the disturbance did not develop into a full slug, or the observed slug showed signs of re-laminarization.[1]

---

[1] This was the case in the result for the fourth measurement location, presented in the original symposium proceedings; the present result is a different measurement, but for identical injection conditions.

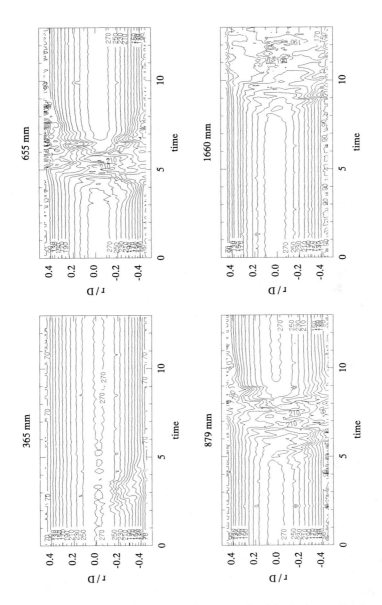

Figure 4: The temporal evolution of the locally-averaged axial velocity profiles for at four measurement locations behind the injection point (from top to bottom: 365, 655, 870 and 1660 mm).

The digital PIV measurements allow us to investigate the detailed structure of the disturbance. Figure 5 shows the velocity maps of the lower halves of four consecutive recordings. At Re=5,800 the bulk velocity is about 135 mm/s; hence, the expected mean displacement of structures between two consecutive recordings is roughly 0.3 pipe diameters. With this knowledge we can identify coherent structures in the subsequent maps in Fig. 5. In particular, note a vortical flow pattern close to the lower pipe wall in Fig. 5; by overlaying identical flow patterns in subsequent vector maps it is possible to compile a quasi-instantaneous picture of the full turbulent slug in the pipe. The problem that arises is that flow structures near the pipe wall advect at a smaller rate than flow structures close to the centerline. Ideally, one would like to use a pattern matching algorithm that takes into account the advection velocity of every individual flow structure. However, such an algorithm could not be developed at this moment; instead we matched the patterns by eye.

To eliminate any discontinuities related to the edges of individual PIV data fields, the overlapping data were merged using a weight value for each data point which decreases linearly as a function of the distance from the middle of the image.

The result in Figure 6 is a compilation of 60 PIV results, covering an elapsed time of 6 seconds, recorded at a distance of 655 mm behind the injection position. The remaining data set contains 59×1392 data points.

At the top of Fig. 6 we plotted the axial velocity at centerline; this would compare to the time trace of a passing slug as measured by a single-point probe (e.g., a laser-Doppler anemometer). Initially the velocity at the centerline is equal to two times the bulk velocity (i.e., 2×135=270 mm/s). One can observe how instabilities grow in size and amplitude, until they cover the full width of the pipe. The velocity at the centerline drops and fluctuates around the dash-dotted line which represents the mean centerline velocity for a fully-developed turbulent pipe flow (i.e., 1.3×135=176 mm/s; the constant of 1.3 is only valid for fully-developed turbulent pipe flow at low Re, see Eggels *et al.* 1994). At the trailing interface the flow returns to a laminar flow state. Note that the re-laminarization sets in at x = -540 mm by the centerline velocity becoming equal to the value for a laminar flow for a very brief instant; this event is perhaps related to a so-called 'incursion' (Darbyshire & Mullin 1995).

The total length of the slug appears to be more than 20 times the pipe diameter. The results of Wygnanski & Champagne (1973) show that the growth rate of a turbulent slug is proportional to the bulk velocity (U_b), which implies a length of $16.4D$; taking into account the duration of the injection (i.e., 0.5 s), yields an expected length of $18D$. This value compares quite well with the length of the slug in Fig. 6.

The compiled data of the slug as shown in Fig. 6 allow us to study the

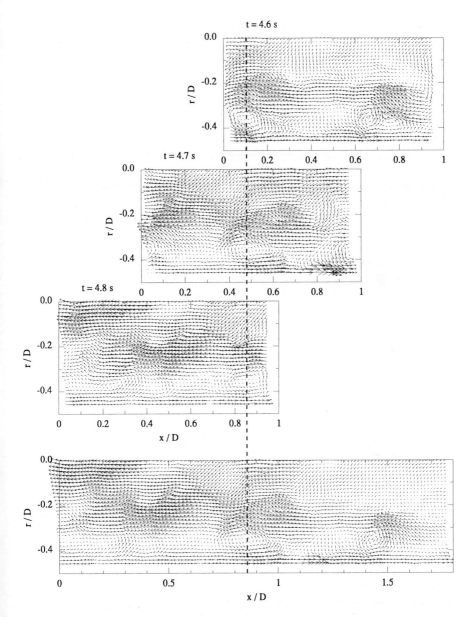

Figure 5: The instantaneous velocity relative to the mean velocity of the laminar flow state for three consecutive PIV measurements (from top to bottom: 4.6, 4.7 and 4.8 s after the end of the injection) at a distance of 655 mm behind the injection location. The bottom plot is the result after merging the PIV data sets.

flow structures that appear in the front, middle and rear sections of the slug. In Figure 7 we show the vector maps at different locations. Note that the instantaneous velocity fields still contain some spurious data (in particular near the upper pipe wall); this is due to the effect described in Section 3.

Before discussing the results shown in Figs. 6 and 7 we would like to emphasize that so far we have only been able to carry out such a compilation for a single image sequence; evidently, observations from these data may not be generally valid, and require validation with results from other measurements.

In Fig. 6 the position $x = 0$ mm denotes the first image in which a deviation from the average laminar velocity profile appears, which occurs 3.1 s after the end of the injection. We use negative x-positions to indicate that this part of the slug can be envisaged to be located upstream of the beginning of the disturbance. From $x = 0$ mm to $x \approx -300$ mm the disturbance increases in strength gradually. Initially, the disturbance seems to display some periodic behavior with a period in x of approximately 60 mm. The vortices shown in Fig. 7a (also visible in Fig. 5) correspond to the white spots in the radial velocity around $x = -286$ mm and $x = -310$ mm. A little further upstream, i.e. $x \approx -325$ mm, the centerline velocity quickly approaches the turbulent value and the flow becomes turbulent over the entire cross section of the pipe. The fluctuations in the centerline velocity are much larger than in a fully developed turbulent flow (Fig. 7b). Re-laminarization sets in at $x \approx -520$ mm (Fig. 7c) and seems to occur in phases. The centerline velocity profiles shows an almost step-wise increase towards the laminar value separated by sudden decreases (Figs. 7d,e). Again, some periodicity appears to be present, now with a spacing of roughly 100 mm. Clearly, the turbulence decays very slowly near the wall (Fig. 7f).

5 Conclusions

A transition of the laminar flow into a turbulent flow state was induced by injection of a small amount of fluid into the pipe, and the development of the transition along the pipe was investigated with digital PIV.

The present results demonstrate that digital PIV is a useful tool for the investigation of transient flow phenomena. It is possible to study a large variety of flow conditions with high accuracy and detail within a relatively short time (provided one has access to a high-capacity storage facility).

At the Reynolds number studied here, a turbulent slug should grow in length (Wygnanski & Champagne 1973). Flow visualization experiments using dye in the same facility (Draad et al. 1995) suggest a transition to turbulence under the same flow and disturbance conditions. The results presented here show the existence of a turbulent slug at 655, 879 and 1660 mm downstream from the point of injection. However, in at least one occasion we observed signs of re-laminarization at the last measurement station, which may indicate that the slug may not always be stable. It might be possible that — due to the skewed laminar profile — the development and stability of the slug depends on the direction of the injection.

We believe that the most prominent result is the compiled data set shown in Fig. 6. This data set allows us to observe the instantaneous transitional flow structure as a whole (which covers more than 20 pipe diameters in the flow direction). The compilation procedure is now done partly by hand, and needs to be optimized.

At this moment we have only compiled one image sequence out of the total of 45; we anticipate that the image sets that remain to be investigated will provide additional information. So far, with only one of these data sets it is impossible to make any definite conclusions, although we feel that the present data set reveals a number of potentially significant structures.

The PIV recordings for the other injection conditions are currently being analyzed; the results will be presented in a forthcoming paper.

Acknowledgments

The research of dr.ir. J. Westerweel has been made possible by a fellowship of the Royal Netherlands Academy of Arts and Sciences. Dr.ir. A.A. Draad received financial support from Shell Research.

References

DARBYSHIRE, A.G., & MULLIN, T. 1995. Transition to turbulence in constant-mass-flux pipe flow. *J. Fluid Mech.*, **289**, 83–114.

DRAAD, A.A. 1996. *Laminar-turbulent transition in pipe flow for Newtonian and non-Newtonian fluids.* Ph.D. thesis, Delft University of Technology.

DRAAD, A.A., KUIKEN, G.D.C., & NIEUWSTADT, F.T.M. 1995. Transition to turbulence in pipe flow. *Pages 103–110 of: et al.*, R. KOBAYASHI (ed), *Proc. IUTAM Symp. "Laminar-Turbulent Transition"*.

EGGELS, J.G.M., UNGER, F., WEISS, M.H., WESTERWEEL, J., ADRIAN, R.J., FRIEDRICH, R., & NIEUWSTADT, F.T.M. 1994. Fully developed turbulent pipe flow: a comparison between direct numerical simulation and experiment. *J. Fluid Mech.*, **268**, 175–209.

KEANE, R.D., & ADRIAN, R.J. 1991. Optimization of particle image velocimeters. Part II: Multiple-pulsed systems. *Meas. Sci. Technol.*, **2**, 963–974.

WESTERWEEL, J. 1994. Efficient detection of spurious vectors in particle image velocimetry data sets. *Exp. Fluids*, **16**, 236–247.

322

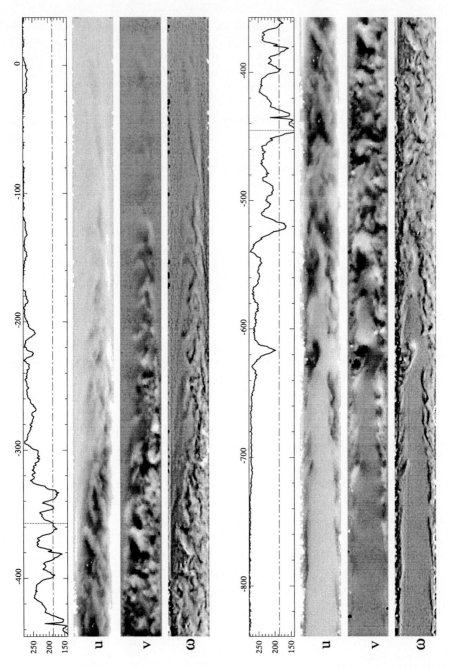

Figure 6: The (quasi-) instantaneous structure of a turbulent slug as compiled from about 60 overlapping PIV measurements. In this figure are shown the axial velocity at the centerline (in mm/s), and maps of the fluctuating axial and radial velocities (u and v respectively), and of the fluctuating vorticity (ω) as a function of the axial position (in mm). For reference the dash-dotted line represents the mean centerline velocity for a fully-developed turbulent flow.

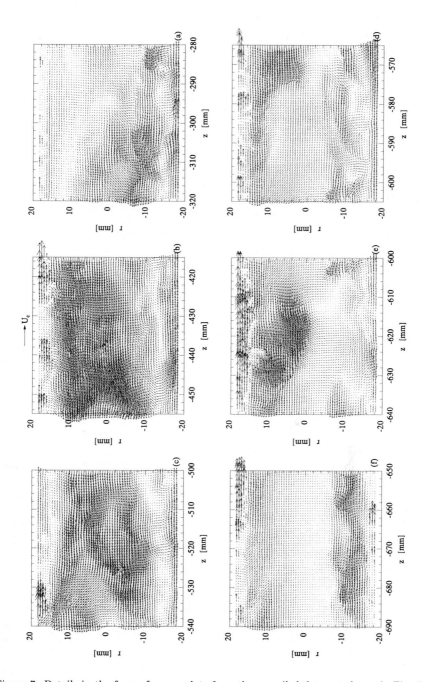

Figure 7: Details in the form of arrow plots from the compiled data set shown in Fig. 6. The arrows represent the velocity relative to the laminar velocity profile. The arrow above (b) represents the direction and magnitude of the centerline velocity in the laminar flow state.

324

WESTERWEEL, J., DRAAD, A.A., VAN DER HOEVEN, J.G.TH., & VAN OORD, J. 1994. A fast data acquisition system for digital PIV: application to fully-developed and transitional turbulent pipe flow. *In: Proc. 7th Int. Symp. on Applications of Laser Techniques to Fluid Mechanics.*

WESTERWEEL, J., DRAAD, A.A., VAN DER HOEVEN, J.G.TH., & VAN OORD, J. 1996. Measurement of fully-developed turbulent pipe flow with digital particle image velocimetry. *Exp. Fluids*, **20**, 165–177.

WYGNANSKI, I.J., & CHAMPAGNE, F.H. 1973. On transition in a pipe. Part 1. The origin of puffs and slugs and the flow in a turbulent slug. *J. Fluid Mech.*, **59**, 281–335.

III.3 Mixing in Gravity Current Heads Flowing over Rough Surfaces

W.D. Peters[1], J.E.S. Venart[2]and S.R. Cogswell

[1] Department of Engineering, University of Prince Edward Island
[2] Fire Science Centre, University of New Brunswick
Fredericton, New Brunswick Canada
[3] EMR Microwave Technology Corporation Fredericton, New Brunswick, Canada

Abstract

A two-dimensional full-field quantitative visualization technique is used to assess the mixing that occurs in gravity current heads flowing over rough surfaces. This technique combines small-scale saltwater modelling, laser-induced fluorescence and digital image processing to provide information about the mixing behaviour and the associated concentration fields that exist in this density-differentiated fluid mixing system. The effects of surface roughness and initial fluid density difference on flow dynamics and head dilution are examined with potential contributions to the area of heavier-than-air gas (HTAG) dispersion.

1. Introduction

Gravity currents belong to a class of fluid flows in which buoyant forces, generated by density variations, produce motions predominantly in a direction normal to the gravity vector. This flow can be created, for instance, when a heavier fluid is introduced into a reservoir of less dense fluid after which it slumps to the bottom and spreads horizontally as a gravity current. Since the fluids forming this flow system are generally miscible, mixing plays an important role in the flow dynamics.

The primary feature of a dense gravity current, flowing over a smooth surface, is a well defined raised head formed at the downstream edge of a shallower flowing source layer. The leading point of this head exhibits a slightly elevated nose which allows less dense ambient fluid to be overrun by the structure as it advances. Behind the head, a stable and quiescent interface separates the source and ambient fluids.

Two basic gravity current mixing mechanisms have been defined by the large amount of research performed on the subject. Firstly, Kelvin-Helmholtz instabilities, initiated by velocity and density gradients at the intruding head/ambient fluid interface, lead to mixing in this region. Secondly, mixing internal to the head region arises as a layer of lighter ambient fluid is overrun by the advancing head. The velocity shear between this overrun fluid and the current head causes further Kelvin-Helmholtz instabilities at this level. Additionally, Taylor instabilities, due to packets of lighter fluid rising through the structure, further contribute to, and, perhaps, even dominate, the internal dilution process.

2. Review of Past Work

Many fluid flow problems exist in which laser-induced fluorescence can provide valuable information about the interior fluid regions that would not be possible with conventional dye techniques. Mapping of species concentration, density or temperature in a planar field, or along a line, is the most frequently cited use of this technique

Research by Hesselink (1988), Nash et al. (1995) and Cruyningen et al. (1990), for instance, adds to the credibility of this widely accepted visualization method. As the success of this technique is closely linked to the fluorescent dye that is used, it is necessary to completely understand its behaviour. Walker (1987) provides a very thorough and extensive quantitative examination of one of the more popular dyes, fluorescein sodium.

Salt-water modelling has been shown to be very useful in the study of gravity current flows (Peters et al., 1996). Researchers, such as Chen (1980), Chobotov et al. (1986), Zukoski and Kubota (1988), Simpson (1987), Simpson and Britter (1979) and Winant and Bratkovich (1977) have used this technique extensively to examine the same class of flows yielding a valuable resource of information. However, none of this work has used laser-induced fluorescence as the visualization technique.

One conclusion of the Simpson and Britter work, for instance, was that the shear between the gravity current head and the ambient fluid was the main mixing process at the front. Consequently, they stated that the overrun less dense fluid may be neglected in terms of the overall mixing. Winant and Bratkovich, however, observed large density fluctuations in the head region. They concluded that these variations seemed to be a result of the gravitational instability between discrete parcels of unmixed lighter fluid entrained at the leading edge of the structure, rather than any active turbulent fluctuations.

Of particular importance to this work is the research concerned with the influence of surface roughness on gravity current propagation and mixing. The amount of salt-water modelling in this area is, unfortunately, small and, unlike the present study, generally limited to flows over single obstacles (Chobotov et al., 1986, Simpson, 1987, Lane-Serff et al., 1995). In terms of heavier-than-air gas dispersion, the effects of surface roughness are generally considered through wind tunnel studies. Petersen and Ratcliff (1989) and Roberts et al. (1990) used wind tunnel simulations to consider flows over a variety of roughness arrays. They reported that cloud concentrations in a dispersion over a roughness scale typical of an urban area were reduced by as much as 25 times over those found for a similar advance over a grassy plain.

3. Flow Parameters and Scaling Laws

The gravity current flow parameters for the salt-water modelling used here are defined in Figure 1. The ambient fluid depth, h_1, and initial fluid densities, ρ_A and ρ_B, are controlled and set prior to the experiment. The controlled upstream source condition is quantified by the source flowrate per unit channel width, Q, and the initial fluid density difference, $\Delta\rho/\rho_A$.

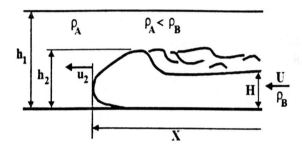

Figure 1. Definition of the experimentally controlled and observed gravity current flow parameters.

Based on continuity, Q is the product of the source layer depth, H, and the source layer velocity, U. At any downstream channel position, the source layer velocity, U, can be derived with good approximation based on the observed source layer depth, H, where U = (Q/H). Other observed parameters are the current head height, h_2, the current head frontal velocity, u_2, and the current nose position, X.

When investigating full-scale heavier-than-air gas dispersion, for instance, much of the scaling difficulty arises when near-source modelling is attempted. In far-field modelling, however, many of these scaling concerns are eliminated as the flow is, then, stabilized as a negatively buoyant layer that propagates away from the source solely under the influence of buoyancy. Under these conditions, the primary dimensionless groups characterizing the flow are the Reynolds and Froude numbers, as well as, the initial fluid density difference. Dilution of the dense gas cloud at this point has, typically, reduced the head density sufficiently so that the Boussinesq approximation is valid.

When modelling large-scale flows, however, equality of both the full-scale Reynolds and Froude numbers is generally not possible for small-scale lab experiments. When a choice has to be made, though, Froude number scaling is more important (Hoot and Meroney, 1974). Fortunately, the requirement of exact

Reynolds number equality can be relaxed as long as highly turbulent characteristics are exhibited in the flows. It has been shown that these turbulent characteristics exist in gravity current flows with Reynolds numbers, based on the current head height and velocity, as low as 1000 (Keulegan, 1957, Simpson, 1987, Chobotov et al. 1986).

Some confusion seems to exist in the literature over the definition to be used for the gravity current Reynolds number. It has been defined in terms of (1) the source layer depth, H, and the current head frontal velocity, u_2, (Simpson and Britter, 1979) and (2) the source layer depth, H, and velocity, U (Zukoski and Kubota, 1988). Fundamentally, the second of these definitions seems to be more appropriate as it is based on the controlled source condition for the current flowrate. Consequently, a Reynolds number defined by the source flowrate, Q, is used here where $Re = Q/v = UH/v$. This definition is, also, independent of the shape of the gravity current head. These same variables are, also, used to define the Froude number.

A balance between buoyant and inertial forces for an inviscid gravity current produces characteristic scales that can be used to normalize lengths, times and velocities. The scales used by Chen (1980) and Chobotov et al. (1986) are adopted, i.e.,

$$d_{REF} = (Q^2/g')^{1/3}$$

$$t_{REF} = (Q/g'^2)^{1/3}$$

$$u_{REF} = (g'Q)^{1/3}.$$

The term g' is the reduced gravity term resulting from the Boussinesq approximation.

4. Experimental Facility

Figure 2 illustrates the water channel apparatus and circulation system used to create small-scale gravity current flows. The facility consists of an open plexiglass channel, 2.4 m long and 0.2 m wide with a maximum possible ambient fluid depth of 0.3 m. It is filled with the less dense ambient fluid prior to an experiment and allowed to become quiescent.

The saline fluid that is injected to form the gravity current flow enters the channel through an entry box section located on the channel bottom at the upstream end. This box is filled with porous plugs that provide a uniform flow entry condition. The source injection flowrate is controlled by a pair of variable area flowmeters. The depth of the less dense ambient fluid in the channel is controlled by means of an adjustable overflow weir located at the upstream end of the channel.

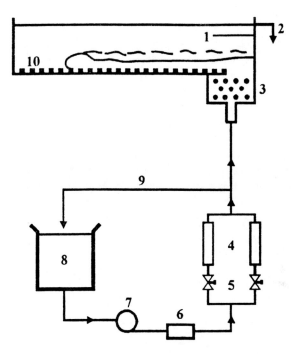

Figure 2. Schematic diagram of the water channel apparatus and circulation system used to generate small-scale gravity current flows.

1) splitter plate	6) filter
2) overflow weir	7) circulating pump
3) inlet entry box	8) saline storage reservoir
4) flowmeters	9) recirculation line
5) flow control valves	10) roughness array

Surface roughness is created using two-dimensional arrays of square cross-section elements positioned on the channel floor perpendicular to the flow direction. These elements span the full channel width over the entire length of the channel. The first is positioned flush with the inlet edge. Four sizes of roughness elements were used with nominal dimensions of 6, 13, 19 and 25 mm on a side. Elements were positioned with a spacing equal to the side dimension. Including the smooth floor, five surface conditions were, thus, examined.

The flow visualization system is represented in Figure 3. Fluorescein sodium dye is added to the gravity current fluid at a known concentration such that its emitted fluorescent intensity is linearly proportional to its concentration in the fluid (Walker, 1987). It is, then, possible to map concentration fields based on sampled intensity fields. Following Walker's recommendations, the ambient and source fluids are buffered to a pH above 8 to maximize dye efficiency.

The 488 nm blue line of a four W argon ion laser is used to excite the dye. The laser beam is reflected off a high speed rotating front-surface mirror scanning at a

330

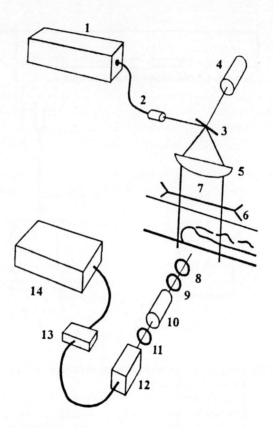

Figure 3. Schematic diagram of the gravity current flow visualization system.

1) argon ion laser	8) high-pass colour filter
2) fibre optics	9) 25 mm lens
3) scanning mirror	10) image intensifier
4) drive motor	11) 18 mm lens
5) collimating lens	12) video camera
6) slit assembly	13) time signal encoder
7) visualization sheet	14) video recorder

frequency of 500 Hz. It is then passed through a plano-cylindrical lens and slit mask to generate a collimated laser sheet approximately 20 cm wide and 1.5 mm thick. Finally, the sheet is passed through the longitudinal vertical mid-plane of the water channel to illuminate the flow.

The laser sheet generating and video recording systems are mounted on a computer controlled, stepper motor driven traversing carriage that can track the gravity current head as it progresses down the channel. With this system, accurate control over the tracking parameters can be maintained, while feedback on carriage position and velocity is provided.

The illuminated flows are sampled at a fixed rate of 30 images/second with a standard B/W CCD video camera (Panasonic WV BD-400) using an 18 mm lens. The camera is coupled to a low light level image intensifier (Astrolight 9100) equipped with a 25 mm lens and a sharp cut-off high-pass Schott colour filter (Ealing OG-515). A S-VHS video cassette recorder (Panasonic AG-6720) is used to record images for future processing and analysis. A time signal encoder (For-A VTG-22) is used to superimpose a reference time signal on the video record.

Video images are analyzed using a PC-based digital image processing system equipped with a Matrox MVP-AT/NP frame grabber board. The board acquires images with a resolution of 512 x 480 pixels representing a spatial resolution of approximately 0.5 mm x 0.5 mm for the field of view utilized. This system, along with custom and commercial software packages for image processing, allows single snapshots and/or time-averaged segments to be grabbed and analyzed.

5. Experimental Matrix

Three series of tests were performed with fluid density differences, $\Delta\rho/\rho_A$, of 0.010, 0.030 and 0.050. Injection flowrates, Q, of 10.3, 10.6 and 10.8 cm^2/s, corresponding to the above density differences, were used to maintain a Reynolds number, Re = Q/ν, of 1000. If the Reynolds number is based on the observed current head height and velocity, i.e., Re = $u_2 h_2/\nu$, a value much greater than 1000 is achieved, ensuring that turbulent characteristics exist in the flows. The depth of the ambient fluid, h_1, was set to 30.0 cm for each test. Five surface roughness conditions were examined for each of the three test series for a total of fifteen tests.

6. Analysis and Discussion of Results

In general, gravity current advance rates are governed by two flow regimes. The first is one in which viscous forces are negligible compared to the inertial and buoyant forces. A balance between these latter two forces yields a constant velocity current advance. The second regime is one in which viscous forces dominate inertial forces in the balance with buoyancy, producing a deceleration due to increasing floor shear stresses with downstream position.

Chen (1980) suggested that the transition from a constant velocity to a decelerating flow occurs at a downstream position given by

$$X_t = 0.1 \frac{Q^2}{u_{REF}\,\nu}$$

When normalized with d_{REF}, this transition length becomes $X_t^* = 0.1Re = 0.1Q/\nu$, so that the normalized transition length is a function of the injection rate and the fluid viscosity. In this form, there is no dependence on the initial fluid density difference except in how it affects the viscosity.

Figure 4 shows the normalized nose position versus normalized time for the fifteen tests performed. The slope of each curve represents the normalized frontal velocity, u_2^*, a form of the Froude number, for each flow. The linear line shown reflects the constant velocity advance in the observed inertially dominated region of the smooth surface flows for each fluid density difference.

Examination of the smooth surface flows shows that the slope of the line is independent of $\Delta\rho/\rho_A$ and has a constant value of 0.89, indicating that pseudo-steady conditions exist. This was, also, the value observed by Chan et al. (1993). If the data for each fluid density difference with a constant surface roughness is considered, it can be further concluded that u_2^* is independent of the fluid density difference over the range of surface roughness used.

Based on Chen's work, the experimental data points for the smooth surface flow cases should deviate from the straight line at a transition length, $X_t^* = 100$. This appears to be supported by the data. It should be noted, however, that flow deceleration can, in part, be attributed to end wall effects (Chobotov et al., 1986).

As Chen's work was developed for smooth surface flows, there is no indication of how the transition from a constant velocity to a decelerating flow regime is affected by surface roughness. It is reasonable to assume, however, that the transition length would decrease with increased roughness. Surface roughness should act, in addition to viscous effects, to intensify the floor shear stresses so that advance rates exhibit greater deceleration for increasing surface roughness.

The experimental data for the rough surface flows strongly support this hypothesis. As roughness increases, further deviation from the smooth surface data is apparent and the flows exhibit a strong transient nature. It is evident, then, that increased boundary shear forces, induced by the roughness, significantly influence the advance rate. The inertially dominated, constant velocity regime is no longer evident as the transition length is observed to have been reduced to zero such that a transient deceleration exists throughout the entire spread.

In addition to the gravity current advance rates, the variation in head concentration with downstream position was, also, examined. Here, concentration refers to saline content and, thus, fluid density. Data extracted from one second (30 frame) time-averaged images were used to produce maps of the normalized head density profile at various positions down the channel. This information was collected for each experiment at nominal downstream positions of 25, 60, 100, 135, 175 and 210 cm.

Figure 5 illustrates the normalized concentrations obtained for three channel positions for a flow with an initial fluid density difference, $\Delta\rho/\rho_A$, of 0.010 and a surface roughness scale, $h_R^* = 0.86$. Contours are shown for 20% intervals and represent the local fluid density difference as a fraction of the initial maximum fluid density difference, $(\Delta\rho/\rho_A)_{MAX}$.

In this example, the roughness element array has a height of approximately 2 cm. The gravity current structure is observed to flow above it, interacting with the

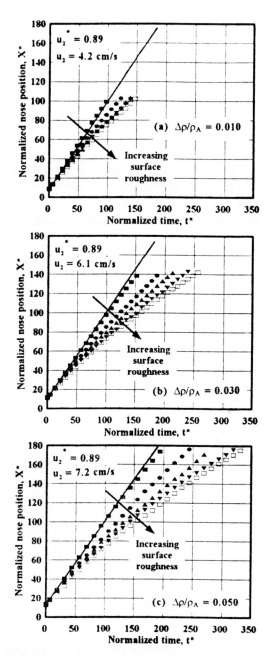

Figure 4. Normalized nose position versus normalized time.
(a) $\Delta\rho/\rho_A = 0.010$, $h_R{}^* = 0.00$ ■ 0.27 ● 0.59 ▲ 0.86 ▼ 1.13 □
(b) $\Delta\rho/\rho_A = 0.030$, $h_R{}^* = 0.00$ ■ 0.39 ● 0.83 ▲ 1.22 ▼ 1.60 □
(c) $\Delta\rho/\rho_A = 0.050$, $h_R{}^* = 0.00$ ■ 0.45 ● 0.97 ▲ 1.42 ▼ 1.87 □

334

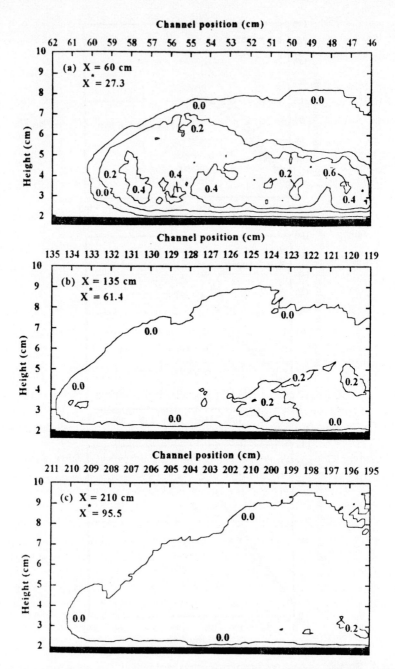

Figure 5. Concentration maps showing normalized head fluid density difference profiles for $\Delta\rho/\rho_A = 0.010$ and $h_R^* = 0.86$ at downstream positions of (a) $X^* = 27.3$, (b) $X^* = 61.4$ and (c) $X^* = 95.5$.

lighter fluid in the spaces between the roughness elements as it progresses down the channel. Individual roughness elements are not apparent in these views as the maps are time-averaged.

As the current advances down the channel, the normalized density within the head reduces significantly due to mixing with the ambient fluid. Surprisingly, though, the head remains intact, even for fluid density differences of 10% of the maximum. This represents a density within the head of only 0.1% greater than ambient.

A spatially determined mean head fluid density difference can be estimated from the concentration maps. This is done by first calculating the volume under the density profile surface using discretized sampling with a resolution of approximately 0.5 x 0.5 cm. The mean head fluid density difference is, then, determined by dividing this volume by the area of the current head that is projected onto the concentration map. The variations of normalized mean head fluid density difference with normalized downstream channel position are given in Figure 6. As before, the normalized mean fluid density difference is presented as a fraction of the maximum fluid density difference, $(\Delta\rho/\rho_A)_{MAX}$.

The pseudo-steady, stable smooth surface flows exhibit very little dilution over the channel length. The dilution rate for these tests is approximately zero such that the mean head fluid density difference is essentially constant. This is indicative of the inertially dominated flow condition that exists, characterized by a constant advance velocity and insignificant mixing.

The immediate conclusion, however, is that dilution rates increase significantly with roughness scale; the normalized mean head fluid density difference appears to decrease asymptotically to zero with channel position. For the test described in Figure 5, the normalized mean head concentration near the end of the channel is 80-90% lower than that for the corresponding smooth surface flow.

On examination of the data and the video records, it is concluded that, for a particular roughness array, the lightest gravity current flows are observed to ride up and over the roughness arrays without significant mixing with the fluid in the array spaces. Heavier flows, however, will more vigorously interact and mix with this fluid resulting in significantly increased dilution over the channel length.

This increase in dilution can be seen particularly between the tests with fluid density differences, $\Delta\rho/\rho_A$, of 0.010 and 0.030. This result strengthens our premise that the contribution of packets of overrun less dense fluid to the overall head dilution is much more significant than presented by other authors such as Simpson and Britter (1979). In addition, this observation is in agreement with the conclusions of Winant and Bratkovich (1977).

7. Conclusions

For the smooth surface flows considered in this work, both the inertial and viscous flow regimes were evident. The observed locations for the transition from

336

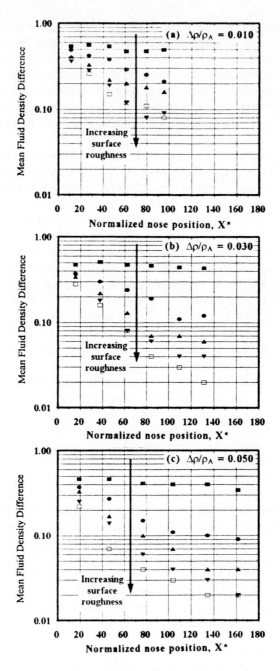

Figure 6. Normalized mean fluid density difference versus normalized nose position.
(a) $\Delta\rho/\rho_A = 0.010$, $h_R^* = 0.00$ ■   0.27 ●   0.59 ▲   0.86 ▼   1.13 □
(b) $\Delta\rho/\rho_A = 0.030$, $h_R^* = 0.00$ ■   0.39 ●   0.83 ▲   1.22 ▼   1.60 □
(c) $\Delta\rho/\rho_A = 0.050$, $h_R^* = 0.00$ ■   0.45 ●   0.97 ▲   1.42 ▼   1.87 □

inertially dominated to viscous dominated flows were in good agreement with Chen's work. It is not possible, however, to extend his conclusions to rough surface flows due to the significantly increased floor shear stresses. The roughness scale range used in these tests tends to reduce the transition to zero such that a transient decelerating flow exists for the full channel length. As a result, the normalized frontal velocities experience a much greater deceleration over rougher surfaces. When normalized, the frontal advance velocity appears to be independent of the fluid density difference.

Closely connected to the frontal advance rate is the dilution rate of the gravity current head as it flows down the channel. Significantly increased dilution rates are observed to be related to flow deceleration and increased surface roughness. Near the end of the channel, normalized mean head concentration in rough surface flows are observed to be at least 20 times less than those of comparable smooth surface flows.

The head dilution rate does not appear to be independent of the fluid density difference. Increased interaction and mixing with the less dense ambient fluid trapped in the spaces between roughness elements is observed for heavier gravity current flows. This is a significant conclusion as the role of the overrun less dense ambient fluid in the total head dilution process has been given little attention in the past.

Nomenclature

d_{REF}	characteristic length scale
g'	reduced gravity term due to Boussinesq approximation
H	source layer depth
h_R	roughness element scale
h_1	ambient fluid depth
h_2	current head height
Q	source injection flowrate per unit channel width
Re	gravity current Reynolds number
t_{REF}	characteristic time scale
U	source layer velocity
u_{REF}	characteristic velocity scale
u_2	current head frontal velocity
X	current nose position
ρ_A	ambient fluid density
ρ_B	source fluid density
$\Delta\rho/\rho_A$	fluid density difference
ν	kinematic viscosity
*	indicates normalized data using d_{REF}, t_{REF} or u_{REF}

338

Acknowledgements

Support for this work was received from Emergency Preparedness Canada, the Natural Sciences and Engineering Research Council of Canada and the Society of Fire Protection Engineers.

References

Chan, W.R., Zukowski, E.E. and Kubota, T., 1993, Experimental and Numerical Studies on Two-Dimensional Gravity Currents in a Horizontal Channel, Report. No. NIST-GCR-93-630, Center for Fire Research, National Institute of Standards and Technology, Gaithersburg, MD., 238 p.

Chen, J.C., 1980, Studies on Gravitational Spreading Currents, Report No. KH-R-40, W.M. Keck Laboratory of Hydraulics and Water Resources, California Institute of Technology, Pasadena, CA.

Chobotov, M.V., Zukoski, E.E. and Kubota, T., 1986, Gravity Currents with Heat Transfer Effects, Report No. NBS-GCR-87-522, Center for Fire Research, National Bureau of Standards, U.S. Department of Commerce, Gaithersburg, MD, 158 p.

Cruyningen, I. van, Lozano, A. and Hanson, R.K., 1990, Quantitative Imaging of Concentration by Planar Laser-Induced Fluorescence, Experiments in Fluids, Vol. 10, pp. 41-49.

Hesselink, L., 1988, Digital Image Processing Flow Visualization, Annual Review of Fluid Mechanics, Vol. 20, pp. 421-485.

Hoot, T.G. and Meroney, R.N., 1974, The Behaviour of Negatively Buoyant Stack Gases, Proceedings of the 67th Annual Meeting of APCA, Denver, CO, Paper 74-210, 20 p.

Keulegan, G.H. 1957, An Experimental Study of the Motion of Saline Water from Locks into Fresh Water Channels, Report No. 5168, National Bureau of Standards, U.S. Department of Commerce, Washington, D.C.

Lane-Serff, G.F., Beal, L.M. and Hadfield, T.D., 1995, Gravity Current Flows Over Obstacles, Journal of Fluid Mechanics, Vol. 292, pp.39-53.

Nash, J.D., Jirka, G.H. and Chen, D., 1995, Large Scale Planar Laser Induced Fluorescence in Turbulent Density-Stratified Flows, Experiments in Fluids, Vol. 19, pp. 297-304.

Peters, W.D., Cogswell, S.R. and Venart, J.E.S., 1996, Dense Gas Simulation Flows Over Rough Surfaces, Journal of Hazardous Materials, Vol. 46, pp.215-223.

Petersen, R.L. and Ratcliff, M.A., 1989, Effect of Homogeneous and Heterogeneous Surface Roughness on Heavier-Than-Air Gas Dispersion, Report No. 4491, American Petroleum Institute, Washington, D.C., 354 p.

Roberts, P.T., Puttock, J.S. and Blewitt, D.N., 1990, Gravity Spreading and Surface Roughness Effects in the Dispersion of Dense Gas Plumes", Proceedings of the AIChE 1990 Health and Safety Symposium, Session IIB: Modelling of Aerosol Clouds, Orlando, FL, March 18-22, 1990.

Simpson, J.E., 1987, Gravity Currents: In the Environment and the Laboratory, Ellis Horwood Limited, West Sussex, United Kingdom, 244 p.

Simpson, J.E., and Britter, R.E., 1979, The Dynamics of the Head of a Gravity Current Advancing Over a Horizontal Surface, Journal of Fluid Mechanics, Vol. 94, Part 3, pp. 477-495.

Walker, D.A., 1987, A Fluorescence Technique for Measurement of Concentration in Mixing Fluids, Journal of Physics, E, Scientific Instruments, Vol. 20, pp. 217-224.

Winant, C.D., and Bratkovich, A., 1977, Structure and Mixing Within the Frontal Region of a Density Current, Proceedings of the 6th Australasian Hydraulics and Fluid Mechanics Conference, Adelaide, Australia, December, 1977.

Zukoski, E.E. and Kubota, T., 1988, Experimental Study of Environment and Heat Transfer in a Room Fire, Report No. NIST-GCR-88-554, Center for Fire Research, National Institute of Standards and Technology, Gaithersburg, MD, 26 p.

III.4 Quantitative Velocity Measurements in Turbulent Taylor-Couette Flow by Flow Tagging

M. Biage*, S.R. Harris[†], W.R. Lempert[£] and A.J. Smits[§]

Department of Mechanical & Aerospace Engineering
PRINCETON UNIVERSITY, Princeton, New Jersey 08544, U.S.A.

Abstract. Quantitative velocity measurements have been performed in Taylor-Couette flow using the PHANTOMM flow tagging technique. The results illustrate the expected three-dimensional features of the flow including the presence of well defined Taylor cells at low Taylor numbers and gradual transition to turbulence as the Taylor number is increased. The spectral density and auto correlation functions were computed over the Taylor number range 21.1×10^3 to 1.10×10^{11}. For Taylor numbers less than approximately 6×10^7, the flow is dominated by large toroidal eddies with a wavelength the same order as the gap size. As the Taylor number is increased above this value, the flow displays more irregular motions, associated with the Görtler instability or instability of the second kind.. Nevertheless, even for the highest Taylor number studied in this work, the structure is far from random, and the flow exhibits an almost periodic behavior. The results show the potential of the PHANTOMM technique to investigate spatial instabilities in confined flows.

Keywords. PHANTOMM technique, Taylor-Couette flow, flow transition.

1 The PHANTOMM Technique

The ability to obtain accurate, high spatial resolution velocity data over a wide range of Reynolds numbers poses a significant challenge to existing optical diagnostic techniques. Flow tagging is a relatively new optical diagnostic in which a laser beam is used to "write" a spatially continuous pattern into a

---

[†]Graduate Student, Department of Mechanical and Aerospace Engineering, Princeton University

*Associate Professor, Department of Mechanical Engineering, Federal University of Uberlândia-MG, Brazil, 38400-902; Visiting Fellow (1995), Department of Mechanical and Aerospace Engineering, Princeton University.

[£] Research Scientist, Department of Mechanical and Aerospace Engineering, Princeton University.

[§] Professor, Department of Mechanical and Aerospace Engineering, Princeton University; Associate Fellow - AIAA.

specific region of the flow field. The temporal evolution of the initial pattern is then tracked ("interrogated") through Laser-Induced Fluorescence (LIF) imaging. The displacement within the elapsed time interval constitutes a measurement of velocity, with an absolute accuracy limited only by the ability to determine position and time.

Flow tagging based on caged dye Photo-Activated fluorophore (PAF) (McCray1989) tracers has been termed Photo-Activated Nonintrusive Tracking of Molecular Motion (PHANTOMM). Caged dye PAF's are organic dye molecules in which a chemical caging group has been attached in order to quench the normally bright laser fluorescence. The caging group is photolytically cleaved upon exposure of the molecule to ultraviolet (UV) light, typically, but not necessarily, from a laser. Upon photolysis, the original dye is recovered which can be tracked indefinitely using ordinary laser sheet fluorescence imaging approaches (Dahm et al. 1990). As was discussed by Lempert (1995), the uncaged dye exhibits an extremely intense fluorescence, with the result that low concentrations (less than 10^{-6} M) are required. As in ordinary dye visualization, the fluorescence is Stokes-shifted so that simple long-pass colored glass filters can be used to attenuate elastic scattering from the laser. This produces high contrast images which are conducive to measurements at high Reynolds number. In many respects, the technique is similar to Laser-Induced Photochemical Anemometry (LIPA), in which time lines are written into a flow using photochromic or phosphorescent materials (Falco and Nocera 1993).

The capabilities of the technique to measure velocity are directly linked to the optical and chemical properties of the PAF tracers. In particular, it should be pointed out that, while there is no limit to the maximum time between cage-breaking (tag) and interrogation (since the photochemical change is permanent), there is a minimum time dictated by the kinetic rate for the cage-breaking photolysis process. The finite rate results in a time lag between the firing of the tagging laser and the evolution of sufficient dye in its fluorescent form to be interrogated with sufficient signal-to-noise. The magnitude of the signal-to-background ratio is a function of the caged dye purity, as well as the sheet thickness. This ratio needs to be adjusted by experiment (Lempert 1995).

2 The Taylor-Couette Flow and Objectives

The flow between concentric cylinders has proved to be a remarkably rich problem in fluid dynamics. Interesting phenomena have been observed and examined since the early work of Taylor (1923) who found that for certain combinations of the rotation rates and geometry, the flow in the gap between the two cylinders became unstable. The first instability is in the form of counter-rotating vortex pairs with their axes in the circumferential direction, and as the rotation rate increases, a series of progressively more complicated flow patterns are observed. Part of the fascination with the Taylor-Couette problem is how such a simple flow geometry can produce such complexities. One particularly interesting aspect is the well-ordered transition from laminar to turbulent flow as the rotation rate increases, which may help to produce a general model of transition.

Some of the studies in Taylor-Couette flow have suggested that the structure of the flow at higher rotation rates is composed of large Taylor cells throughout the gap, typically due to the Taylor instability, and a thin boundary layer on the wall of the inner cylinder where the instability has been found to be characteristic of the Görtler instability. The disturbances in the boundary layer are of the form of counter-rotating vortex pairs with axes in the circumferential direction, with motion similar to that of the Taylor cells (Wei et al., 1992; Smith and Townsend, 1982). The Taylor-Görtler instability may also be an important mechanism for transition in more general boundary layer flows over curved surfaces, for both compressible and incompressible flows.

For a cylindrical geometry (r,θ,z), where r, θ and z represent the radial, azimuthal and vertical axial directions, the basic-state velocity vector is $\vec{V} = (v_r, v_\theta, v_z)$, where v_r, v_θ and v_z represent the radial, tangential and axial velocity components. Rayleigh (1916) showed that the necessary and sufficient condition for the existence of an inviscid axisymmetric instability is:

$$\frac{d(\Gamma^2)}{dr} < 0 \tag{1}$$

anywhere in the flow. Γ is the circulation defined as $\Gamma = rV_\theta$.

The instability in Taylor-Couette flow is governed by the Taylor Number, Ta, defined by

$$Ta = Re^2 \delta, \qquad Re = \frac{V_i d}{\nu} \tag{2,3}$$

$$V_i = R_i \Omega_i, \qquad d = R_o - R_i \tag{4,5}$$

$$\delta = \frac{d}{\bar{r}}(1 - \mu^2) \tag{6}$$

$$\bar{r} = \frac{R_i + R_o}{2}, \qquad \mu = \frac{\Omega_o}{\Omega_i} \tag{7,8}$$

where ν is the kinematic viscosity, R_i and R_o are the radii of the inner and outer cylinders, respectively, Ω_i and Ω_o and are the angular velocities of the inner and outer cylinders. When the outer cylinder is fixed and the inner cylinder is rotating about its axis, $\mu = 0$. This is the case which was studied in this work.

The gap size is at the higher end of the medium gap size range ($R_i/R_0 = 0.38$, where the large gap range starts at $R_i/R_0 = 0.33$).

The progressive complexity of Taylor-Couette flow has not been investigated in detail by quantitative velocity measurements. Intrusive methods such as hot wire anemometry are not suitable for this study because they can interfere with the flow structure. Also, single point techniques can not provide information on the global features of the large Taylor cells. On the other hand, the PHANTOMM technique is a non-intrusive method that provides spatially resolved measurements along a line to illustrate the principal features of the flow.

The primary objective of the current study is the characterization of the performance of the PHANTOMM technique for measuring velocity in liqiud flows at high Reynolds numbers. Additionally, the results give new insight in the overall structure of the flow as a function of Taylor number, in the range where the flow goes from laminar to turbulent.

3 Experimental System

3.1 Taylor-Couette Experimental Apparatus

The inner and outer cylinders of the Taylor-Couette apparatus are made of Plexiglas. The inner cylinder was machined to a radius of R_i = 3.14 cm. The nominal inside radius of the outer cylinder is R_0 = 8.18 cm and the variation in the inner radius around the circumference at fixed height was determined by visual inspection to be less than 1%. The height of the cylinders is L = 102 cm; giving an aspect ratio $\Theta = L/d = 20$ (d is the gap length), which is large enough to minimize end effects. The top and bottom endplates are stationary. The temperature of the system was not controlled but the ambient temperature of the room was constant between 22 and 25°C. The experimental system is mounted on a vibration-isolation table. The cylinders are driven by variable speed DC motors that allow uniform and stable rate of rotation for both cylinders. The rotation rates of the inner and outer cylinders can be chosen independently and co-rotating and counter-rotating cases are possible. For the results presented here, the outer cylinder was fixed, so that $\mu = 0$ (see Eq. 8), and the Taylor number was varied by changing the rotation frequency of the inner cylinder. A small He-Ne laser was directed on to a small black tape spot on the wall of the inner cylinder, and the signal was captured with a photo diode. The rotation frequency of the inner cylinder was determined using a digitizing oscilloscope. Since 0.355 micron light is attenuated by Plexiglas, the outer cylinder was fitted at its midpoint with a small quartz window, approximately 1 cm in diameter, to facilitate tagging. The working fluid was a water solution with Dextran Carboxy fluorescein as a tracer material. A typical concentration is 0.5 mg/l.

3.2 Optical Apparatus

In the optical configuration used in this study, the tagging was performed using the third harmonic of a Q-switched Nd:YAG laser at 0.355 micron. Single-pulse energies between 40 mJ and 60 mJ were used, depending upon the experiment. Lines were tagged along the radial axis (the r-direction), where the UV tagging beam from the laser was loosely focused with a 30 cm focal length lens, resulting in a beam waist of approximately 1 mm. The interrogation was performed with a flashlamp pumped dye laser using LD490 dye in methanol with no interactive line-narrowing optics. The dye laser is capable of pulse outputs between 50-400 mJ with a pulse duration of approximately 2 microseconds. The interrogation beam was formed into a sheet that, for the lateral recording, emerged from the top of the experimental apparatus. The sheet, approximately 2 cm thick, was formed by mirrors and cylindrical lenses with positive and negative focal length. This sheet was located in the (r,z) plane in the annular space between the cylinders, where the z- direction is along the axis of the cylinders. For recording from the top view, similar optics were used to place the sheet in the (r, θ) plane. It covered the entire gap and also had a thickness of approximately 2 cm. A colored glass filter (OG530) was used to block stray elastic scattering from reaching the camera. Fluorescence was imaged onto a standard CCD camera and recorded on a VHS VCR. The delay between the tagging and interrogation lasers was adjusted with a digital delay generator. The maximum repetition rate was 10 Hz.

4 Results

We have applied the PHANTOMM technique to measure the instantaneous velocities in the (r,z) and (r,θ) planes. Images from both views were recorded at 40 different Taylor numbers in the range $21 \times 10^3 < Ta < 0.11 \times 10^{12}$ and for several different delays between tagging and interrogation.

Figures 1 to 5 show representative PHANTOMM line images. Figs. 1 and 2 show images in the (r,z) plane for the lowest value of the Taylor number, $Ta = 21 \times 10^3$. The lines indicate flow structure typical of the Taylor cells. The flow has a rotational laminar structure with a positive velocity close to the outer cylinder wall and a negative velocity from the center to the inner cylinder wall. This flow configuration is completely steady, and the Taylor cells cover the complete gap.

Figures 2 and 3 illustrate the flow at low Taylor numbers. The corresponding Reynolds number is about 154, which in narrow gap experiments corresponds to the wavy vortex regime, but for this experiment with a much larger gap appears to be in the regime where simple Taylor vortex flow is established. The rotation of the line about an equilibrium point is seen in the (r,z) plane (Fig. 1), and the higher azimuthal velocity is seen near the middle of the gap in the (r,θ) plane (Fig. 2). The three-dimensional characteristics of the flow are seen in the way the lines close to the outer cylinder wall are affected by the axial flow in the vertical direction.

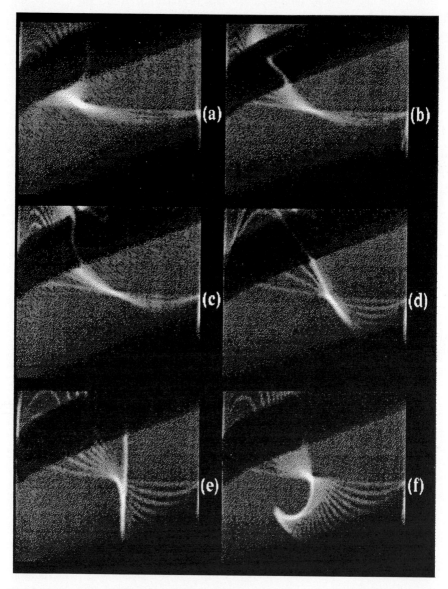

Fig. 1. Representative PHANTOMM line images for Ta=0.21×10^5 (Re=150), in the (r,z) plane. Seven seconds of delay between successive pictures.

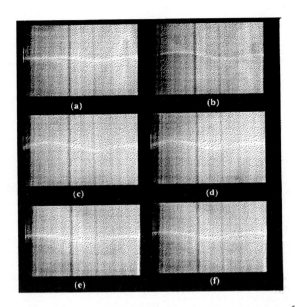

Fig. 2. Collage of representative PHANTOMM line images for Ta=0.21×10^5 (Re=154), in the (r,z) plane. $\Delta t=1$ sec between tag and interrogation. Two seconds of delay between successive pictures.

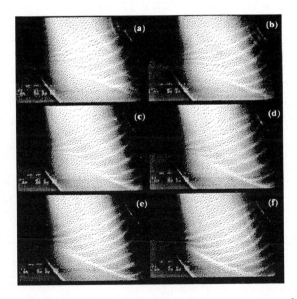

Fig. 3. Collage of representative PHANTOMM line images for Ta=0.21×10^5 (Re=154), in the (r,θ) plane. $\Delta t=20$ msec between tag and interrogation and $\Delta t=1$ sec between sucessive lines. Two seconds of delay between successive pictures.

348

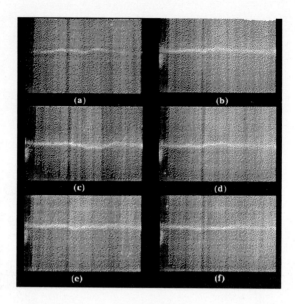

Fig. 4. Collage of representative PHANTOMM line images for Ta=2.45x10⁸ (Re=1.66x10⁴), in the (r,z) plane. Δt=20 msec between tag and interrogation. One second of delay between successive pictures.

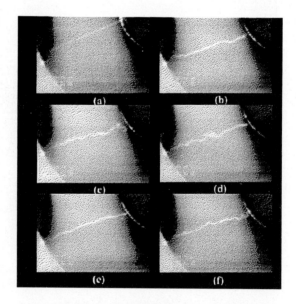

Fig. 5. Collage of representative PHANTOMM line images for Ta=2.45x10⁸ (Re=1.66x10⁴), in the (r,θ) plane. Δt=2 msec between tag and interrogation. One second of delay between successive pictures ((a) baselineimage).

As the Taylor number increases, the regularity of the flow in the (r,z) plane changes and the coherence of the Taylor cells diminishes. At the same time, fluctuations with higher frequencies appear in the flow. Figs. 4 and 5 show the flow line structures observed at higher Taylor number, for the (r,z) and (r,θ) planes, respectively. By comparing Figs. 2 and 4 and Figs. 3 and 5, it appears that the laminar rotational flow has begun to transition to turbulent flow. In Figs. 4 and 5, the delay-time between tagging and interrogation, Δt, is equal to one and two msec, respectively, much less than for Figs. 1 and 3 (Δt = one second). Nonetheless, a significant level of turbulence, indicated by shot-to-shot variation, is clearly illustrated.

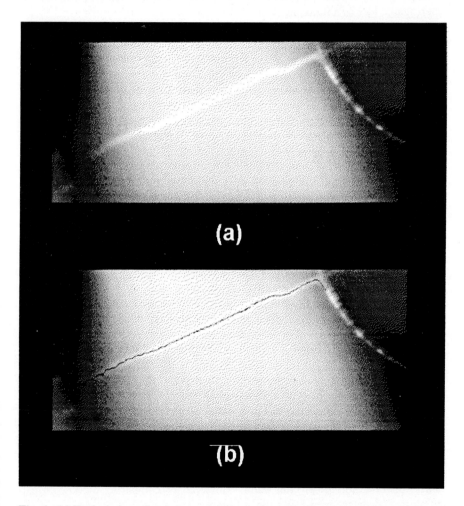

Fig. 6. (a) Typical view of an interrogated line in the (r,θ) plane, 15 ms after tagging, for Ta=5.96x10^7 (Re=8176) and (b) the corresponding fitted line position.

350

Fig. 6 shows a typical interrogated line, overlaid with an estimate of the center of the line estimated using the gray scale intensity as a function of r. The center of intensity was found using a peak searching algorithm. For the measurements in the (r,z) plane, the algorithm simply located the pixel corresponding to the maximum gray scale intensity over each vertical "slice" of data. For the measurements in the (r,θ) plane, the center of intensity was found in two steps: as a first approximation, the maximum intensity pixel in a vertical "slice" was located, as above for the (r,z) plane. Since the principal flow direction is along theta, a set of 30-40 pixels about each "vertical" maximum, normal to the radial direction, was, subsequently, evaluated. The maximum intensity pixel of each of these sets was taken as the center point of the interrogated line as a function of r.

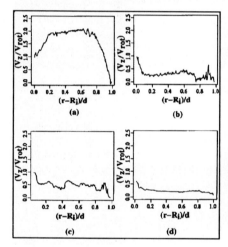

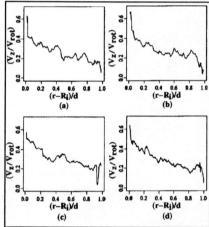

Fig. 7. Dimensionless instantaneous velocities along the gap, in the plane (r,θ). (a) Ta=0.21x10^5, (b) Ta=6.02x10^6, (c) Ta=5.96x10^7 and (d) Ta=2.45x10^8.

Fig. 8. Dimensionless instantaneous velocities along the gap, in the plane (r,θ). (a) Ta=1.63x10^9, (b) Ta=4.76x10^9, (c) Ta=2.0x10^{10} and (d) Ta=0.11x10^{12}.

The calibration of the image magnification was performed external to the experimental system because of the difficulty of performing this in-situ. We utilized an image of a grid placed in the same optical geometry used for the experiments. In this calibration, we did not apply corrections for the distortion caused by viewing the tagged lines through the cylindrical surface in the (r,z) plane. Also, we did not correct the distortion caused by the mismatch of the indices-of-refraction for air and water. However, by comparing the known gap length to that determined using the calibration measurements, we estimated the error to be less than 3%. The principal effect of the cylindrical surface is to compress the radial scale of the image very near the outer wall. The z-axis is not affected, except for a very small translation. This translation was minimized by positioning the camera at the same height as the tagging laser. A conservative estimate of the absolute velocity uncertainty introduced by these effects is less than ±5%.

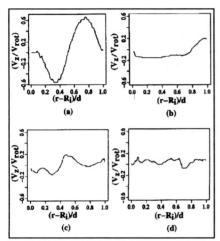 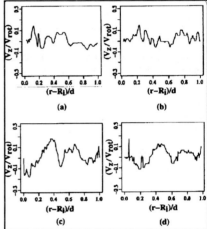

Fig. 9. Dimensionless instantaneous velocity along the gap in the (r,z) plane. (a) Ta=0.21x10^5, (b) Ta=6.02x10^6, (c) Ta=5.96x10^7 and (d) Ta=2.45x10^8.

Fig. 10. Dimensionless instantaneous velocity along the gap in the (r,z) plane. (a) Ta=4.76x10^9, (b) Ta=2.0x10^{10}, (c) Ta=0.48x10^{11} and (d) Ta=0.11x10^{12}.

Figures 7 to 10 show the instantaneous velocity profiles, non-dimensionalized by the inner cylinder wall linear velocity, for the planes (r, θ) and (r,z) for several Taylor numbers in the range of $21x10^3 < Ta < 0.11 x 10^{12}$. These figures were plotted on the same scale to illustrate the relative fluctuations in the two planes as a function of Taylor number.

Figures 7 and 8 show the velocity profiles for several Taylor numbers in the (r,θ) plane. In Fig. 7, which corresponds to Ta = 21x10^3 (Re = 154), we observe that the velocity in the center of the gap achieves a velocity two times greater than the linear velocity of the inner cylinder wall. This occurs because of the three-dimensionality and the resulting streamwise convergence, as indicated earlier. The mean velocity in the (r, θ) plane becomes less affected by the three-dimensional characteristic of the flow as the Taylor number increases. Moreover, the outflow velocity profile evolves with the Taylor number, and it achieves an almost constant slope (about 0.42) when the Taylor number becomes larger than approximately 6×10^7. This tendency is clearly seen in the velocity profiles shown in Figs. 7 and 8, suggesting that the velocity profile is tending toward an asymptotic state at high Taylor number.

Almost all the variation of the angular momentum occurs within the two wall boundary layers since surface stresses are expected to be nearly proportional to the 7/5 power of the rotational velocity, according to Smith and Townsend (1982). At a Taylor number of 21×10^3 the boundary layer in the inner cylinder wall is concentrated in a region smaller than 0.07 of the gap length. The thickness of the boundary layer shrinks when the Taylor number increases. For larger Taylor numbers, this thickness is less than 0.03 of the gap length. The boundary layer thickness on the outer cylinder wall is larger than that of the inner cylinder wall. However, the flow in this outer boundary layer appears to be more disturbed, fluctuating more intensely close to the wall. This fact suggests that flow in the boundary layer on the outer cylinder is more affected by curvature, which is expected since the curvature there is destabilizing. It must be noted, however, that optical distortions are maximum near the outer wall, and it is therefore difficult to quantify the boundary layer thickness.

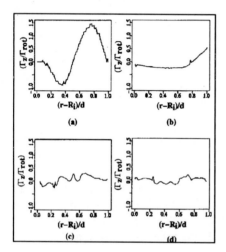

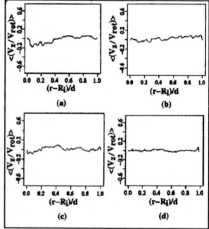

Fig. 11. Angular-momentum ratio in the plane (r,z). (a) Ta=0.21×10^5, (b) Ta=6.02×10^6, (c) Ta=2.45×10^8 and (d) Ta=1.63×10^9.

Fig. 12. Angular-momentum ratio in the plane (r,z). (a) Ta=4.76×10^9, (b) Ta=2.0×10^{10}, (c) Ta=0.48×10^{11} and (d) Ta=0.11×10^{12}.

The velocity profiles shown in the Figs. 9 and 10 correspond to the (r,z) plane measurements. The velocity profile in Fig. 9a is characteristic of a laminar rotational flow, typical of Taylor cells with a very slow circulation velocity.

When the Taylor number increases (above Ta=5.96x10^7, where Re=8,183), the characteristic laminar rotational flow disappears and is replaced by flow with larger fluctuation with irregular aspects. At all Taylor numbers, the mean velocity is zero in the (r,z) plane.

It is useful to consider the angular-momentum ratio, defined by

$$\left(\frac{\Gamma}{\Gamma_{rot}}\right) = \left(\frac{v_i r}{v_{rot} R_i}\right) \quad \text{with} \quad i = \theta \text{ or } z \tag{9}$$

where, v_i is the velocity in either the θ or z direction, v_{rot} is the linear velocity of the inner cylinder wall, and R_i is the radius of the inner cylinder.

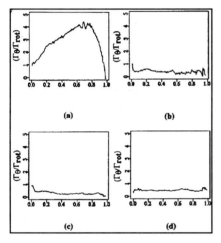

Fig. 13. Angular-momentum ratio in the plane (r,θ). (a) Ta=0.21x10^5, (b) Ta=6.02x10^6, (c) Ta=2.45x10^8 and (d) Ta=1.63x10^9.

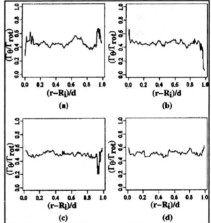

Fig. 14. Angular-momentum ratio in the plane (r,θ). (a) Ta=4.76x10^9, (b) Ta=2.0x10^{10}, (c) Ta=0.48x10^{11} and (d) Ta=0.11x10^{12}.

Figs. 11 to 14 show the angular-momentum ratio across the gap for the planes (r,θ) and (r,z) for several Taylor numbers in the range of 21x10^3 < Ta < 0.11 x 10^{12}. The mean angular-momentum ratio is seen to be approximately constant across the gap for larger Taylor numbers. For the measurements in the (r,θ) plane, the mean angular-momentum ratio is close to 0.50, for Taylor numbers larger than approximately 6x10^6 (this is the same value found by Smith and Townsend, 1982). For the measurements in the (r,z) plane, the mean angular-momentum ratio is approximately zero for Taylor numbers larger than approximately 6×10^6. These observations suggests that for high Taylor number the core flow is not affected by the curvature effect associated with the cylinder surfaces, and in this it is similar to boundary layer flow in channels.

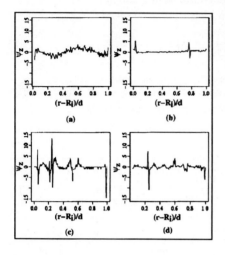 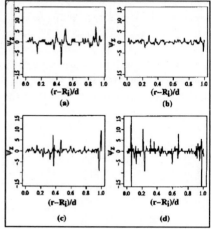

Fig. 15. Dimensionless radial velocity gradient in the plane (r,z). (a) Ta=0.21x10^5, (b) Ta=6.02x10^6, (c) Ta=2.45x10^8 and (d) Ta=1.63x10^9.

Fig. 16. Dimensionless radial velocity gradient in the plane (r,z). (a) Ta=4.76x10^9, (b) Ta=2.0x10^{10}, (c) Ta=0.48x10^{11} and (d) Ta=0.11x10^{12}.

Since the angular momentum ratio is approximately constant and equal to 0.5, we have, from the Equation 9, for Taylor numbers larger than approximately 6×10^6:

$$v_\theta = 0.5 v_{rot}\left(\frac{R_i}{r}\right) \tag{10}$$

The dimensionless radial velocity gradient is given by:

$$\Psi_j = \frac{\left(\dfrac{\partial v_j}{\partial r}\right)}{\left(\dfrac{v_j}{r}\right)} \qquad \text{where} \quad j = z \text{ or } \theta \tag{11}$$

Figs. 15 to 18 show the instantaneous dimensionless radial velocity gradient across the gap in the (r,θ) and (r,z) planes, for several Taylor numbers in the range of $21 \times 10^3 < Ta < 0.11 \times 10^{12}$. The average dimensionless radial velocity gradient is almost constant in the central flow along the gap, for both planes of measurements. The more important information obtained from Figs. 15 and 16 is the presence of Görtler vortices. One observes in these figures that, at certain locations along the gap, sharp peaks occur, accompanied by a rapid change in sign. From the results of a companion visualization study, presented in Biage et al. (1996), we can identify these peaks as being due to Görtler vortices. Figs. 17 and 18 show similar peaks in the central flow, however, no evidence of vortex pairs (i.e., Görtler vortices), can be seen.

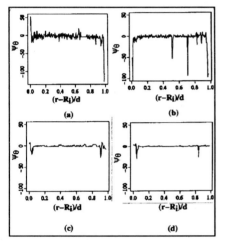

Fig. 17. Dimensionless radial velocity gradient in the plane (r,θ). (a) Ta=0.21x10^5, (b) Ta=6.02x10^6, (c) Ta=2.45x10^8 and (d) Ta=1.63x10^9.

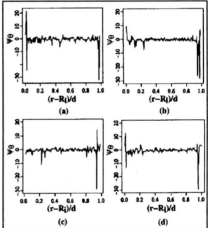

Fig. 18. Dimensionless radial velocity gradient in the plane (r,θ). (a) Ta=4.76x10^9, (b) Ta=2.0x10^10, (c) Ta=0.48x10^11 and (d) Ta=0.11x10^12.

The spatial one-dimensional and one-sided spectral density function gives further information about the flow behavior. For example, the spectral density function of a periodic signal gives a maximum in the characteristic wave number of the signal. In the case where the dynamic system has a quasi-periodic behavior, then the spectral density function exhibits several maxima that correspond to a discrete set of characteristic wave numbers. On the other hand, the spectral density function for a random signal has a continuous spectral representation over a large bandwidth of wave numbers that characterizes a disordered behavior. The degree of disorder of a system is associated with the wave number bandwidth of the spectral representation. A generalized spectral density function may be evaluated by (Bendat and Piersol, 1985, Tennekes and Lumley, 1972):

$$G_{xx}(K_i) = \frac{2}{n_d N \Delta d} \sum_{j}^{n_d} \left| X_j(K_i) \right|^2 \qquad i = 0,1,\dots, \frac{N}{2} \qquad (12)$$

where $X(K_i)$ is the K_ith Fourier component of the fluctuating part of a chosen variable, $x(r)$; j is an index corresponding to a particular instantaneous image; and n_d is the total number of frames analyzed, which was of order ten in the present work.

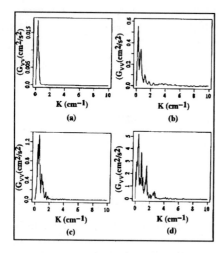

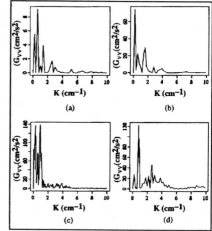

Fig. 19. One-dimensional and one-sided spectral density function in the (r,z) plane. (a) Ta=0.21x10^5, (b) Ta=6.02x10^6, (c) Ta=5.96x10^7 and (d) Ta=2.45x10^8.

Fig. 20. One-dimensional and one-sided spectral density function in the (r,z) plane. (a) Ta=1.63x10^9, (b) Ta=4.76x10^9, (c) Ta=2.0x10^10 and (d) Ta=0.11x10^12.

For the (r,z) plane, the spectral density function was evaluated using the instantaneous velocity signal, so that $x(r)=v_z(r)$. The spectral density function was also evaluated, for (r,θ) plane, but in this case the angular momentum, $x_j(r)=rv_z(r)$, was used. In the (r,θ) plane, the average velocity profile changes significantly along the gap. One fundamental wave number with large amplitude that represents the average velocity variation along the gap appears in the spectral density function. The amplitude of the fundamental wave number is much larger than the amplitudes corresponding to the wave numbers of the fluctuations in the instantaneous velocity. In this case, the qualitative feature of the spectral density function is better analyzed using the instantaneous angular momentum rather than the velocity.

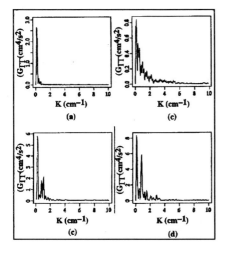 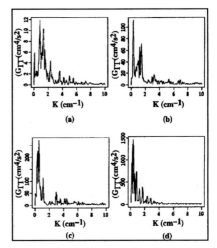

Fig. 21. One-dimensional and one-sided spectral density function in the (r,θ) plane. (a) Ta=0.21x10^5, (b) Ta=6.02x10^6, (c) Ta=5.96x10^7 and (d) Ta=2.45x10^8.

Fig. 22. One-dimensional and one-sided spectral density function in the (r,θ) plane. (a) Ta=1.63x10^9, (b) Ta=4.76x10^9, (c) Ta=2.0x10^{10} and (d) Ta=0.11x10^{12}.

Figs. 19 to 22 show the spectral density function across the gap in the (r,z) and (r,θ) planes, for several Taylor numbers in the range $21x10^3 < Ta < 0.11x10^{12}$. The flow is dominated, at low Taylor numbers, by eddies with a regular wave number ($K = 0.20$ cm^{-1}), corresponding to the Taylor cells, which have a characteristic scale with the same dimension of the gap. The spectral density function displays additional maxima with significant energy as the Taylor number is increased. As can be seen in Figs. 19 and 20, the flow becomes more irregular, with important contributions at characteristic wavenumbers greater than the fundamental. We conclude that the flow is almost perfectly periodic, for Taylor number less than order $6x10^7$. For larger Taylor numbers, the flow structure is quasi-periodic, with contributions from several characteristic wavelengths. As seen in Fig 20d, this flow structure appears to persist, even at the very highest Taylor numbers studied in this work. The measurements in the (r,θ) plane leads to similar conclusions, as can be seen in the Figs. 21 and 22.

In conclusion, we have used the PHANTOMM flow tagging technique to obtain instantaneous velocity profiles in Taylor-Couette flow over a Taylor number range from $21x10^3$ to $0.11x10^{12}$. From the computed spectral density function, and in combination with a previous visualization study, a Taylor number of 6 x 10^7 was found to characterize the transition from flow dominated by Taylor cells to one dominated by Gortler vortices located near the walls. However, even at the highest Taylor number studied, a reasonably well-defined coherent flow structure was found to persist.

5 Acknowledgments

We acknowledge the U.S Air Force Office Scientific Research and U.S. National Science Foundation for support of this research. Also, we acknowledge the CNPq (National Office of Sciene and Technology Development of Brazil) for providing support for Mr. Biage.

6 References

BARCILON, A. and BRINDLEY, J., 1984, Organized structrures in Taylor-Couette flow. J. Fluid Mech., 143,29.

BENDAT, J. S. And PIERSOL, A. G., 1986, RANDOM DATA: Analysis and measurement procedures, John Wiley & sons.

BIAGE, M.; HARRIS, S. R.; LEMPERT, W. R. and SMITS, A. J., 1996, Visualization study of Taylor-Couette flow: A description of the transition to turbulence, AIAA Paper, 96-1989.

DAHM, W.J.A. and DIMOTAKIS, P.E., 1990, Mixing at large Schmidt number in the self-similar far field of turbulent jets, J. Fluid Mech.,. 217, p. 299.

FALCO, R.E. and NOCERA, D., 1993, Quantitative Multi-Point Measurements and Visualization of Dense Liquid-Solid Flows Using Laser-Induced Photochemical Anemometry (LIPA), in: Particulate Two-Phase Flow (ed. M.C. Roco) Boston: Butterworth-Heinemann.

HARRIS, S. R., LEMPERT, W. L., HERSCH, L., BURCHAM, C. L., SAVILLE, D. A. And MILES, R. B., 1995, Flow tagging measurements of internal circulation in droplets. AIAA paper, 95-0168.

LEMPERT, W. R., MAGEE, K., GEE, K. R. and HAUGHLAND, R. P.,1995, Flow tagging velocimetry in incompressible flow using Photo-Activated Nonintrusive Tracking of Molecular Motion (PHANTOM). Exp. in Fluids., 18, 249-257.

McGRAY, J. A. and TRENTHAM, D. R. Properties and uses of photoreactive caged compounds. Annu. Rev. Biophys. Chem. 18, 239-270, 1989.

SMITH, G. P. and TOWNSEND, A. A. 1982. Turbulent Couette flow between concentric cylinders at large Taylor number. J. Fluid Mech. 123, 187.

RAYLEIGH, l. 1916, On the dynamics of revolving fluids. Scientific Papers. 6, 447-53.

TAYLOR, G. I. 1923. Stability of a viscous liquid contained between two rotating cylinders. Phil. Trans. R. Soc. Lond. A 223, 289.

TENNEKES, H and LUMLEY, J. L. A First Course in Turbulence. The MIT Press, 1972.

WEI, T.; KLINE, E. M.; K-LEE, S. and WOODRUFF, S., 1992, Gortler vortex formation at the inner cylinder in Taylor-Couette flow. J. Fluid Mech., 245, 47.

III.5 LDV-Measurements on the Chaotic Behaviour in Wide Gap Spherical Couette Flow

P. Wulf, C. Fechtmann, C. Egbers and H.J. Rath

Center of Applied Space Technology and Microgravity (ZARM), University of Bremen, 28359 Bremen, F.R.G.

Abstract. We report on a concurrent study of LDV-measurements on bifurcation scenario in rotating spherical Couette flow. As an example for a bifurcation scenario a complex route into chaos can be observed in the wide gap spherical Couette flow experimentally. By increasing the Reynolds number with the angular velocity of the driving inner sphere the flow bifurcates from laminar axisymmetric basic flow to the periodic motion of non-axisymmetric secondary spiral waves for relative large aspect ratios (Egbers, 1994). In the present study the relative wide gap width of ß = 0.50 is chosen. The spiral waves exist over a wide range of the Reynolds number. In this range a change in shape and periodicity can be detected by visualization with small aluminum flakes and also measured by Laser Doppler velocimetry (LDV-technique). At high Reynolds numbers, the flow undergoes a bifurcation to low-dimensional chaotic motion before it eventually becomes turbulent. The dynamic behaviour is discussed by spectral bifurcation diagrams, reconstructed attractors and their Lyapunov exponents as a quantitative parameter. These quantities are calculated from the measured LDV time series.

1 Introduction

The subject of hydrodynamic instabilities and the transition to turbulence has an importance for the understanding of non-linear dynamic systems. Progress in understanding instabilities, bifurcations and routes into chaos has been made primarily by focussing attention on a small number of relatively simple hydrodynamic systems like Rayleigh-Be'nard convection and the flow between two concentric rotating cylinders (Taylor-Couette flow). Furthermore, a considerable progress in understanding the first instability in the form of Taylor vortices of a viscous incompressible fluid flow between two concentric rotating spheres for small and medium gap widths has been achieved over the last two decades. The three examples just reviewed are examples of transition to turbulence through a repeated finite number of symmetry-breaking bifurcations. Especially the study of instabilities and turbulence in spherical Couette flow is of basic importance for the understanding of global astrophysical and geophysical motions. Much of the universe is filled with fluids in turbulent motion, and instabilities are quite common in planetary at-

362

mospheres. But the study of spherical Couette flow is also important for general theory of hydrodynamic stability since this flow is a natural combination of circular Couette flow at the equator and the flow between rotating disks at the poles. Another important feature of the spherical geometry is that the basic flow involves two types of symmetry, the reflection symmetry with respect to the equator and the translational symmetry with respect to the axis of rotation. Depending on the aspect ratio, both types of symmetry-breaking bifurcations can exist in the spherical Couette flow.

In this work, we consider the flow between two concentric spheres with the inner sphere rotating and the outer one at rest as illustrated in figure 1. This flow can be characterized by the following control parameters: The aspect-ratio $\beta = (R_2 - R_1)/R_1$ and the Reynolds number $Re = (R_1^2 \Omega)/v$, where R_1 and R_2 are the inner and outer radii, Ω is the angular velocity of the inner sphere and v is the kinematic viscosity. Another control-parameter coming into account is the acceleration rate $d\Omega/dt$, where t is time, because the occuring flow pattern during the transition to turbulence are also determined by the history of the flow, i.e. it depends on whether the Reynolds number is increased or decreased, quasistationary or fast.

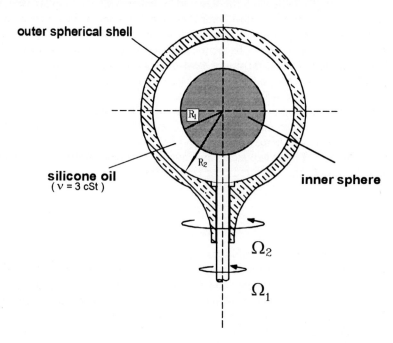

Fig. 1: Principle sketch of the spherical Couette flow model

Most of previous experimental investigations on spherical Couette flow were restricted to small and medium gap widths, where Taylor vortices exist as reported by Sawatzki & Zierep (1970), Munson & Menguturk (1975), Wimmer (1976), Yavorskaya et al. (1980), Wimmer (1981), Bühler (1985) and Bühler & Zierep (1984,

1986). Although spherical Couette flow is more relevant to astrophysical and geo-physical applications, in comparison with the Taylor-Couette flow system (e.g. Fenstermacher et al., 1979, Andereck et al., 1986, Buzug et al., 1992, 1993, v. Stamm et al., 1993, Mullin, 1993) the dynamic behaviour during the laminar-turbulent transition of the flow between two concentric spheres has been studied less (e.g. Belyaev et al., 1984, Nakabayashi & Tsuchida, 1988, Egbers, 1996).

Some new aspects of the dynamic behaviour of the spherical Couette flow during the transition to turbulence for a wide range of Reynolds numbers and for the case of the wide gap width (ß = 0.5) are presented in this study. The instabilities arising are in contrast to Taylor-instabilities (Taylor, 1923). They occur in the form of non-axisymmetric secondary waves with spiral arms, which break the spatial symmetry-behaviour of the basic flow. With increasing the Reynolds number, the number of secondary waves with spiral arms decreases, before the flow loses its stability and the flow seems to become chaotic with increasing the Reynolds number (Egbers, 1995). This transition was earlier thought to be a direct transition into turbulence without the existence of instabilities (e.g. Munson & Menguturk, 1975).

With the means of time series analysis of the LDV-measurements we want to give some additional quantitative estimates for the existence of a transition region from laminar basic flow to chaotic motion and for the onset of turbulence in the wide gap width. Experimental results connected with the problems of stability, bifurcation, non-axisymmetry, periodicity, quasi-periodicity, chaotic and turbulent motions in the spherical Couette flow will be discussed.

The common theoretical models of the transition from a laminar basic state to the turbulent one are summarized in chapter 2. The experimental apparatus and experimental methods used for this investigation are described in chapter 3. Experimental results on simultaneous flow visualization studies and the LDV-measurements on the dynamic behaviour during the laminar turbulent transition are presented in chapter 4. In chapter 5 we give some conclusions.

2 Theoretical Models of the transition to turbulence

In this chapter, we summarize previous theoretical models of the transition from laminar basic to turbulent flows: The first model was postulated by Landau (1944) presenting the following idea. If the fluid flow is in some laminar state and the stress parameter - the Reynolds number - is increased then a simple periodic oscil-lation will be excited in the flow above a critical value of the Reynolds number. Further increase in Reynolds excites another oscillation whose frequency is incom-mensurate with the first one and the process continues so that more and more mo-des are excited sequentially as the Reynolds number is increased. Each of the new modes is introduced at smaller and smaller increments of the Reynolds number so that an accumulation point is rapidly reached where a continuum of discrete modes is excited. This sequence of events for the Landau scenario is shown in figure 2. In this model, turbulence may be thought of in these terms as an infinite sum of dis-crete incommensurate frequencies with random phase. These ideas are consistent

364

with a statistical description of turbulence, which has been successfully used in the development of turbulence models.

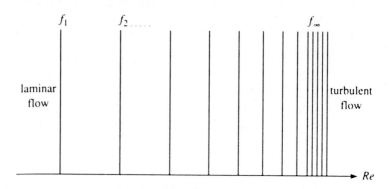

Fig. 2: Principle sketch of the Landau model of the transition from laminar to turbulent flow (from Mullin, 1993)

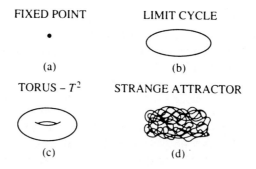

Fig. 3: Principle sketch of the Ruelle-Takens model of the laminar turbulent transition (from Mullin, 1993)

More recently, Ruelle & Takens (1971) suggested that the laminar basic state can be thought of as an attracting fixed point in the phase space. Thus, if the system is disturbed from this state, then any resulting trajectory will head back towards the fixed point, as shown in figure 3(a). If the Reynolds number is increased in this case, a time-periodic mode will be excited in the flow which can be represented by a limit cycle in phase space (figure 3(b)). A further increase in the Reynolds number will lead to the appearance of a second frequency which is incommensurate with the first one. This will give a torus in phase space, as shown in figure 3(c). As the Reynolds number is increased furthermore, a third frequency begins to appear which would form a three frequency torus. Ruelle & Takens (1971) suggested, however, that this is a non-robust situation and it is more typical for a so-called strange attractor to be in the solution structure at this stage, as illustrated in figure 3 (d). This model is said to be a typical description for classes of equations for which the Navier-Stokes equations are believed to belong. This has not yet been proved for all fluid-dynamic problems but evidence is mounting that, at least in certain confined flows - as i.e. Taylor Couette flow is - it might be a better descriptor than the Landau picture. But the question is, which route into chaos will happen in the spherical Couette flow and which model is suited to describe this transition?

3   Experimental methods

3.1   Spherical Couette flow apparatus

An experimental set-up for spherical Couette flow was constructed, consisting of an inner sphere rotating concentrically inside another rotating outer spherical shell. A principle sketch of the experimental apparatus is illustrated in figure 4: The outer sphere (R_2 =40.00±0.02 mm) is composed of two transparent acrylic plastic hemispheres. The upper hemisphere has a spherical outer surface of about $0^0 < \Theta < 110^0$ to investigate whether the occuring flow patterns are symmetric with respect to the equatorial plane or not. The inner sphere is made out of aluminium having the five various radii R_1, to variate the aspect ratio ß from 0.08, 0.18, 0.25, 0.33 to 0.5.

Generally, both spheres can be rotated independently by means of two belt-drives. The eccentricity between outer and inner sphere could be minimized to ±0.015 mm using high precision bearings and shafts. Two different synchronous motor drives were adapted, which provide a uniform and stable rate of rotation up to n = 850 rev/min with fluctuations of less than 1.5%. They variate the Reynolds numbers of the inner and outer sphere respectively in a range from 0 to Re = 10^5. The revolutions were calibrated using an optical tracking system (optical coupling). Most of our experiments were carried out by increasing the Reynolds number quasistationary from zero. However, because the occuring flow structures depend also on initial conditions (Egbers 1994), the acceleration rate for both spheres could be variated. Experiments with the spherical Couette flow system were carried out in a laboratory condition, where the temperature could be kept uniformly up to ±0.3 C. Since the viscosity of the silicone oils, which were used as working fluids, vary by approximately 2% / ^{0}C, the temperature must be precisely controlled and measured in order to have a well-defined Reynolds number. A temperature accuracy of ±0.15 C was achieved for all six temperature sensors (PT 1000) used in our experiment: For measuring the fluid temperature, three temperature sensors are installed just below the outer surface of the inner sphere at latitudes $\Theta = 10^0$, $\Theta = 80^0$ and $\Theta = 160^0$. These temperature data were transferred from the rotating sphere to the stationary part by a slip-ring system. For investigations withonly the inner sphere rotating three other temperature sensors are installed at the inner surface of the outer sphere at latitudes $\Theta = 0^0$, $\Theta = 45^0$ and $\Theta = 90^0$. Our measurements were carried out with the silicone oil M3 at a constant temperature of 25^0 C. As tracer particles small aluminium flakes were used. The concentration by weight was 0.05%. The viscosities of the working fluids with tracer particles were measured with a VOR-rheometer (BOHLIN-Reologi AB, Sweden). An effect on the viscosity was not detected.

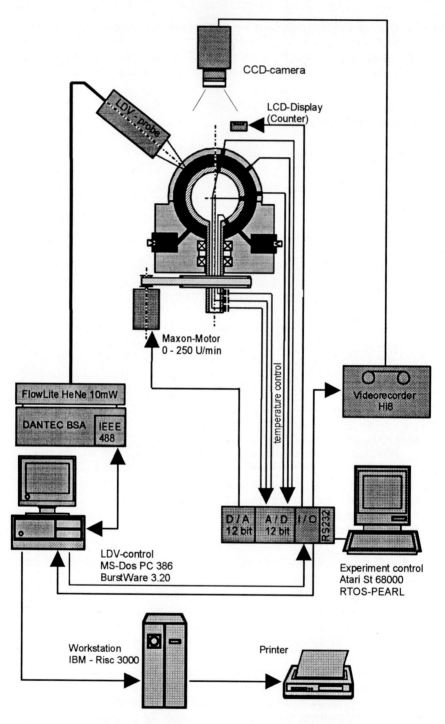

CCD-camera

LCD-Display
(Counter)

LDV - probe

Maxon-Motor
0 - 250 U/min

temperature control

FlowLite HeNe 10mW

DANTEC BSA | IEEE
488

Videorecorder
Hi8

D / A
12 bit | A / D
12 bit | I / O | RS232

LDV-control
MS-Dos PC 386
BurstWare 3.20

Experiment control
Atari St 68000
RTOS-PEARL

Workstation
IBM - Risc 3000

Printer

Fig. 4: Principle sketch of the experimental apparatus with the applied LDV-measuring
technique and the observation technique with CCD-camera

3.2 LDV measuring system and visualization methods

Because the investigated flow structures appearing in the spherical Couette flow during the laminar-turbulent transition are non-axisymmetric, it was necessary to use an observation-technique, which provides a simultaneously flow visualization of both the azimuthal and the meridional flow. Therefore, a combination of the following two visualization methods was used: To investigate the flow structures occuring in the meridional cross-section of the spherical annulus, a slit illumination technique is employed. In addition, a system with a fiber-optic is applied to visualize the polar region with the azimuthal waves. In this way, the cellular structure of the occuring vortices in the meridional plane as well as the azimuthal and polar behaviour of the arising flow pattern can be obtained (Egbers, 1994). Photographs or prints from video-records were taken.

For the application of laser-Doppler-velocimetry (LDV) on the spherical Couette flow experiment, a special traversing system has been constructed to mount the LDV optic probe on the spherical Couette flow experiment as depicted in figure 4, while the laser and the transmitter are mounted apart from the experiment on a mounting bench. The traversing system consists of a high-precision bow with a traversing sledge and a traversing table. The traversing sledge is capable of moving the optic probe in meridional direction ($0^0 < \Theta < 110^0$) and the traversing table is capable of moving in radial direction over a range of 60 mm in order to determine the meridional dependence of the velocity and to obtain velocity profiles. The laser-Doppler-velocimeter system used in our experiments, consists of a 100 mW Ar-laser, a transmitter unit and a 1-D fiber flow optic probe (DANTEC-Electronics, Denmark). A frequency shift is added by the Bragg-cell to one of the beam pair to allow for measurements of reversing flows. The optic probe with a fibre optic cable is connected to the transmitter via manipulators. The backscattered light is focussed on a photomultiplier tube. Data records of each measurement in the centre of the gap consisted of 32000 validated Doppler bursts. More details on the calculation of the time series is contained in the work of Fechtmann et al. (1996). Parameters characterizing our LDV-system and the traversing data are listed in table 1.

Ar-laser	100 mW
focal length	160 mm
wavelength	514.5 nm
beam spacing	38 mm
beam diameter	2.2 mm
number of fringes	22
probe volume:	
length	0.4 mm
diameter	0.05mm
fringe spacing	2.18 mm
Bragg-cellfrequency	40 MHz
meridional traversing angle	$0^0 < \Theta < 110^0$
radial traversing distance	$0 < r < 60$ mm

Table 1: Parameters characterizing the LDV-measuring system and traversing data

368

As tracer particles for the LDV-measurements we use polysterene spheres (Bangs-Lab.) with a diameter of 1.6 μm. The concentration of tracers in the working fluid was 0.01 Vol.%. The application of the LDV-technique on the spherical Couette flow experiment requires an optical correction for the accurate determinations of the probe volume locations and for the interference fringe spacing due to refraction effects of the spherical outer surface. Because the probe is adjusted in radial directi-on, the optical axis of the front lense of the probe passes perpendicular through the spherical outer surface. Thus, the correction for the two laser beams, which are in the same plane, could be calculated for a cylindrical surface. However, the fact, that a small probing volume is needed to produce sufficient spatial resolution, which could be obtained only by a large intersection angle, the small-angle approximation cannot be used in this case. The correction method used in this work for the case of large intersection angles was derived in our previous work (Egbers & Rath, 1996).

4 Experimental results

4.1 Visual observations of wide gap instabilities

In our previous investigation on the stability of spherical Couette flow (Egbers & Rath, 1995), it was found, that the laminar basic flow between two concentric sphe-res, where only the inner sphere rotates and the outer one is at rest, loses its stabili-ty not only in the form of Taylor-instability. The instabilities occuring depend strongly on the aspect ratio. For wide gap widths (ß > 0.33) considered here, the first instability manifests itself as a break of the spatial symmetry and a new non-axisymmetric secondary wave mode was observed, which spreads from the pole to the equator. With increasing the Reynolds number, the number of secondary waves with n spiral vortices decreases. For ß = 0.33, secondary waves with n = six, five and four spiral arms were found, while in the gap with an aspect ratio of ß = 0.5 waves with n = five, four or three spiral arms exist. Flow visualization studies on the development of the spiral wave flow up to chaotic motion were carried out du-ring this work. The results are depicted in figure 7(a)-7(f) in chapter 4.3. The flow regimes, their spatial states with non-axisymmetric and non-equatorial symmetric secondary waves were characterized by measuring the phase velocities and the displacement between the northern and the southern hemisphere of the arising secondary waves (Egbers 1994). Furthermore, the spherical Couette flow in wide gap widths shows the well-known transition phenomena of hysteresis and non-uniqueness, depending on the acceleration rate of the inner sphere (Liu et al., 1996). In order to give some additional quantitative estimates on the dynamic be-haviour of these secondary waves up to chaotic motions, the dynamic behaviour of the meridional velocity component (time series) was recorded in the center of the gap. From these records, the autocorrelation functions, power spectra, reconstructed attractors, Lyapunov exponents and the fractal dimensions were calculated. Furthermore, by counting the minima and maxima of the oscillating velocities in meridional plane we obtain the spectral and frequency bifurcation diagrams. The results will be presented in the following chapters.

4.2 Bifurcation diagrams

The flow dynamics are obtained by LDV-technique with a sampling rate more then ten times higher then the highest significant frequency of the dynamic system. The obtained times series are filtered by an Optimal filter in the frequency domain to preserve the dynamic information from being eliminated by. An additional filtering is done by a Savitzky-Golay filter with an order of four filtering (Fechtmann et al., 1996). From the time series the power spectrum is calculated. The peaks in the power spectrum quantify the periodic behaviour of the flow. The frequencies of the peaks with a high significance are collected and presented in spectral bifurcation diagram as a function of the Reynolds number. The "traces" in the diagram give a good representation on how the motion changes from basic flow over different forms of periodic motion to chaos. The bifurcation scenarios are illustrated in the form of a velocity bifurcation diagram (Fig. 5) and in the form of the a frequency bifurcation diagram (Fig. 6). As can be seen from both diagrams, the flow under-goes a Hopf-bifurcation from laminar basic state to non-axisymmetric periodic motion in the form of five secondary waves at Re = 1170 (Egbers,1994). The nor-malized frequency is $f_1 = 1.28$. At Re=1360 the shape of the periodic flow pattern changes to five rosettes with a normalized frequency of $f_1 = 1.02$. At Re=1550 the rosettes begin to pulsate from the equator to the poles. Finally the rosette modes decay to a spiral four arm structure at Re=1700. This structure is also periodic with a normalized frequency of $f_1 = 1.28$ and begin pulsation at Re=1700. For Re $\geq$ 2080 only a week periodic motion can be observed by visualization of the fluid motion. However, the bifurcation diagram shows that there is still periodic motion in the flow. The motion becomes clearly chaotic after Re $\geq$ 2300.

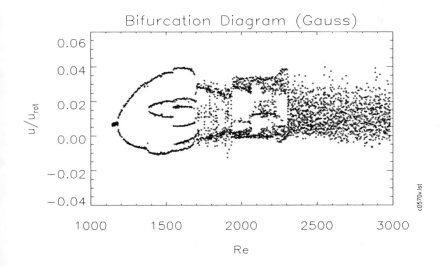

Fig. 5: Bifurcation diagram for the position of $\Theta = 70°$ as a function of the Reynolds number. The meridional velocities are normalized by the velocity of rotation ($\beta = 0.5$).

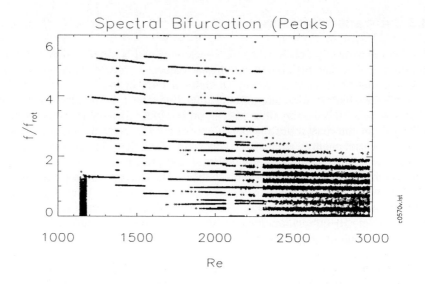

Fig. 6: Spectral bifurcation diagram for the position of $\Theta = 70°$ as a function of the Reynolds number. The frequencies are normalized by the frequency of rotation ($\beta = 0.5$).

4.3 Nonlinear dynamic behaviour

To detect the dynamic behaviour using linear signal processing techniques the most common and very usefull way is to analyse the time series of a representative velocity component, which is in our experiment the meridional one, and to construct the power spectra and the autocorrelation function. The spectrum gives a measure of the amount of power in a given frequency band over a selected frequency range. The autocorrelation function for a peridic signal for example is itself periodic and can often give a less confusing representation of the data than the power spectrum. It therefore provides a useful complementary representation of the data. Irregularity in the time series give rise to a decay in the autocorrelation function and the rate of decay gives a measure of the degree of irregularity. However, this method is linear and therefore fails when chaotic motion occurs. The extraction of quantitative estimates from the autocorrelation function can be problematical and it is usually used as a complement to the power spectrum.

But to detect the nonlinear behaviour during the route into chaos, in addition it is necessary to use nonlinear signal processing techniques to analyse the time series. Therefore, an alternative way to investigate the bifurcations and the nonlinear dynamics is to reconstruct the underlying attractors which represent the dynamic behaviour. This is done by the method of time delayed coordinates introduced by Takens (1980). The appropriate time delay τ and the embedding dimension dimE are estimated by the calculation of the fill factor and the integral local deformation (Buzug & Pfister, 1992) and singular value decomposition (Broomhead & King,

1986). The reconstructed attractor is projected onto an orthonormal basis that is calculated from the covariance matrix of the time-delayed time series. This can be interpreted as an optimal stretching of the attractor on the primary axes of an dimE -dimensional hyper ellipsoid containing the averaged spatial extension of the attractor in the phase space. It is then possible to apply an estimation calculation of the Lyapunov exponents onto the attractor. The sign and magnitude of the greatest exponent is providing information about the dynamic behaviour of the system. The Lyapunov exponent spectrum is calculated by a linear approximation of the local flow in the phase space (Holzfuss, 1987).

The results of our measurements on the nonlinear dynamic behaviour during the laminar turbulent transition in the wide gap spherical Couette flow are illustrated in figure 7(a)-7(f) as a function of the Reynolds number: The diagrams depicted contain the following informations on the dynamic behaviour during the transition from laminar basic flow to chaotic motion:

o velocity time series   o pictures of flow visualization
o autocorrelation function   o power spectra
o Lyapunov exponents   o reconstructed attractors
o correlation integral   o correlation dimension

4.4 Route into chaos

The summary of the analysis of the nonlinear dynamic behaviour of wide gap spherical Couette flow - the route into chaos - is illustrated in figure 8: With increasing the Reynolds number quasistationary in small steps with $\Delta Re = 5$, the spherical Couette flow bifurcates from a laminar basic state via the first Hopf bifurcation ($Re_{crit1} = 1170$) into a single periodic mode with n = 5 spiral waves having a frequency $f_1 \approx 1.33 \ f_{rot}$ (limit cycle). The second bifurcation ($Re_{crit2} = 1490$) even is a Hopf bifurcation, but in this special case no quasi-periodic motion was found. In this case, the flow shows n = 4 stable spiral waves and a second limit cycle with frequency ratio $f_2 \ / \ f_1 = 4/5$ occurs, which seems to yield to a Neirmark-Sacker bifurcation (Fechtmann et al, 1996). Thus, this special bifurcation scenario is in contrast to those models derived in chapter 2. With further increasing the Reynolds number over $Re_{crit3} = 1640$, the third Hopf bifurcation was obtained connected with period-doubling phenomena. The flow consists of n = 3 spiral and pulsating waves and a new frequency f_{31} occurs, which has a frequency ratio of $f_{31} \ / \ f_2 = 4/3$, but also a new frequency f_{32} was obtained with $f_{32} \approx 0.38 \ f_{rot} = 0.5 \ f_{31}$ (period-doubling). The next Hopf bifurcation occurs at $Re_{crit4} = 1760$ in the form of spiral asymmetric and pulsating waves. the basic frequency f_4 corresponds to the frequency f_1 of the mode with n = 5 spiral waves pulsating with the frequency f_{32}. In analogy to the frequency change from the stable mode with n = 5 to the mode with n = 4 the next Hopf bifurcation occurs at $Re_{crit5} = 2120$ showing additional pulsating frequencies. With a further increase of the Reynolds number ($Re_{crit6} = 2500$), the flow shows the typical behaviour of chaotic motion and a strange attractor is generated (chaos). The corresponding autocorrelation function decreases to zero, the power spectra decrease with 1/f and the Lyapunov exponents are positive.

372

Laminar basic flow

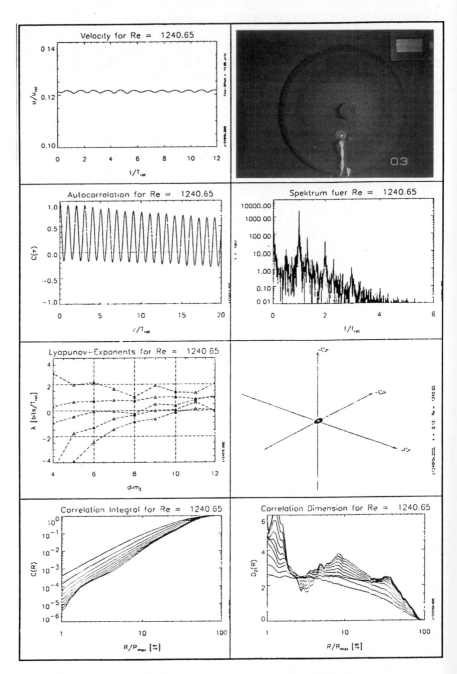

Fig. 7(a): Velocity, flow visualization, autocorrelation, spectrum, Lyapunov-exponents, reconstructed attractor, correlation integral and dimension (Re = 1240)

Spiral wave flow (n = 5, stable)

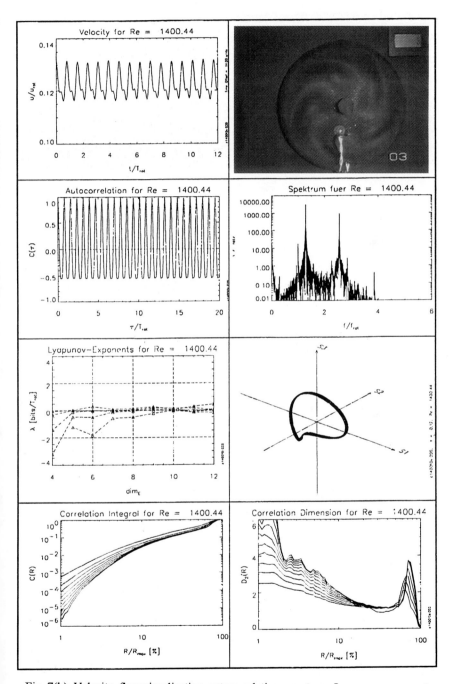

Fig. 7(b): Velocity, flow visualization, autocorrelation, spectrum, Lyapunov-exponents, reconstructed attractor, correlation integral and dimension (Re = 1400)

374

Spiral wave flow (n = 4, stable)

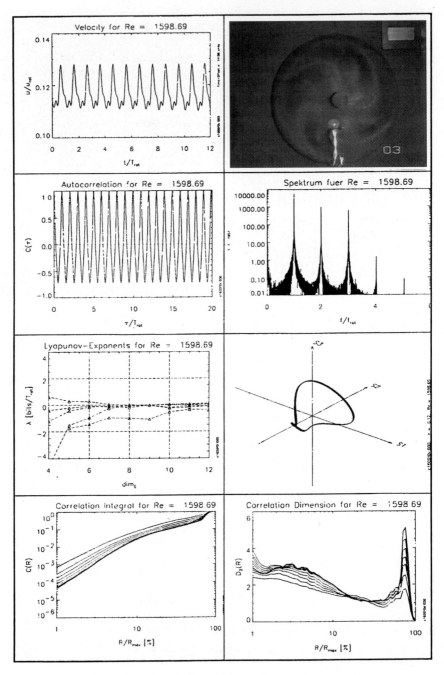

Fig. 7(c): Velocity, flow visualization, autocorrelation, spectrum, Lyapunov-exponents, reconstructed attractor, correlation integral and dimension (Re = 1598)

Spiral wave flow (n = 3, pulsating)

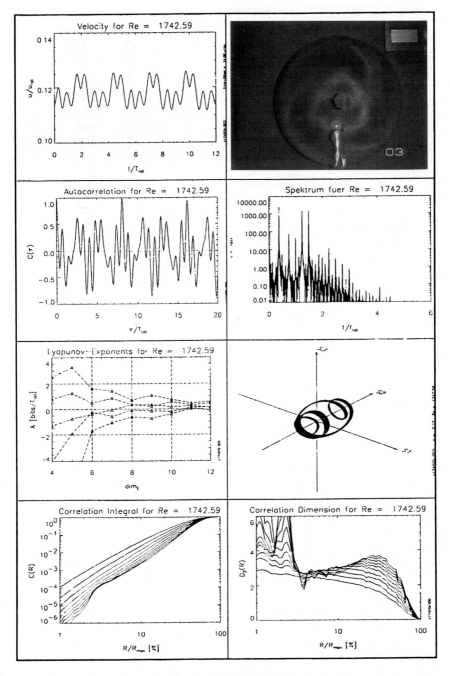

Fig. 7(d): Velocity, flow visualization, autocorrelation, spectrum, Lyapunov-exponents, reconstructed attractor, correlation integral and dimension (Re = 1742)

Spiral wave flow (n = 4, pulsating, asymmetric)

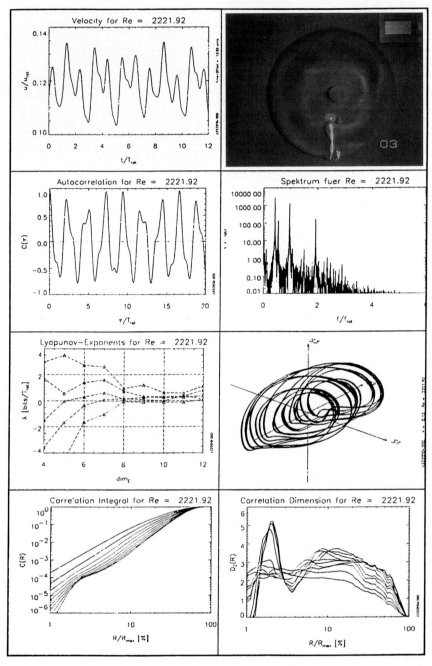

Fig. 7(e): Velocity, flow visualization, autocorrelation, spectrum, Lyapunov-exponents, reconstructed attractor, correlation integral and dimension (Re = 2222)

Chaotic flow

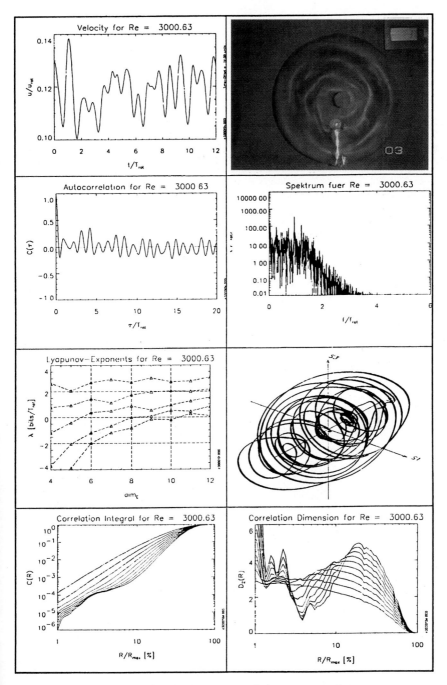

Fig. 7(f): Velocity, flow visualization, autocorrelation, spectrum, Lyapunov-exponents, reconstructed attractor, correlation integral and dimension (Re = 3000)

Route into chaos

spiral waves (n=5, stable)

1. limit cycle
   $f_1/f_{rot} \approx 1.33$

$\Rightarrow$ $1244 < Re < 1490$

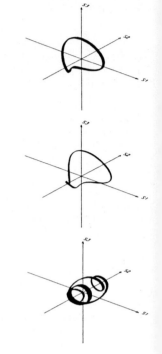

spiral waves (n=4, stable)

2. limit cycle
   $f_2/f_{rot} \approx 1.03$ $f_1/f_2 \approx 5/4$
$\Rightarrow$ $1490 < Re < 1640$

spiral waves (n=3, pulsating)
period doubling
   $f_{31}/f_{rot} \approx 0.76$ $f_{31}/f_2 \approx 4/3$
   $f_{32}/f_{rot} \approx 0.38$ $f_{31}/f_{32} \approx 2$
$\Rightarrow$ $1640 < Re < 1760$

spiral waves (pulsating, asymmetric)

1760 < Re < 2120   2120 < Re < 2500

n=5   n=4

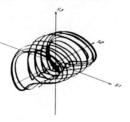
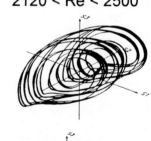

chaos
Re > 2500

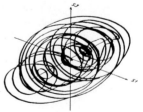

Fig. 8: The route into chaos in wide gap spherical Couette flow as a function of the Reynolds number

5 Conclusions

We have considered the nonlinear dynamic characteristics of the laminar-turbulent transition of the spherical Couette flow for the wide gap width ß = 0.5. In contrast to previous investigations, we could detect a transition region, where instabilities in the form of secondary waves with spiral arms exist, which break the spatial symmetry. The LDV-technique could be applied successfully for measurements of time series of the meridional velocity component. In this way, the time-dependent behaviour of the flow with spiral waves during the transition to turbulence was obtained. The system shows a rich variety of bifurcations to chaos, e.g. the periodic, quasi-peridic and chaotic regimes were identified. Simultaneous LDV-measurements and flow visualization studies give a good impression of the spatial-temporal development of the flow during the transition to weak turbulence. The route into chaos could be identified as a new scenario of Hopf bifurcations. The analysis of autocorrelation function, power spectra, reconstructed attractors and fractal dimensions shows that the laminar basic flow loses its stability to chaotic motion in a way, which is neither identical with the picture of Landau (1944) nor with the model of Ruelle & Takens (1971). For the spherical Couette flow, we found a new route into chaos. For a complete description of these transitions, however, we need to conduct simultaneous high-resolution LDV-measurements of the fully three-dimensional velocity vector. Furthermore, we need to extend our investigations on chaotic behaviour to other wide gap aspect ratios. Nevertheless, further investigations must be carried out to give more informations about the development of turbulence as a function of the gap width and the meridional coordinate.

Acknowledgements. This work is supported by the Deutsche Forschungsgemeinschaft (DFG). The authors thank Dipl.-Ing. Peter Wulf, Dipl.-Ing. Wolfgang Beyer and cand. phys. Carsten Fechtmann for their assistence during this work.

References

Andereck, C.D., Liu, S.S. and Swinney, H.L. (1986): Flow regimes in a circular Couette system with independently rotating cylinders. J. Fluid Mech., vol. 164, 155-183

Belyaev, Yu.N., Monakhov, A.A., Scherbakov, S.A. and Yavorskaya, I.M. (1984): Some routes to turbulence in spherical Couette Flow. in: Kozlov,V.V. (ed.): Laminar-Turbulent Transition. IUTAM-Symp. Novosibirsk/USSR, Springer

Broomhead, D.S. & King, G.P. (1986): Physica D 20, 217

Bühler, K. (1985): Strömungsmechanische Instabilitäten zäher Medien im Kugelspalt. VDI-Berichte, Reihe7: Strömungstechnik Nr.96

Bühler, K. and Zierep, J. (1984): New secondary instabilities for high Re-number flow between two rotating spheres. in: Kozlov,V.V. (ed.): Laminar-Turbulent Transition. IUTAM-Symp. Novosibirsk/USSR, Springer, Berlin, Heidelberg, New York, Tokyo

Bühler, K. and Zierep, J. (1986): Dynamical instabilities and transition to turbulence in spherical gap flows. in: Comte-Bellot, G. and Mathieu, J.: Advances in turbulence. Proc. 1st Europ. Turb. Conf., Lyon, France, 1-4 July, Springer, Berlin,Heidelberg, NewYork,London, Paris, Tokyo

380

Buzug, Th. & Pfister, G. (1992): Physical Review A 45, 10

Buzug, Th., v. Stamm, J. and Pfister, G. (1992): Fractal dimensions of strange attractors obtained from the Taylor-Couette experiment. Phys.A, 191, 559

Buzug, Th., v. Stamm, J. and Pfister, G. (1993): Characterization of period-doubling scenarios in Taylor- Couette flow. Physical Review E, 47, no. 2, 1054-1065

Egbers, C.(1994): Zur Stabilität der Strömung im konzentrischen Kugelspalt. Dissertation, Universität Bremen

Egbers, C. and Rath, H.J. (1995): The existence of Taylor vortices and wide-gap instabilities in spherical Couette flow. Acta Mech. 111, 3-4, 125-140

Egbers, C. & Rath, H.J. (1996): Developments in Laser Techniques and Applications to Fluid Mechanics (Eds.: R.J. Adrian, D.F.G. Durao, F. Durst, M.V. Heitor, M. Maeda, J.H. Whitelaw), Springer,45-66

Fechtmann, C., Wulf, P., Egbers, C. & Rath, H.J.; 1996; LDA-Messungen zum chaotischen Verhalten der Strömung im konzentrischen Kugelspalt; Lasermethoden in der Strömungsmeßtechnik, 5. Fachtagung der GALA, 11.-13. Sept. Berlin, vol. V, pp 17.1-17.7

Fenstermacher, P.R., Swinney, H.L. and Gollub, J.P. (1979): Dynamic instabilities and the transition to chaotic Taylor vortex flow. J. Fluid Mech., 94, part1, 103-128

Holzfuss, J. (1987): Doctoral thesis, Universität Göttingen,

Landau, L.D. (1944): On the problem of turbulence. C. R. Acad. Sci. USSR 44, 311

Liu, M., Blohm, C., Egbers, C., Wulf, P. & Rath, H. J.; 1996; Taylor vortices in wide spherical shells.; Phys. Rev. Lett., vol. 77, No. 2, pp 286-289

Mullin, T. (1993): The nature of chaos, Oxford Science Publications

Munson, B.R. and Menguturk, M. (1975):Viscous incompressible flow between concentric rotating spheres. Part 3: Linear stability and experiments. J. Fluid Mech., vol. 69, 705-719

Nakabayashi, K. and Tsuchida,Y. (1988): Spectral study of the laminar-turbulent transition in spherical Couette flow. J. Fluid Mech., vol. 194, 101-132

Sawatzki, O. und Zierep, J. (1970): Das Stromfeld im Spaltzwischen zwei konzentrischen Kugelflächen, von denen die innere rotiert. Acta Mech. 9, 13-35

v. Stamm, J., Buzug, Th. and Pfister, G. (1993): Frequency locking in axisymmetric Taylor-Couette flow. Physical Review E., submitted

Takens, F. (1980): Lecture Notes in Mathematics, 898, Springer

Taylor, G.J. (1923): Stability of a viscous liquid contained between two rotating cylinders. Phil.Trans. A223, 289-293

Wimmer, M. (1981): Experiments on the stability of viscous flow between two concentric rotating spheres. J. Fluid Mech., vol.103, 117-131

Yavorskaya, I.M., Belyaev, Yu.N. and Monakhov, A.A. (1975): Experimental study of a spherical Couette flow. Sov. Phys. Dokl., vol. 20, 4, 256-258

Yavorskaya, I.M., Belyaev,Yu.N., Monakhov, A.A., Astaf'eva, N.M, Scherbakov, S.A. and Vvedenskaya, N.D. (1980): Stability, non-uniqueness and transition to turbulence in the flow between two rotating spheres. IUTAM-Symposium, Toronto

Chapter IV

Free Flows and Flames

Chapter IV

IV.1 Three-Component Laser-Doppler Measurements of the Confined Model Flow Behind a Swirl Nozzle

B. Lehmann[1], C. Hassa[2], J. Helbig[1],

Deutsche Forschungsanstalt für Luft- und Raumfahrt (DLR) e.V.
Institut für Antriebstechnik
[1]Abt. Turbulenzforschung, D-10623 Berlin
[2]Abt. Aerothermodynamik und Verbrennung, D-51147 Köln
Germany

ABSTRACT. Results are reported of three-component laser-Doppler measurements in the highly complex flow field behind a model nozzle as used for the atomization of liquid fuel in gas turbines. The flow field consists of two concentric air streams leaving a concentric dual nozzle system both with slightly different but equally directed swirls. The outer ring flow forms a 1 mm thick ring film which interacts strongly with the inner flow and the surrounding flow field.

Mean and rms velocities were measured as well as double and triple cross-correlations of the three velocity components. Frequency analysis exhibited a strong coherent fluctuation of the radial velocity component near the nozzle exit. The measured data are needed to improve atomization and combustion effectiveness and for numerical modelling of the flow.

1 Introduction

For aeroengine combustors the method of fuel preparation is of importance for all performance aspects related to mixing and flow homogeneity in the combustor primary zone: combustion efficiency, the required pressure loss of the combustor, smoke production and gaseous emissions of unburnt hydrocarbons, carbon monoxide and nitrogen oxides. Since the installation of international emission regulations in 1981, the use of airblast atomizers, which were particularly successful in limiting smoke production, became customary throughout the industry. In these devices, the pressure difference between the customer casing and flame tube is used to produce a swirling air flow. It forms in some extent the recirculation zone for the aerodynamic flame stabilisation, but also disrupts the low velocity fuel film due to the shearing forces between liquid and air and subsequently atomizes and disperses the liquid.

To obtain mathematical models of airblast atomizers which can be used for design purposes, physical submodels for the turbulent swirling air flow, the atomization and the turbulent dispersion of the polydisperse spray have to be found and validated. Then their interaction can be investigated in a practical combustor. For that purpose, an airblast atomizer nozzle was built (Blümcke, E., Eickhoff, H. and Hassa, C., 1989) to provide good experimental conditions. Axial symmetry of the swirling flow, optical access to the atomizer lip and the swirl channels had to be realised whereas the essential features of practical nozzles had to be retained.

Such features were a good atomization and mixing, a wide range of flow and combustion stability and low emissions. Investigations of the cold two-phase flow (Hassa, C., Blümcke, E., Brandt, M. and Eickhoff, H., 1992) with low and high liquid loading (Brandt, M., Hassa, C. and Eickhoff, H., 1992) and with combustion (Hassa, C., Deick, A. and Eickhoff, H., 1993) have already been made. Hassa (1994) gave a detailed analysis of the production and the dynamics of liquid fuel spray behind a combustion nozzle. Finally, Hassa and Arold (1995) reported on periodical droplet diameters' variations which resulted in a related frequency spectrum of the droplet events.

This contribution reports on the use of a 3-D laser-Doppler measurement system for the detailed investigation of the aerodynamics, especially near the nozzle exit. As it is well known, such flows exhibit anisotropic turbulence and often diverge from ideal axisymmetry near the geometric centerline. Therefore one- or two-component laser Doppler measurements are problematic to be applied for the measurement of the three velocity and six Reynolds-stress components by means of subsequently rotate the systems. Such kind of two-component measurements have been reported by Hoffmeister, Hertwig, Kreul, Kretschmar and Erler (1996) who calculated stress-tensor data for the total Reynolds stress tensor from their dual-component results.

2 Experimental Facility

2.1 Nozzle and Flow Chamber

The experimental facility consisted of the nozzle device blowing into a cylindrical plexiglass flow chamber with 128 mm inner diameter and a length of 300 mm. The air was taken from a high-pressure relief system and expanded to almost the ambient pressure before entering the nozzle. At the end of the nozzle chamber, a circular end plate left open a 30 mm wide slit between the plate and the cylinder body so that the air flow could leave it in a preferably radial direction. The pressure drop in the nozzle structure was about 3 percent of the flow's

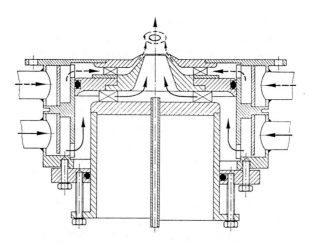

Fig. 1. Plane view into the co-swirling dual-flow atomization nozzle

total pressure. Two partial flows were introduced into the nozzle structure by means of two hoses each.

The two air-supply hoses for each partial air stream led to the nozzle plenum chambers and were coupled together and connected with two separate rotameter measurement devices. In between they were connected to the common exit of a seeder device which served to introduce solid tracer particles into the nozzle flows. The seeder-air flow rate was kept low enough to be neglected compared with the main nozzle flow rates.

The nozzle structure is shown schematically in Fig. 1. It consisted of an inner convergent nozzle mouth with a diameter of 17 mm being concentrically enveloped with a ring-film nozzle the air-film flow of which was nearly one mm thick. The thickness of the concentric wall separating the two different jets reduced at knife-edge dimensions at the nozzle exit plane. Both jets obtained a uniquely directed swirl introduced by means of two separately arranged swirl vanes being passed by the two partial air streams in the radial direction. The flow volume rates were different. Nearly 34 m^3/h passed through the central nozzle and the ring-film flow was controlled to be 20 m^3/h. This resulted in nearly 75 m/s maximum axial and circumferential velocities forming a swirl number of nearly s=1 for the different jets.

2.2 Optical Devices and Signal Processing

A three-component laser-Doppler measurement facility was built up of a DISA 55X dual-colour dual-component anemometer with a common 40 MHz Bragg cell for the two components. An additional single-component system consisting of the very first generation DISA 55L „flow direction adapter" was rectangularly directed onto the common measurement volumes of the dual system. Only one of the two Bragg cells of this adapter system was exited for a frequency shift.

For the dual-component system the green and blue light colours of a water-cooled 5 Watt Argon laser were applied. They served to measure the axial w and the circumferential v velocity components (Fig. 2). The radial u component was measured with the green light of an air-cooled 300 mW Argon laser. Signal separation of the two green component devices could be easily attained by means of working in forward-scatter modes for all three components.

The light-beam separation angles were adjusted between about 4.5 and 7.5 degrees and the focus lengths of the emission optics were 500 mm. The signal-receiving photomultipliers were positioned at about 300 mm away from the measurement volumes. All emitting and receiving optical systems for the three velocity components were fixed on a common support which could be vertically shifted together with the big Argon laser by means of a computer control. The light for the third component was led from the stationary small laser to the 55L system by means of an optical glass fibre device. Adjustment devices enabled the probe volumes to be sensitively adjusted for maximum effective intersection as required for the correlation measurements.

Most of the discussed data have been collected and analysed by means of three BSA enhanced FFT processors for Doppler signals. They were controlled for coincidence conditions under the exactly working hardware coincidence mode in most of the discussed cases. Some data had been acquired and analysed by means of a self-developed FFT processor the possibility of which to run the FFT with up to 2048 data points was helpful in certain cases.

2.3 Optical Access to the Flow Chamber

The flow chamber with an inner diameter of 128 mm was fixed vertically on a horizontal two-coordinate traversing device. Together with the vertical traverse of the optics, the total space of interest could be covered.

An undisturbed optical access for the light beams of the two-component device and for scatter-light reception was achieved by means of two 250 mm long and about 25 mm wide glass windows pasted in opposite positions on the related slits of the cylinder wall. In this way the inner wall contour of the chamber exhibited only small deviation from circular shape.

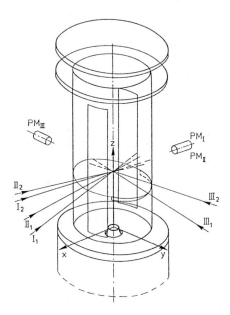

Fig. 2. Nozzle and flow chamber with optical arrangements and coordinates

Thus, the two-component optics could be traversed in a parallel direction with the chamber axis and with the windows as well as rectangularly to them in parallel with a chamber diameter. The necessity of also traversing the radial u component device in parallel with a chamber diameter would imply severe optical deadjustments due to the 3 mm thick plexiglass wall of the flow chamber.

This problem could be overcome by replacing the chamber wall by a piece of overhead projector foil in those regions where the optics had to be traversed. Because of its low thickness, the foil exerted only small disturbances to the measurement volume even if the light beams passed it under rather oblique orientation.

In order to keep the mechanical stability of the wall's contour, the chamber wall was provided with circumferential slits in the regions where the light beams had to penetrate the wall, and the slits were covered by the foil pasted on the inner wall of the chamber. The signal scatter light, however, was received under the forward direction through the original wall thickness and needed only small re-adjustments of the receiving optics in the course of traversion.

2.4 Seeding Technique

Different seeding materials had been tested for the discussed application. These materials ranged from liquid glycerine particles to solid particle powders of SiC, TiO_2 and SiO_2. In all cases, the most numerous particle diameters were below 1 micrometer, and the maximum diameter was about 2 micrometers. The most suitable choice was the use of the SiO_2 particles. This product, named „monospher 800" of the German Merck company is a nearly monodisperse powder around the particle diameter of 800 nm +/- 20 % and the particle shape is nearly spherical. Similar powders with other particle diameters are available. It may be supposed that the monodispersity would be the main reason of suitability.

This material was dispersed into a low volume rate air flow with a self-developed seeder. The seeder enables small portions of the powder to be separated almost continuously from a supply vessel and its dispersion by means of a high-velocity air jet from a 0.6 mm diameter wide air nozzle. The primarily seeded air was introduced into the two main air streams of the experimental nozzle through two pressure apertures whose diameters were adapted to the different flow rates of the main streams. In this way an optimum continuity of seeding concentration for the two nozzle air streams should be attained.

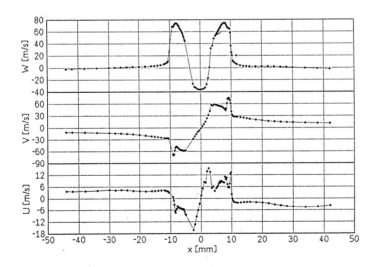

Fig. 3. Mean velocity profiles at z=2 mm behind nozzle exit

3 Measurement Results

3.1 Results from the Nozzle Exit Plane

3.1.1 Mean Velocities

Though no fuel atomization took place in the present case, the velocity field near the nozzle was of primary interest. Therefore, and also as the initial flow condition for numerical flow-field modelling, the velocity profiles in close proximity to the nozzle had been measured. This was at about 2 mm downstream of the nozzle exit.

In Fig. 3 the measured mean exit velocity profiles are plotted. The mean axial component W shows maximum velocities of about 75 m/s and in the center a backflow region with nearly 40 m/s maximum negative velocity. This backflow is caused by the swirl which forms a low-pressure region along the nozzle axis. It is demanded for flame stabilisation.

The mean tangential (circumferential) velocity V exhibits a quite characteristic difference between the kernel and the ring-film flow. The ring flow has a higher swirl than the kernel flow, whereas the mean axial w component of the film flow is obviously somewhat smaller than for the central flow. Thus we can estimate from the W and V velocities (Fig. 3) that the swirl number is about s=1.15 for the film flow and s=0.75 for the inner flow. The mean radial velocity U is almost an order of magnitude lower than V and W.

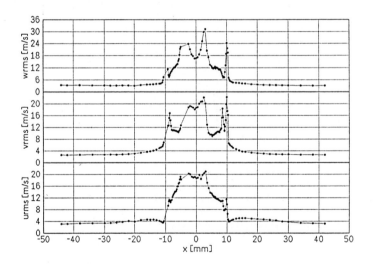

Fig. 4. Turbulence intensities at z=2 mm

It must be mentioned that there is a lack of velocity data around the positions x=-10 mm and x=-4 mm. Due to a 1 mm thick pin which supported the slit chamber wall, the measuring light beams were blocked, and under coincidence conditions no data were collected for the three components. Therefore, at z=2 mm, the profiles are more complete and more detailed for the positive x coordinate being, therefore, the subject of further discussion.

The linear slope of the U profile across the nozzle axis and the slight non-linear slope of the V profile point out a deviation of the flow field from solid-body rotation behaviour. For increasing radial displacement, the flow field is more and more influenced by the shear layer along the inner nozzle wall. An intermediate maximum of the U profile accompanied by a minimum of the V profile are due to the wake of the knife-edged wall that separates the inner and the ring-film air streams.

A more detailed and finer structured shape can be recognised in the U profile compared with the V and W profiles in the region near x=10mm where the film flow interacts with the center flow. This might be due to the orientation of the optical probe volumes with regard to the film flow's contour and dimensions. In the course of the optical traverse, the u component measurement volume passed the film flow contour under tangential orientation.

The common optical probe volume of the v and w system with the length of more than 1.5 mm penetrated the film under rectangular orientation and thus could suffer from an integration effect along its own length. One might deduce therefore that the spatial resolution of the u component in respect to the film thickness must be better than for the v and w components.

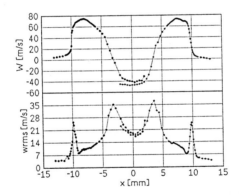

Fig. 5. Test of the optical resolution under differently outligned probe volume at z=1.5 mm. Half profiles at left-hand side: probe volume penetrates the film flow perpendicularly, at right-hand side: probe volume tangentiates the film

This interpretation contradicts the plots in Fig. 4 where the vrms and the wrms profiles seem to have at least the same spatial resolution as the urms profile. The sharp double peaks of the vrms plot exhibit a much higher spatial resolution than the length of the probe volumes would allow.

The only explanation for this is the fact that the discussed measurements have been collected under triple coincidence conditions for the three components. The discrimination effect of coincidence condition is a reduction of the probe volumes to the location and the dimensions of the common intersection area. That means that the effective probe volume has the dimensions of the optical probe volumes' diameters, which were of the order of 0.15 mm.

Fig. 5 shows the result of a check concerning the orientation of the probe volume in respect to the circular film flow. Due to the geometric conditions of the experimental devices it was possible to carry it out only for the w component. This component was measured at z=1.5 mm for the tangential (right-hand side half profile) and for rectangular (left-hand side half profile) orientation of the probe volume with respect to the flow film. No coincidence condition was applied. The half profiles of the mean velocities as well as of the wrms values do not show any characteristic difference concerning the spatial resolution of the measurement.

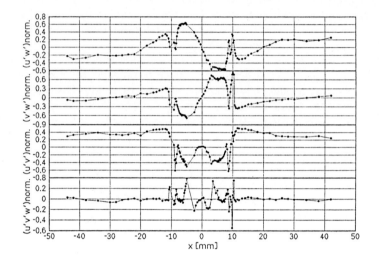

Fig. 6. Normalised cross-correlation profiles at z= 2mm

3.1.2 Correlation Results

Cross-correlation coefficients were measured and calculated of the velocity fluctuations u', v', w' and named e.g. u'v', meaning the time average of the product u'v'. Normalisation was executed by dividing them with the products of the related components' rms values.

In Fig. 6 the three normalised double cross-correlation coefficients of the velocity components and the triple correlation coefficients are plotted. All the evaluated correlation coefficients are zero in the center of the nozzle. Maximum u'v' correlation is attained just at the outer edge of the ring film flow at x=10 mm. All double correlation coefficients attain maximum variation in the region of the inner nozzle flow and near the film flow. Correlation values are obtained of 0.9 maximum and frequently are of the order of +/-0.6. The v'w' correlation has a positive maximum at the outer edge of the film flow and a negative peak at its inner edge where also the u'v' values peak to considerable correlation.

Antisymmetry can be stated for the normalised u'w' and v'w' correlations, whereas the u'v' correlation coefficient has a symmetrical profile referred to the x=0 position. Far off the nozzle area the normalised u'w' and u'v' tend to absolute values between 0.2 and 0.4, since the v'w' correlation coefficients tend to zero.

The normalised triple correlation u'v'w' slopes are more complicated and more variable. A certain symmetry of the variations with respect to the x=0 position can be recognized.

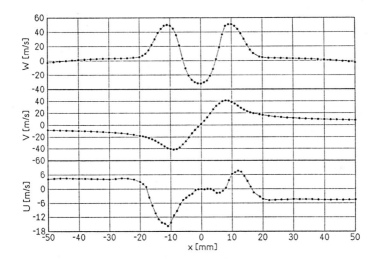

Fig. 7. Mean velocities at z=15 mm

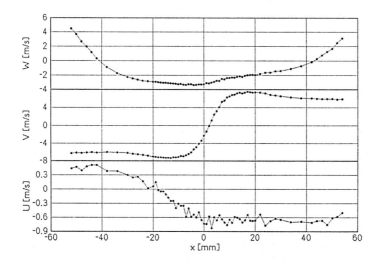

Fig. 8. Mean velocities at z=100 mm

392

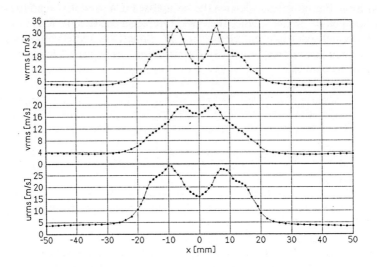

Fig. 9. Turbulence intensities at z=15 mm

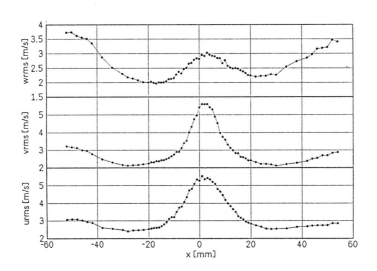

Fig. 10. Turbulence intensities at z=100 mm

3.2 Further Development of the Flow Field

Additional flow field data have been measured in the chamber cross-sections z=15 mm and z=100 mm downstream of the nozzle. Fig. 7 and Fig. 8 show the plots of the mean velocities at the two axial positions. The velocity values reduce by a factor on the order of 10 up to the z=100 mm position. The backflow region exists even at that position along a diameter of 80 mm. The tangential velocity V basically keeps its shape and spreads its maxima for increasing z.

In the near-nozzle region up to z= 15 mm, a divergence angle of the flow can be estimated from the maxima of the U and W profiles of Fig. 3 and Fig. 7. It is about 22 degree if taken from the W profiles but the V profiles make no divergence to be recognised.

The U profiles in Fig. 7 and Fig. 8 show a considerable deviation from having the expected antisymmetrical slope. It is caused by a small deviation of the mean flow's rotation center from the origin of the x-y-coordinate system which is bent to the flow chamber.

The estimated delay of both centers was less than 1mm in the z=100 mm plane, which is obviously enough for the relatively large shift and deformation of the U velocity profile. If this 1mm value is interpolated to the z=15 mm plane, only the order of 0.15 mm of center deviation seems to be responsible for the strong deformation of the U profile near x=0 mm of Fig. 7.

Further experiments showed that it was almost impossible, to adapt the traverse for a reproducible check of the flow rotation center. It is assumed that the observed unstable behaviour of the data is caused by the length of the measurement probe volume together with the low u velocity to be measured.

Fig. 9 and Fig. 10 exhibit the fluctuation rms values of the velocity components. At the z=15 mm position the wrms and the urms values dominate the v velocity component but are adapted to it or reduced to relatively lower values at z=100 mm.

This differently intensive exchange of turbulent energy between the components is also expressed by the correlation coefficients in Fig. 11 and Fig. 12. The u'w' correlations are higher than the other ones which expresses a stronger interaction between these two components than between the others.

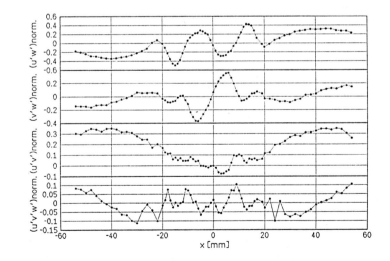

Fig. 11.Cross correlation profiles at z=15mm

The structure of the triple correlation profiles seems to be still influenced by the mixing process of the different jets at z=15mm but considerably equalised at z=100 mm. In spite of the low triple correlation values at z=100 mm, its profile's shape scales with the chamber diameter and shows that the flow field has attained a structure at this position which is dominated by the chamber dimensions and no longer by the features of the nozzle.

Due to the considerable deviations of the U profiles from antisymmetry (Fig. 7 and Fig. 8), a check was to be made in order to consider the influence of the flow center delay to the correlation profiles. The result is also shown in Fig. 12. At z=100 mm only small deviations were exerted to all four correlation profiles in the region near x=0 if the traverse was delayed by 2 mm away from the geometrical center. The correlation data far off the flow rotation center were practically not influenced.

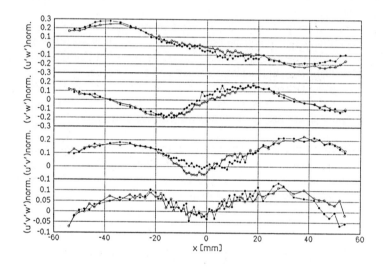

Fig. 12. Cross correlation profiles at z=100 mm. Traverses of closed and open symbols' profiles differ with about 2 mm of flow rotation-center distance, closed symbols profile nearly passing the center

3.3 Detection of Instability Effects

An inspection for flow-field instabilities was done in the course of the measurements by means of frequency analysis of the velocity fluctuations. In fact, there were found very distinct fluctuation frequencies of the spectra in the vicinity of the nozzle.

Fig. 13 shows the superimposed plots of ten linear frequency spectra of the u component as could be measured at the z=15 mm and x=10 mm position. The strong and reproducible frequency peak describes a strong fluctuation of only the u component, whereas similar results for the v and the w components could not be obtained. If a frequency peak could be observed at all for these components, it was highly smeared and diffuse in the spectrum. Obviously similar frequency spectra have been measured by Hoffmeister et. Al. (1996) by means of hotwire techniques.

These observations may be related to the results of Hassa and Arold (1995), who detected a periodicity of the droplet arrival events. They showed by means of Fig. 14 periodic characteristics of the detected droplet rates as well as of the droplet diameters and in Fig. 15 a related frequency spectrum of the droplet arrival sequences and tried to explain it as a result of the original liquid film instability in the course of the atomization process. The velocity spectra of Fig. 13 give also probability that the droplet statistics are caused by the aerodynamic flow field instability the origin of which is not yet clear.

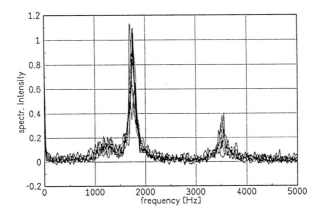

Fig. 13. Frequency spectra of the radial u fluctuations, measured at z=15 mm and x=10 mm, 10 spectra superimposed

A similar frequency spectrum as in Fig. 13 was measured without the flow chamber and with the nozzle flows blowing as free flows. This check was done in order to investigate whether the fluctuation coherence would possibly be a resonance effect of the chamber. Compared with the chamber conditions and for the same volume flow rates, the frequency peaks were shiftet to somewhat higher frequencies. In addition the experiment showed clearly that the second frequency peak is not a higher harmonic of the first and most intensive one.

In the course of further experiments the partial volume flows were varied for frequency analysis in the freely blowing nozzle flow and at the same position as chosen for Fig. 13. If the inner partial flow was reduced and the outer ring flow volume was kept constant, both frequency peaks are shifted to lower frequencies. At the same time the higher frequency peak reduced its intensity and vanished totally at 50% inner flow volume rate whereas the lower frequency peak reduced its bandwidth and became very sharp.

No remarkable effect was exerted to the frequency spectrum if the inner volume flow rate was kept constant and the one of the ring film was reduced. If both partial flows were reduced by the same amount, both frequency peaks reduced their frequencies in proportion to the volume flow rates.

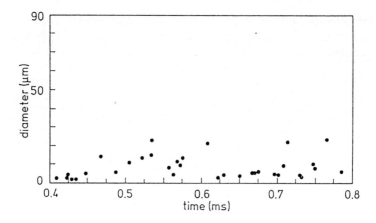

Fig. 14. Droplet arrival history at z= 10.5 mm, x=10.5 mm, measured, Hassa and Arold (1995)

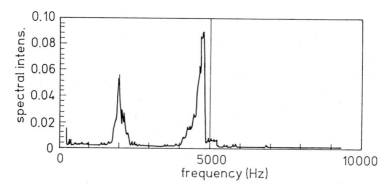

Fig. 15. Frequency spectrum deduced from the droplet arrival history, Hassa and Arold (1995)

It must be concluded from the experiments, that the inner partial flow of the nozzle dominates the fluctuation behaviour especially of the radial velocity component. The lower frequency peak seems to be caused by a primary instability of the inner flow, whereas the interaction with the outer ring-film flow produces obviously the higher frequency peak.

Frequency spectra were also measured at z=100 mm away from the nozzle. At this position, no risidual KHz frequencies were detected but low frequency peaks were found near 20 Hz.

4. Conclusions

The three component laser Doppler measurements of the flow field behind the atomization nozzle with two concentrical, separately swirled air streams with equal swirl directions enable us to conclude that

- in spite of relatively large optical probe volumes with more than 1.5 mm of length and working in forward-scatter mode, measurements with a high spatial resolution were possible to be made also in the shear region of a not much more than 1 mm thick film flow area,
- it is is supposed that such a high resolution could only be attained due to the applied triple-coincidence technique, which causes a reduction of the effective probe-volume dimensions since the coincidence check would meet exclusively the same tracer particle for all three components of the velocity events,
- the radial mean U profile is very sensitive to a small deviation of the traverse from the actual rotational flow center,
- the double and triple cross-correlation coefficients of the three velocity components' fluctuations are widely insensitive in respect to the afore-mentioned deviation,
- a very coherent instability was detected with a frequency of about 1.7 KHz measured from the radial u velocity component at z=15 mm behind the nozzle exit, whereas a similarly strong fluctuation behaviour of the v and the w velocity components could not be detected. At z=100 mm only low frequencies on the order of 20 Hz were analysed,
- the frequency results may cause supposition of a still not analyzed aerodynamic instability mechanism especially connected with a swirl flow,
- the coherent velocity fluctuations seem to influence the arrival history of the droplet events and their diameters, as observed by Hassa and Arold (1995).

The measurement results are of interest for the development of more effective atomization nozzles and for numerical modelling of the flow.

5. References

Blümcke, E., Eickhoff, H., Hassa, C. 1989, Untersuchungen zur turbulenten Partikeldispersion an einer Luftstromzerstäuberdüse, VDI-Berichte Nr 765, pp. 635-644, VDI-Verlag GmbH, Düsseldorf, 1996, Germany

Brandt, M., Hassa, C., Eickhoff, H. 1992, An Experimental Study of Spray-Gasphase Interaction for a Co-Swirling Airblast Atomizer, Proc. Eighth Annual European Conference on Liquid Atomization and Spray Systems, Sept./Oct. 1992, Amsterdam, The Netherlands.

Hassa, C., Blümcke, E., Brandt, M., Eickhoff, H. 1992, Experimental and Theoretical Investigation of a Research Atomizer/Combustion Chamber Configuration, Proc. International Gas Turbine and Aeroengine Congress and Exposition, June 1992, Cologne, Germany.

Hassa, C., Deick, A., Eickhoff, H. 1993, Investigation of the Two-Phase Flow in a Research Combustor under Reacting and Non-Reacting Conditions, AGARD Conference Proc. 536 on 'Fuels and Combustion Technology for Advanced Aircraft Engines, May 1993, pp. 41-1 to 41-12, Fiuggy, Italy.

Hassa, C., 1994, Experimentelle Untersuchung der turbulenten Partikeldispersion in Drallströmungen, Thesis, Research Report No. 94-20, Deutsche Forschungsanstalt für Luft- und Raumfahrt, Institut für Antriebstechnik, Cologne, Germany

Hassa, C., Arold, M., 1995, Investigation of Droplet Concentration Fluctuations in a Research Spray Combustion Chamber, <u>Proc. 11th European Conference of ILASS-Europe on Atomization and Sprays, Nürnberg Messe GmbH, D-90471 Nürnberg, Germany</u>, edited by K. Bauckhage, Univ. of Bremen, Germany

Hoffmeister, M., Hertwig, K., Kreul, K.-J., Kretschmar, H., Erler, K., 1996, Modellierung von turbulenten Drallstrahlen für Brennkammern von Flugtriebwerken, VDI-Fortschritt-Bericht, Reihe 7: Strömungstechnik, Nr. 291, VDI-Verlag GmbH, Düsseldorf, 1996, Germany

IV.2 On the Relationship Between the Formation Number and Passive Scalar Pinch-Off in Starting Jets

H. Johari, D. Dabiri, A. Weigand and M. Gharib

1 Mechanical Engineering Department, Worcester Polytechnic Institute, Worcester, MA 01609, USA

2 Center for Quantitative Visualization, Graduate Aeronautical Laboratories, California Institute of Technology, Pasadena, CA 91125, USA

Abstract. A series of experiments were conducted in a water tank to investigate the behavior of starting vortex ring in starting jets at Reynolds numbers up to 2×10^4. The starting vortex separates from the rest of jet in the near-field. The velocity vectors were measured via the digital particle image velocimetry while the passive scalar concentrations were observed by the laser sheet technique. The measured Formation number was in the range of 4.5 to 5.5, consistent with previous low Reynolds number results. The Formation number appears to be a robust parameter for characterizing the pinch-off of vorticity in starting jets.

Keywords. Starting Jets, Formation Number, Particle Image Velocimetry

1 Introduction

Steady, round jets have been investigated extensively in the past because of their wide ranging applications in industrial settings as well as their utilization in scientific studies. The starting process for jets is much less well-understood and only a few studies have attempted to characterize impulsively-started jets. Abramovich and Solan (1973) examined the laminar starting jet experimentally and modeled the flow as a laminar jet combined with a spherical vortex cap. Their experimental data corroborated the model predictions and the speed of the jet front was found to be about one half of the steady jet at the same location. Numerical studies conducted by Kuo et al. (1986) of laminar starting jets produced a Reynolds number scaling different than that of Abramovich and Solan. A model for the turbulent starting jet was developed by Witze (1980); it was based on Turner's (1962) model for a starting plume. The model, in agreement with measurements, indicated that the jet tip in the turbulent case also advances at one half speed of a steady jet at the same location. Lahbabi et al. (1993) have verified this scaling using non-intrusive image processing techniques applied to images of a passive scalar being carried by the jet.

In a number of recent experiments on starting jets, it has been observed that the starting vortex pinches off and separates from the rest of the flow. It is interesting that this phenomenon has not been reported previously in the literature on starting jets, presumably due to the lack of detailed data on the velocity (or vorticity) field. The pinch-off and subsequent separation of the starting vortex from the rest of the jet has been characterized by two different means. Gharib *et al.* (1994) have utilized the Digital Particle Image Velocimetry (DPIV) method to quantify the total circulation ejected from the source as well as that contained within the vortex core, and to follow these quantities in time. At a certain point in time, the circulation associated with the vortex core reaches a constant value while the total circulation in the flow field is increasing (due to the continued ejection of fluid at the source). Since the pinch-off time cannot be very accurately estimated from the vorticity contours extracted from the DPIV data, the time τ that corresponds with the total ejected circulation equal to the constant starting vortex circulation has been selected as the characteristic pinch-off time. The (running) average velocity of the source $\bar{V}(t)$ divided by the source diameter d has been used to non-dimensionalize the characteristic time τ, as indicated below.

$$\frac{\bar{V}\tau}{d} = \frac{1}{\tau}\int_0^\tau V(t)dt \frac{\tau}{d} = \frac{1}{d}\int_0^\tau V dt$$

This dimensionless time is referred to as the Formation number F. For a variety of nozzle velocity time-histories, the measured Formation number has a mean value of approximately 4 (Gharib *et al.* 1994). For an impulsively-started flow of velocity V_o, the Formation number will be $V_o\tau/d$. The time evolution of circulation along with the vorticity contours during and after pinch-off are shown Fig. 1.

In a separate effort, the planar laser-induced fluorescence (LIF) technique was utilized by Kouros *et al.* (1993) and Johari *et al.* (1995) to investigate the dispersion of a passive scalar in starting jets. For the case of impulsively-started flow, nozzle Reynolds numbers $V_o d/\nu$ ranging from 5×10^3 to 2×10^4 were achieved in a pressure-driven system with nozzle diameters of 0.64, 1.3, 2.5 cm. In these experiments, the starting vortex, as marked by a passive scalar (fluorescent dye), also separated from the ensuing jet. LIF images of the starting vortex just after the pinch-off and separation are shown in Fig. 2. The separation time was chosen at the instant that pure ambient fluid existed across the entire width of the flow and the starting vortex was completely isolated from the rest of the jet. The normalized time for the separation of the starting vortex, derived from the passive scalar images, had values ranging from 10 to 40 and appeared to vary significantly among the different set of runs in the these experiments (Johari *et al.* 1995).

Interestingly, the normalized separation time derived from passive scalar experiments does not agree with the Formation number obtained from the circulation data. In fact, the former exhibits strong (jet) Reynolds number dependence, whereas the Formation number does not appear to vary strongly with the Reynolds number, based on the vortex core circulation, in the range of a few

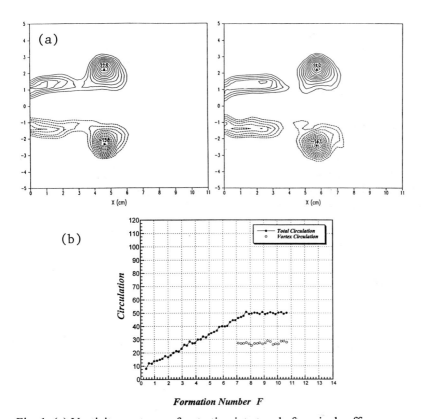

Fig. 1 (a) Vorticity contours of a starting jet at and after pinch-off
(b) Vortex core and total circulation as a function of Formation number

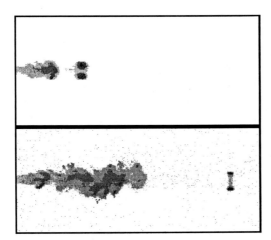

Fig. 2 LIF images of an impulsively-started jet at a nozzle
Reynolds number of 1.5×10^4

thousands (Gharib *et al.* 1994). There are two plausible reasons why the observed Formation numbers differ substantially from the normalized pinch-off time of passive scalar. First, the Formation number may exhibit strong Reynolds number dependence once it is varied over a considerable range. Second, the two distinct definitions used for the Formation number and the passive scalar pinch-off time may make any direct comparison between the two inappropriate.

To resolve these issues, a piston-driven apparatus was designed which was capable of achieving higher Reynolds numbers, i.e., comparable to those in the impulsively-started pressure-driven case. At the same time, the exit diameter was made large enough such that accurate DPIV measurements could be attained. In this manner, both DPIV and LIF measurements could be performed on the flow generated from the same apparatus. The primary objectives of this project were: (i) to investigate the Formation number resulting from high Reynolds number starting jets, and (ii) to examine any possible relationship between the Formation number and the pinch-off of a passive scalar in starting jets.

2 Experimental Setup

2.1 Apparatus

The experiments were conducted in a horizontal free-surface water tank 90 cm wide and 60 cm deep. The dimensions of the tunnel were much greater than the vortex ring so that wall effects were negligible. The flow generating apparatus is shown in Fig. 3 and consists of three main sections, a piston-cylinder assembly, a plenum, and an ejection tube. The piston has a 10.2 cm diameter and is driven by a stepping motor. The maximum piston travel is also 10 cm; however, only about

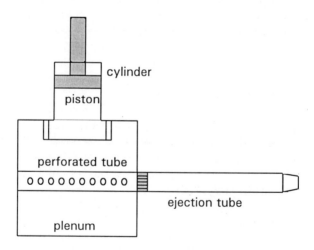

Fig. 3 Schematic of the apparatus

8 cm of travel were utilized in the present experiments. The piston forced the fluid into the plenum; the fluid escaped through a perforated PVC tube. The PVC tube had a fill factor of unity such that the sum of perforated area was equal to the cross-sectional area of the tube. The flow passing through the PVC tube then entered the ejection tube via a honeycomb-screen element. The ejection tube was made of clear acrylic and had a 5.1 cm internal diameter and a length of 52 cm. The last 2.4 cm of the tube was tapered on the outside, creating a 15° angle with the tube wall. This last section will be referred to as the 'nozzle' even though the tube's inside diameter was constant.

The piston motion was controlled by a computer such that a constant acceleration was maintained until the piston reached the end of its travel. A linear variable displacement transducer was used to record the piston motion. Thus, the piston velocity increased linearly with time. Since the ejection tube's diameter was one-half of the piston-cylinder diameter, the velocity at the tube exit was also linear with time and 4 times greater than the piston velocity. Two piston 'ramp' rates of 6.63 and 2.63 cm/s^2 ($\pm$ 2%) were selected in the present study. The piston travel durations were 1.5 and 2.4 s for the two ramp rates, respectively, corresponding to fluid slug lengths of approximately $5.9d$ and $6.0d$.

2.2 Optical Setup

The flow was seeded either with a fluorescent dye for the passive scalar runs or with silver-coated glass spheres with an average diameter of 14 μm. The flow was illuminated by a 1 mm thick light sheet, generated by an argon-ion laser and a cylindrical lens. The laser sheet was coincident with the jet symmetry plane and revealed side views of the starting jet. The laser beam was shuttered to reduce blurring of the images. The image exposure times were 1.5 and 2 ms for the large and small piston ramp rates, respectively. Flow images were acquired by a CCD camera (768×480) and recorded on laser disk. The area imaged was $2.1d$ in the axial direction and $2.7d$ transverse to the tube exit. Either the immediate near-field, from the tube exit to $2.0d$, or farther downstream, $2.2d$ to $4.3d$, were imaged on the camera.

The particle images were interrogated in pairs using the cross-correlation method of Willert and Gharib (1991). The time differences between consecutive exposures were 2 and 5 ms for the large and small piston ramp rates, respectively. A window size of 32×32 pixels and a step size of either 8×8 (75% window overlap) or 16×16 (50% window overlap) pixels were chosen, resulting in 96×60 or 48×30 velocity vectors, respectively. Circulation measurements obtained from the same images processed by the two different step sizes were quite consistent. According to Willert and Gharib (1991), the location of cross-correlation peak can be resolved with an accuracy of 0.01 pixels. Then, velocity vectors and vorticities could be measured with uncertainties of 1 and 3 percent, respectively. The temporal sampling rate of the measurements is 15 Hz for the current system.

The primary quantity of interest in the present study is the circulation Γ ejected from the source and that associated with the starting vortex ring. Circulation was calculated by taking the line integral of velocity along a closed path. Two different closed paths were chosen in order to observe any differences between the circulation values resulting from the choice of path. The first path consists of a rectangle originating from the nozzle center and extending along the axis of symmetry to an axial position beyond the starting vortex. The radial extent of the rectangle was about 6 cm from the origin at nozzle center for the near-field images and 8 cm for the downstream images. As the flow evolved, the axial extent of this path also increased to always include the starting vortex completely.

The second method of determining circulation takes advantage of the calculated vorticity contours. The vorticity contour at 10% of the local peak vorticity was chosen as the path for the small piston ramp rate (8% contour for the large ramp rate). These contour levels provided circulation values which were quite consistent with the values resulting from the rectangular path. The overall accuracy of circulation values is estimated to be better than 5% of the local value. Incidentally, Reynolds number for the present experiments was based on the ratio of starting vortex circulation at $3d$ over the kinematic viscosity, i.e., Re = Γ/ν. The Reynolds number for the small and large piston ramp rates were 2×10^4 and 4×10^4, respectively.

3 Results

3.1 DPIV Data

For each of the two piston ramp rates more than 10 runs were performed and among these, five runs with optimum seeding were selected for further processing. The vorticity contours of a starting jet with the small piston ramp rate are shown in Fig. 4 at three different time instances. The time was measured from the first image pair with non-zero velocity at the nozzle exit. The first contour level is at ±2 s^{-1} and the difference between the contours is also 2 s^{-1}. The solid contours indicate positive (counterclockwise) vorticity and dashed contours negative (clockwise) vorticity. The origin is at the nozzle center and the markings are in centimeter. The first plot is at 1.5 s after the initiation of flow and only the vortex core with negative vorticity can be observed completely. The nearly circular contours with peak vorticity of 20 s^{-1} in the core center are noteworthy. The vortex core is 2.8 cm away from the nozzle. As the starting vortex moves farther away from the nozzle, a secondary (toroidal) vortex in the (round) shear layer bounding the jet is evident. The total circulation ejected from the nozzle at this time is 91 cm^2/s, using either of the two closed paths.

The second plot in Fig. 4 depicts the starting jet at a time of 2.1 s after flow initiation. The starting vortex core is 5.6 cm away from the nozzle and the entire flow is still relatively symmetric. The vorticity within the starting vortex is segregated into a primary zone with peak vorticity of 23 s^{-1} and a secondary zone with peak vorticity of 13 s^{-1}. A possible explanation for the secondary vorticity within the starting vortex at this early time is the engulfment of the first shear layer vortex by the starting vortex. Furthermore, there are now two vortices in the

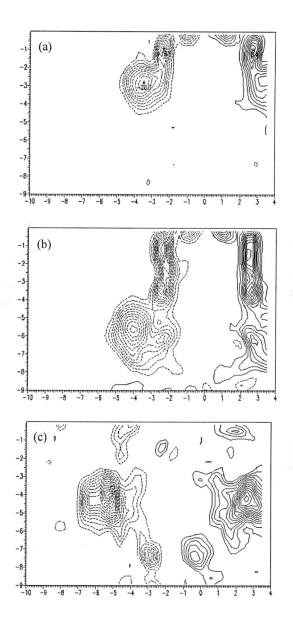

Fig. 4 Vorticity contours of a constant-acceleration jet at 1.5 s
(a), 2.1 s (b), and 3.2 s (c) after flow initiation

jet shear layer. The total circulation, which is still increasing due to the continued fluid release, is 191 cm^2/s. The circulation values computed using either of the two closed paths is within ±1% of this average value.

The last plot in Fig. 4 shows the vortex ring after the nozzle flow had ceased, at a time of 3.2 s. The image corresponds to the downstream position of the camera, where the vortex core is 13 cm away from the nozzle. Clearly the starting vortex has separated from the rest of the jet. Note the appearance of separated vorticity regions ahead and behind the primary vortex core. The circulation computed from the vorticity contour path is 206 cm^2/s, whereas the rectangular path resulted in a value of 230 cm^2/s due to the extra vorticity outside the primary zone. The vortical parcels behind the vortex core are postulated to have been shed from the main vortex ring, similar to those in turbulent vortex ring studies (Maxworthy 1974, Glezer and Coles 1990).

In order to calculate the Formation number for this run, circulation was plotted against time in Fig. 5 for both integration paths. The portion of data indicating the rapid increase of circulation with time comes from the runs with the camera viewing the region immediately downstream of the nozzle, and the data indicating relatively constant values of circulation from the downstream camera position. Note the close correspondence between the circulation values obtained from the two different integration paths, throughout the near-field. To arrive at the Formation number, time has to be normalized with the (running) average velocity and the nozzle diameter. Since the nozzle velocity increased linearly with time at a known rate, the average velocity as a function of time is easily found. Multiplying the average velocity with time is the effective fluid slug length ejected from the nozzle and is denoted by L.

The second graph in Fig. 5 is a plot of circulation against L/d for the near field portion of the run. The horizontal line is at a value equal to the mean circulation of the separated vortex, obtained from the downstream data. The cross-section of this line with the near-field circulation data provides the Formation number for this run. The value for this run is approximately 4.6. It is important to realize that near-field data contain all the circulation ejected from the nozzle, i.e. the starting vortex circulation as well as that in the jet shear layer. The farther downstream data contain only the circulation associated with the separated starting vortex core. Therefore, a Formation number of 4.6 indicates that the circulation of the final vortex ring is equal to that derived from the first 4.6d long slug of fluid ejected from the nozzle.

The Formation number was found for a number of runs with both the small and large piston ramp rates. In all cases, the Formation number was in the range of 4.5 to 5.5. These values are quite consistent with those of earlier studies (Gharib et al. 1994) with Reynolds numbers smaller than the present runs by an order of magnitude. Then, it appears that the Formation number is a robust method of characterizing the separation and pinch-off of vorticity in starting vortices. The experiments with the passive scalar are described next.

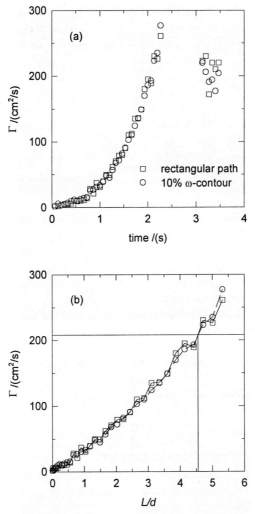

Fig. 5 Total circulation of the flow in Fig. 4 as function of time (a) and
slug length (b)

3.2 Passive Scalar Visualization

The main purpose of passive scalar runs was to investigate whether the pinch-off
of the starting vortex, as revealed by the separation of dyed vortex, can be
corroborated with the Formation number. Sample LIF images of the flow at the
large ramp rate is shown in Fig. 6. The dyed fluid has traveled $1.9d$ after 1.2 s in

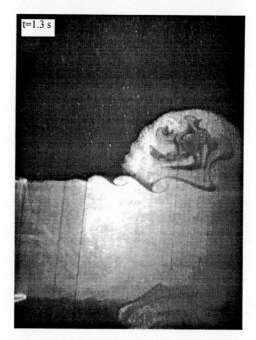

Fig. 6 Close-up LIF images of the large ramp rate flow

the first image. Of particular interest is the development of the Kelvin-Helmholtz instability of the jet shear layer. It is important to note that the velocity in the potential core increased linearly with time. The second image shows the flow at 1.3 s where the tip has extended beyond the field of view. Roll up of the shear layer vortices and their subsequent ingestion into the vortex core are remarkable. Shear layer vortices observed in this manner correspond with the vorticity contours in Fig. 4(b), even though the piston ramp rates are different. The waves on the separating streakline, as in the first image, took about one second to develop for both ramp rates.

The pinch-off and separation of the passive scalar only took place after the piston had reached its final position, i.e., after the supply had ceased. Based on the previous observations, it is believed that the dye within the starting vortex would have pinched-off and separated from the rest of the flow had the supply been continued. Unfortunately, the present apparatus was limited in this manner and the time for passive scalar pinch-off could not be directly measured. However, it was clear that the passive scalar pinch-off time would not correspond with the time associated with the Formation number. The latter time was clearly smaller than the duration of each run, whereas the former time was certainly larger than the run time by an appreciable amount. The discrepancy associated with the passive scalar pinch-off time and the time associated with the Formation number stems from their respective definitions and it is not due to the Reynolds number differences in the past experiments.

4 Summary

The present results indicate that the Formation number appears to be a consistent parameter for characterizing the pinch-off of vorticity not only at low Reynolds numbers but also at vortex Reynolds numbers extending to about 4×10^4. Remarkably, the Formation number was nearly constant with a value between 4.5 and 5.5 for the case of constant acceleration piston motion in the present setup. This range of values is compatible with prior experiments with different nozzle velocity time-histories. At the same time, the definition used for the Formation number appears to be rather incompatible with that employed for the quantification of passive scalar pinch-off. The latter is defined in terms of the local time when the complete separation of the starting vortex is observed. In contrast, the former definition goes back in time to the instant the starting vortex circulation has just been generated by the jet. For this reason, the two normalized times cannot be the same. Lastly, the Formation number can be used to predict the amount of circulation which will accumulate in the separated starting vortex core if the nozzle velocity time-history is known.

References

Abramovich, S. and Solan, A. 1973, The Initial Development of a Submerged Laminar Round Jet, *J. Fluid Mech.*, vol. 59, pp. 791-801.

410

Gharib, M., Rambod, E., Dabiri, D., Hammache, M., Shiota, T. and Sahn, D. 1994, Pulsatile Heart Flow: A Universal Time Scale, Proceedings of the Second International Conference on Experimental Fluid Mechanics, Lovrotto and Bella, Torino, Italy, pp. 34-39.

Glezer, A. and Coles, D. 1990, An Experimental Study of a Turbulent Vortex Ring, *J. Fluid Mech.*, vol. 211, pp. 243-283.

Johari, H., Rose, M., Bourque, S. and Zhang, Q. 1995, Experiments on Impulsively Started Turbulent Jets, Individual Papers in Fluids Engineering, ASME FED-207, pp. 9-16.

Kouros, H., Medina, R. and Johari, H. 1993, Spreading Rate of an Unsteady Turbulent Jet, *AIAA J.*, vol. 31, no. 8, pp. 1524-1526.

Kuo, T-W., Syed, S.A. and Bracco, F.V. 1986, Scaling of Impulsively Started, Incompressible, Laminar Round Jets and Pipe Flows, *AIAA J.*, vol. 24, no. 3, pp. 424-428.

Lahbabi, F.Z., Boree, J., Nuglisch, H.J. and Charnay, G. 1993, Analysis of Starting and Steady Turbulent Jets by Image Processing Techniques, Third Symposium on Experimental and Numerical Flow Visualization, ASME FED-172, pp. 356-368.

Maxworthy, T. 1974, Turbulent Vortex Rings, *J. Fluid Mech.*, vol. 64, pp. 227-239.

Turner, J.S. 1962, The 'Starting Plume' in Neutral Surroundings, *J. Fluid Mech.*, vol. 13, pp. 356-368.

Willert, C. and Gharib, M. 1991, Digital Particle Image Velocimetry, *Exp. Fluids*, vol. 10, pp. 181-193.

Witze, P. O. 1980, The Impulsively Started Incompressible Turbulent Jet, Sandia Laboratories Report SAND80-8617.

IV. 3 Application of PIV to Turbulent Reacting Flows

L. Muñiz, R.E. Martinez and M.G. Mungal

Department of Mechanical Engineering
Stanford, University, Stanford, CA 94305-3032, USA

Abstract. Particle Image Velocimetry (PIV) is used to measure the instantaneous velocity fields in the near and far field of an axisymmetric nitrogen-diluted methane jet ($Re = 6000$) issuing into a co-flowing air stream. Velocity measurements are taken under reacting and non-reacting conditions to determine the effects of heat release on flow structure and mixing. In addition, velocity measurements are made at the base of a lifted methane flame ($Re = 4200$) to examine the velocity criteria for flame stabilization. In each case, the jet and co-flow are seeded with nominally 0.3 μm alumina (Al_2O_3) particles to obtain planar, two-component (axial and radial) velocity data in the jet, flame zone and free stream. Measurement uncertainties are assessed for the current PIV configuration, and beam steering, image distortion and thermophoresis effects on reacting measurements are discussed.

Keywords. PIV, turbulent flames, lifted flames, thermophoresis, beam steering, image distortion

1. INTRODUCTION

Established in the early 1980's, particle image velocimetry (PIV) is now routinely used by the experimental fluid mechanics community to measure the instantaneous two-dimensional velocity fields in a wide variety of complex flows (Lourenco, *et al.* 1989, Adrian, 1991). While refinement of the technique continues through the improvement of accuracy and resolution (*e.g.* Westerweel, *et al.* 1996, Keane, *et al.* 1995), PIV is finding wider applicability through its extension into the combustion community. The non-intrusive laser diagnostics traditionally used in reacting flows have been planar laser-induced fluorescence, Raman scattering, Rayleigh scattering and laser Doppler velocimetry. To complement the planar techniques used to measure temperature and species concentration, PIV is becoming increasingly popular to measure velocity fields in reacting flows.

The non-uniform temperature fields associated with reacting flows add to the complexity of making PIV measurements because as the temperature rises across the flame front, the fluid density (thus seeding density) decreases. Early PIV studies in reacting flows had either no or sparse data in the hot post-flame gases due to the reduced seeding density (*e.g.* Reuss *et al.* 1989, Post *et al.* 1991 and Driscoll *et al.* 1993). Mungal *et al.* (1994, 1995) and Paone (1994) successfully reported velocity in the pre- and post-flame gases of laminar and turbulent premixed Bunsen flames by heavily seeding the pre-flame gases thus providing the full vector fields in the hot and cold regions of the flow simultaneously.

In the present study, a PIV investigation is underway to obtain planar two-component velocity data in reacting and non-reacting flows. The aim of the current study is to determine how turbulent mixing is affected by heat release and provide insights into the stabilization mechanism of lifted jet diffusion flames. Contributions to the assessment of PIV measurement uncertainty due to the high density gradients (such as beam steering, image distortion and thermophoresis) are also discussed.

2. EXPERIMENTAL SYSTEM

2.1 Flow Facility

The facility, shown in Fig. 1, is a vertical indraft wind tunnel with a 30 x 30 x 80 cm test section and 4:1 contraction inlet. Particle Image Velocimetry (PIV) is used to measure the velocity fields in the near and far field of an axisymmetric jet in co-flow at $Re = 6000$ under non-reacting and reacting conditions. The jet, comprised of methane diluted with 60% nitrogen (by volume), has an exit velocity of 19 m/s and is surrounded by a co-flowing stream of air with a velocity of 0.5 m/s. Under reacting conditions, the nitrogen-diluted methane flame is stabilized by a hydrogen pilot with a volumetric hydrogen to methane flow rate ratio of 3%. The jet and pilot issue from two concentric copper tubes, 120 cm long, with a jet inner diameter of 4.8 mm (6.3 mm O.D.) and pilot inner diameter of 8.0 mm (9.5 mm O.D.). PIV measurements are also made at the base of a lifted methane flame; the flow configuration is the same as above except pure methane (99.0%) issues from the 4.8 mm jet and no pilot gas is used in the outer annulus. For both studies, the jet is seeded with nominally 0.3 μm alumina (Al_2O_3) particles, while the co-flow can be seeded with alumina or a micron-sized fog produced from a glycerol-water mixture. The fog, which is non-flammable, evaporates at elevated temperatures (80°C) and thus provides a useful means to mark the room-temperature free-stream fluid surrounding the hot jet flame. The alumina, however, survives flame temperatures and is used when making PIV measurements in flames. The fog is used primarily for jet-flame visualization.

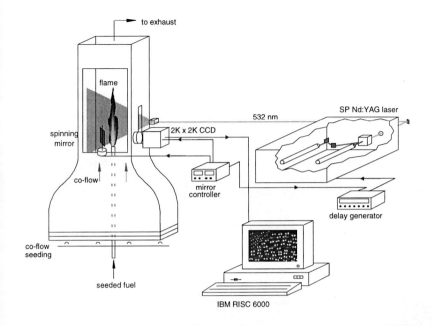

Figure 1: Experimental facility

2.2 PIV System

The seed particles are illuminated by a double-pulsed laser sheet at 532 nm, approximately 0.2 mm thick coincident with the jet centerline. The light source is a Nd:Yag, 400 mJ/pulse, Spectra-Physics PIV-400 laser containing two independent cavities. The pulse separation ranges between 5-50 µs provided by a delay generator (SRS Model DG 535). The scattered light from both laser pulses in a 5 cm x 5 cm region of the test section is collected with a 105 mm Nikkor lens at f#/5.6 onto a 2000 x 2000 pixel CCD array (Kodak Megaplus 4.2). The double-exposed image is processed by an IBM RISC 6000 workstation to yield the velocity field. The MKIV software package (written by FFD Inc., now sold through TSI) uses an autocorrelation technique to find the average particle displacement in each subregion of the image, which is converted into velocity based upon laser pulse separation. The images are processed with 60 x 60 pixel subregions in a 66 x 66 grid so that the overlap between two neighboring subregions is 50%, and each vector represents the average velocity in a 1.5 x 1.5 x 0.2 mm interrogation volume. To remove velocity ambiguity and increase dynamic range, the second particle image is shifted from the first by a known amount using a spinning mirror, and the bias is subsequently removed in post-processing. This CCD-based image acquisition system coupled with high-speed processing provides rapid data collection. The images can be collected at 2 Hz and processed at a rate of 50 vec/sec.

414

3. RESULTS AND DISCUSSION

3.1 Flow visualization

The three flows examined in this study are visualized in Fig. 2 by means of Mie scattering and flame emission. The photographs reveal the instantaneous (10 ns) large-scale jet structure and time-averaged (1/30 sec) flame shape. The jets are seeded with alumina in Figs. 2(a,b) and the co-flow is seeded with fog in Figs. 2(b,c). The non-reacting and reacting jet in co-flow ($Re = 6000$) are shown in Figs. 2(a,b). The flame in Fig. 2b is anchored to the burner rim by a hydrogen pilot. The outer flame jet structure (marked by fog) is similar to that of the non-reacting jet (marked by alumina), however, the flame does influence the inner (fuel) jet structure. The jet widths, characterized by the non-reacting jet visualization and the flame jet's influence on the free stream, are comparable. Figure 2c shows the pure methane lifted flame ($Re = 4200$). Fog, marker of the free-stream fluid, is shown being entrained into the jet upstream of the flame base location, implying premixing.

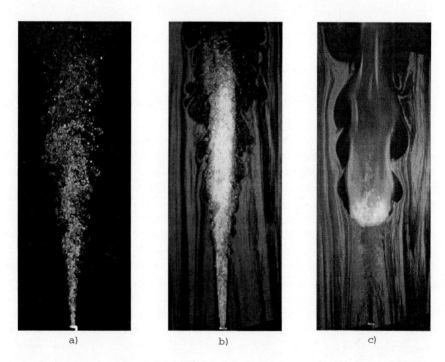

a)   b)   c)

Figure 2: a) non-reacting jet ($Re = 6000$); b) reacting jet ($Re = 6000$); c) lifted methane flame ($Re = 4200$)

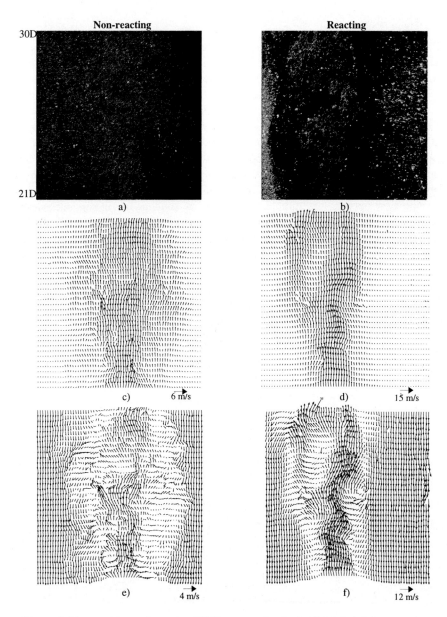

Figure 3: Non-reacting and reacting PIV images and velocity fields in lab frame and convecting frame at $x/d = 25$ and $Re = 6000$

3.2 Heat Release

To examine the effect of heat release on turbulent mixing, PIV measurements are taken in a non-reacting and reacting jet at the same jet exit Reynolds number (Re = 6000). PIV data are taken at three axial locations (from jet lip to flame tip) centered at 7, 25 and 67 diameters downstream (see Muñiz & Mungal, 1995). Figure 3 shows side by side comparisons of the non-reacting and reacting case at 25 diameters downstream where the average flame length is 70 diameters. The upper PIV images (a,b) show the types of images captured on the CCD array. Note, the low seeding density regions on both sides of the centerline indicate the high temperature zones where the flame resides. Figures 3(c,d) show the velocity in the lab frame after removal of the bias velocity imposed by the spinning mirror. The reacting flow appears less turbulent, with the jet showing a zig-zag, meandering appearance. In Figures 3(e,f), large scale structures convecting downstream are visualized by performing a Galilean transformation to the reference frame convecting at one half the centerline minus co-flow velocity. The non-reacting flow shows a broad range of structures while the reacting flow shows a considerable reduction in vortical activity, and the meandering of the jet is readily apparent. Figure 4 shows the instantaneous centerline velocity decay for the reacting and non-reacting case. A factor of two increase in centerline velocity is seen under reacting conditions for axial distances greater than 10 diameters downstream. Since the jet conserves momentum, a less rapid decay in centerline velocity implies reduced entrainment of ambient fluid for the reacting case. Similar results for the hydrogen jet are reported by Takagi *et al*. (1981) using LDV measurements.

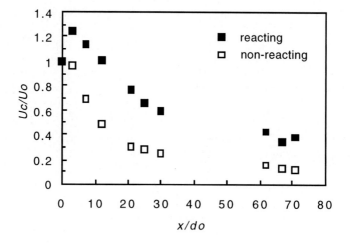

Figure 4: Centerline velocity decay

3.3 Lifted Results

PIV data are taken at the base of a lifted methane flame (Re = 4200) to determine what velocities the flame base can stabilize itself against. Figure 5 show a typical PIV image with the accompanying velocity field. The abrupt change in seeding density marks the thermal boundary of the flame base. This flame boundary is sketched on the velocity field. Since the flow is turbulent and unsteady, the flame base location moves up and down. Since PIV is an imaging technique, the velocity can be extracted at the flame base location even though the flame base is moving within the imaged region (see Muñiz & Mungal, 1996a, 1996b, 1997).

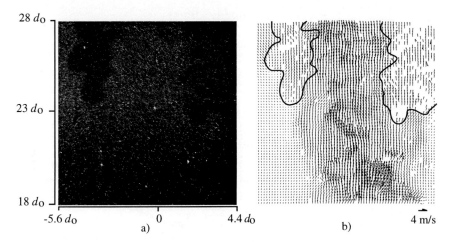

Figure 5: PIV image and instantaneous velocity field of lifted flame base

Figure 6 shows the distribution of velocities at the instantaneous flame base location taken from 66 realizations. Note, the peak in the distribution is near 0.5 m/s and can be compared to the maximum laminar flame speed of 0.38 m/s for methane. Figure 7 contrasts the instantaneous and average behavior. Figure 7a shows the instantaneous axial velocity as a function of axial distance along the centerline (filled symbols) and through the flame base on the left side (open symbols) of the image presented in Fig. 5. The meandering nature of the jet is seen in the centerline profile while the flame base profile shows the velocity decreasing just before being accelerated through the flame front. Similar profiles are seen in triple flame simulations by Ruetsch *et al.* (1995) where the triple flame is characterized by stratified premixed combustion at the base followed by a diffusion flame tail. The mean (from 66 images) profiles are shown in Fig. 7b. The smooth decay in centerline velocity is observed as well as the slow rise in velocity across the flame front.

418

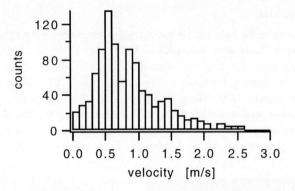

Figure 6: Velocity histogram at instantaneous flame base

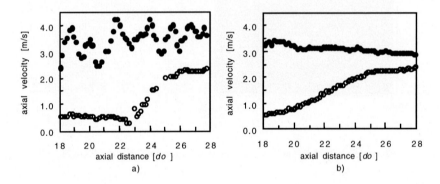

Figure 7: a) instantaneous and b) mean profiles (open symbols -through flame base; filled symbols - along jet centerline)

4. UNCERTAINTY ANALYSIS

PIV measurement uncertainty due to temperature gradients is assessed by addressing beam steering, image distortion and thermophoresis. Before isolating these effects, the algorithm uncertainty is determined excluding the effect of out of plane motion and velocity variations within a subregion which as has been studied by Keane & Adrian (1992).

4.1 Algorithm

The algorithm and imaging contribution to the PIV measurement uncertainty, which applies to non-reacting and reacting flow velocities, is assessed by performing an artificial shift experiment. A single-pulsed image of uniform co-flow seeding is captured on the CCD array. A double-exposed test image is generated by digitally adding the same single-pulsed image to itself shifted to the right by a specified number of pixels. Images of different offsets were generated and processed with different sized interrogation regions to estimate the uncertainty of the present PIV configuration excluding the effect of out of plane motion and velocity variations within a subregion. Figure 8 shows the relative measurement uncertainty for the above cases. Current PIV experiments are designed to have pixel displacements between 10 and 15, are processed with interrogation regions of 60 square pixels and contain 10 or more particle pairs yielding 0.8% pixel displacement uncertainty. These design criteria are consistent with those put forth by Adrian (1991), Lourenco *et al.* (1989) and Keane & Adrian (1992).

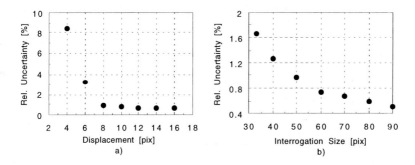

Figure 8: PIV measurement uncertainty in terms of processing parameters

4.2 Beam Steering

To evaluate beam steering, a HeNe beam passes through the flame and terminates on a target 2 m away to measure the angular deflection. The angular deflection is 5.6 x 10^{-4} rad when the beam passes through the center of the Re = 6000 flame and is 2.3 x 10^{-3} rad when passing through the flame tangentially at x/d = 20. The angular deflection in the beam is caused by a density gradient in the direction perpendicular to the beam propagation (Holder & North, 1956) and is given by

$$\varepsilon_x = \frac{L}{\delta} \frac{n(T) - n_o}{n_o}$$

where ε_x is the deflection angle towards the region of highest density, L is the length of gradient along the beam axis and, δ is the width of the density gradient or flame thickness. The index of refraction as a function of temperature is given by

$$n(T) = K\frac{\rho(T)}{\rho_o} + 1 = K\frac{T_o}{T} + 1$$

where T_o equals 300K. Since the index of refraction of air at 300K equals 1.000298, K equals 0.000298 and the index of refraction at flame temperatures (2000K) equals 1.00199.

The measured deflections agree well with those calculated for $L/\delta = 1/3$ and $L/\delta = 1$ for center and tangential propagation of the beam through the flame. For $L/\delta = 1/3$, the calculated angular deflection is 5.5 x 10^{-4} rad. For a laminar flame with circular cross-section, no beam steering would be expected since the density gradient is parallel (or flame surface perpendicular) to the beam propagation. However, for a turbulent flame, it is reasonable to expect the length of the flame surface parallel to the beam propagation to be about one third the flame thickness yielding $L/\delta = 1/3$. When the beam is tangential to the flame, it is reasonable to expect $L/\delta = 1$ yielding a calculated deflection angle of 1.6 x 10^{-3} rad. When making PIV measurements, the laser sheet passes through the center of the jet, and the sheet is expected to be deflected only 28 μm (less than one seventh of the sheet thickness) as it reaches far side of the imaged region. Therefore, the beam steering effect on PIV measurements in this study is negligible.

4.3 Image Distortion

To determine the extent of image distortion caused by the flame, a steel ruler is placed in the flow and imaged with the flow reacting and non-reacting. When in the flame, the ruler is imaged quickly before the ruler thermally expands. The image taken in the reacting flow is identical to the non-reacting ruler image down to the pixel as determined by their addition.

Significant blurring and distortion occurs when the object plane is a substantial distance behind the density gradients, see Fig. 9. The image becomes increasingly blurred as L_1 increases while keeping $L_1 + L_2$ constant. In the present configuration, L_1 is essentially zero, and no blurring is seen. The camera lens (f = 105 mm) is approximately 25 cm away from the object plane (M = 0.36), the flame is 1.5 cm in front of the object plane at the center of the image and is coincident with the object plane on the sides. Since PIV measurements are taken in the flame and not far behind the flame, image distortion effects on the velocity measurements are thus found to be negligible.

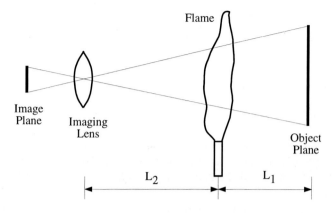

Figure 9: Schematic of imaging system with flame present

4.4 Thermophoresis

Thermophoresis is phenomenon by which a particle suspended in a gas with a temperature gradient, will tend to drift down the gradient. Therefore, the particle velocity perpendicular to a flame surface will lag the fluid velocity as it approaches the flame, introducing an error to the PIV measurement. This lag in particle velocity due to thermophoresis is termed thermophoretic velocity. Thermophoretic velocities are estimated from the experimental and theoretical work of Sung *et al.* (1994) and Gomez & Rosner (1993). Sung *et al.* calculated and measured the thermophoretic velocity to be 15 cm/s for a 0.3 µm alumina particle traveling through a nitrogen-diluted methane/air, counterflow diffusion flame with a strain rate of 240 sec^{-1}. The calculation was done by solving the governing equation of the particle where F_{SD} is the Stokes drag, and F_{TP} is the thermophoretic force.

$$m_P \frac{dv_P}{dt} = F_{SD} + F_{TP}$$

$$F_{TP} \propto \frac{-\nabla T}{T}$$

In the calculations and LDV measurements, the particle velocity lagged behind the flow velocity in the preheat zone where the gradients were high, and the estimated thermophoretic velocity accounted for this lag such that

$$v_F = v_P + v_{TP}$$

Gomez & Rosner (1993) calculated the thermophoretic diffusivity, I_p, by determining the thermophoretic velocity, v_{TP}, from the following expression:

$$v_{TP} = \left(\alpha_T D\right)_P \cdot \left(-\frac{\nabla T}{T}\right)$$

where α_T is the dimensionless thermophoretic diffusion factor, D_P is the particle Brownian diffusivity, T is the local temperature and ∇T its local spatial gradient. The thermophoretic diffusivity of a TiO_2 particle with a diameter in the range between 2 nm and 0.4 μm was found to be $0.5\mu/\rho$. Sung et al. using Gomez & Rosner's results calculated the thermophoretic velocity of 0.3μm TiO_2 particles for the case where k = 240 sec^{-1} with $(\nabla T)_{max} \approx 2000$ K/mm at T $\approx$ 1300K to be 14 cm/s. From this, we took the thermophoretic diffusivity, hence velocity, of TiO_2 and Al_2O_3 to be approximately equal for diameters near 0.3 μm.

To evaluate the thermophoretic velocity in the current study, we needed to estimate ∇T. To this end, we examined the temperature field data of Everest et al. (1995) for a similar turbulent jet flame. The typical temperature gradients for methane/hydrogen jet flames at $Re = 4000$ and 8000, $x/d = 42$ were 1000 K/mm at T = 1300, yielding a thermophoretic velocity of 8 cm/s. In this study, PIV velocities measured on the air-side of the flame are 1 and 0.5 m/s (axial and radial respectively). The magnitude of error due to thermophoresis is estimated as 16 % of the radial component of velocity and 7% of the overall velocity. On the fuel side of the flame front, typical PIV velocities are 6 m/s (axial) and 0.5 m/s (radial) leading to a 1% error on the overall velocity and 16% error on the radial component. While these errors are not overwhelming for the velocity measurements themselves, they could have a serious impact on radial velocity gradient calculations. Since the nature of flame vortex interactions are complex in turbulent reacting flows, no attempt has been made to correct the PIV data for thermophoresis because the instantaneous temperature field is not known.

5. CONCLUSIONS

This experimental work provides instantaneous two-dimensional, two-component velocity fields in turbulent diffusion flames using particle image velocimetry. The non-reacting and reacting jet at $Re = 6000$ are compared and results suggest that heat release does affect jet behavior, namely increased centerline velocity, implying reduced entrainment, and reduced turbulent activity. The results from the lifted methane jet at $Re = 4200$ show the flame base is anchored primarily in the low velocity regions of the jet. Measurement uncertainties related to high temperature gradients are assessed. While effects of image distortion and beam steering are found to be negligible, the radial particle velocity (thus PIV measurement) at the flame surface may be lagging the fluid's radial velocity by as much as 16% due to thermophoresis.

6. ACKNOWLEDGEMENTS

This work is sponsored by the Gas Research Institute, Contract #5093-260-2697, R. V. Serauskas, technical monitor. L. Muñiz is sponsored by AT&T Bell Laboratories Cooperative Research Program.

7. REFERENCES

Adrian R. J. (1991), "Particle-Imaging Techniques for Experimental Fluid Mechanics," Annu. Rev. Fluid Mech., 23, 261.

Driscoll, J. F., Sutkus, D. J., Roberts, W. L., Post, M. E. & Goss, L. P. (1993) "The Strain Exerted by a Vortex on a Flame - Determined from Velocity Field Images," AAIA-93-0362, 31st Aerospace Sciences Meeting, Reno, NV, Jan. 1993.

Everest, D., Driscoll, J. F. & Dahm, W. J. A. (1995), "Images of the 2-D Temperature Field and Temperature Gradients to Quantify Mixing Rates within a Non-premixed Turbulent Jet Flame," Combustion and Flame, Vol 101, n. 1-2, 58.

Gomez, A. & Rosner, D. E. (1993), "Thermophoretic Effects on Particles in Counterflow Laminar Diffusion Flames," Comb. Sci. & Tech., Vol 89, 335.

Holder, D. W. & North, R. J. (1956), Optical Methods for Examining the Flow in High-Speed Wind Tunnels: Part One - Schlieren Methods, AGARDograph, November 1956.

Keane, R. D., Adrian, R. J. & Zhang, Y. (1995), "Super-resolution Particle Image Velocimetry," Measurement Science & Technology, Vol 6, n. 6, 754.

Keane, R. D. & Adrian, R. J. (1992), "Theory of Cross-Correlation Analysis of PIV Images," Applied Scientific Research, Vol 49, 191.

Lourenco, L. M., Krothapalli, A. & Smith, C. A. (1989), "Particle Image Velocimetry," Lecture Notes in Engineering 45, M. Gad-el-Hak (editor), Advances in Fluid Mechanics Measurements, Springer-Verlag, 127.

Mungal, M. G., Lourenco, L. M. & Krothapalli, A. (1994), "Instantaneous Velocity Measurements in Laminar and Turbulent Premixed Lames Using On-Line PIV," Seventh International Symposium on Applications of Laser Techniques to Fluid Mechanics, Lisbon, Portugal, 15.1.

Mungal, M. G., Lourenco, L. M. & Krothapalli, A. (1995), "Instantaneous Velocity Measurements in Laminar and Turbulent Premixed Lames Using On-Line PIV," Combust. Sci. and Tech., Vol 106, 239.

424

Muñiz, L. & Mungal, M. G. (1995), "A PIV Investigation of Turbulent Diffusion Flames," WSS/CI-95F-206, Western States Section of the Combustion Institute, Fall Meeting, Stanford, CA.

Muñiz, L. & Mungal, M. G. (1996a), "PIV Study of Lifted Jet-Diffusion Flames: Low-Speed Stabilization and Evidence for Triple Flames," 26th Symp. (Int.) on Comb., The Combustion Institute, Work-in Progress Poster # 428.

Muñiz, L. & Mungal, M. G. (1996b), "Velocity Measurements in Lifted-Jet Diffusion Flames," WSS/CI-96F-113, Western States Section of the Combustion Institute, Fall Meeting, Los Angeles, CA.

Muñiz, L. & Mungal, M. G. (1997), "Instantaneous Flame-Stabilization Velocitites in Lifted Jet Diffusion Flames", to appear in Combustion and Flame.

Paone, N. (1994), "Velocity Measurements in Turbulent Premixed Flames: Development of a PIV Measurement System and Comparison with LDV," Seventh International Symposium on Application of Laser Techniques to Fluid Mechanics, Lisbon, Portugal, p15.3.

Post, M. E., Goss, L. P. & Brainard, L. F. (1991), "Two-Color Particle-Image Velocimetry in a Diffusion Flame," Central States Section of the Combustion Institute, Spring Meeting, Nashville, TN.

Reuss, D. L., Bardsley, M., Felton, P. G., Landreth, C. C. & Adrian, R. J. (1989), "Velocity, Vorticity, and Strain-Rate Ahead of the Flame Measured in an Engine Using Particle Image Velocimetry," SAE Trans., Vol 99, n 3, 249.

Ruetsch, G. R., Vervisch, L. & Liñán, A. (1995), "Effects of Heat Release on Triple Flames," Physics of Fluids, Vol 7, n 6, 1447.

Sung, C. J., Law, C. K. & Axelbaum, R. L. (1994), "Thermophoretic Effects on Seeding Particles in LDV Measurements of Flames," Comb. Sci. & Tech, Vol 99, 119.

Takagi, T., Shin, H.-D. & Ishio, A. (1981), "Properties of Turbulence in Turbulent Diffusion Flames," Combustion and Flame 40:121-140.

Westerweel, J., Draad, A. A., van der Hoeven, J.G. Th. & van Oord, J. (1996), "Measurement of Fully-Developed Pipe Flow with Digital Particle Image Velocimetry," Experiments in Fluids, Vol 20, n 3, 165.

IV. 4 Whole Field Measurements on an Excited Premixed Flame Using On-Line PIV

L. Lourenco[1], H. Shen[1], A. Krothapalli[1] and P. Strykowski[2]

[1]Florida A&M University and Florida State University
[2]University of Minnesota

1 Introduction

A novel shear layer control technique used to enhance the mixing of heated jets has been the subject of study in our laboratory (Strykowski et al., 1993; 1996). The technique is based on the self excitation of a countercurrent shear layer which is established by the introduction of a reverse flow around the perimeter of an axisymmetric or rectangular jet. Because of its effectiveness, it is suggested that such an active shear layer control technique may also enhance the combustion processes in jet flames. To evaluate this hypothesis, an **ON-LINE PIV** system is used to characterize the velocity field, in both the cold reactant flow and the hot post-flame regions, of an excited, premixed jet flame issuing from a modified round nozzle burner.

2 The Test Facility

The test facility, shown in figure 1, consisted of a modified circular nozzle burner to include a circular collar concentrically installed around the burner exit.

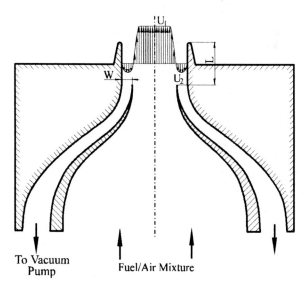

Figure 1: Burner Configuration with Counterflow

To achieve an exit top-hat velocity profile the nozzle has a very large contraction ratio of 31.6:1 and an exit diameter, D=10.16mm. The collar has a diameter of D_c=14.83mm and extends one nozzle diameter from the exit. The collar is connected to a vacuum pump so that a counterflow which is in the opposite direction of the fuel/air mixture flow can be established through it. The local counter-current shear layer is thus created in the near field of the jet just out of the burner exit. Propane is used as the fuel in these experiments. Parameters such as the equivalence ratio (ϕ) of the mixture, the velocity ratio, $r=U_1/U_2$, between the mixture flow (U_1) and counterflow (U_2) were varied during the experiments.

In this experiment, the extension length (L) of the collar above the burner exit, and the gap width (W) between the tube and the collar are kept constant, respectively L/D=1.0 and W/D=0.23. However, these parameters can be easily varied, and their influence on the total performance of the system will be evaluated in a subsequent experimental study. The premixed flames were produced using the arrangement shown in figure 2. Commercial grade propane is supplied from a tank, pressure regulated and metered using a standard rotameter. Shop air is also metered, using both a rotameter and an orifice plate arrangement. In this facility, both the primary jet and the environment air are seeded with Aluminum Oxide particles with a nominal diameter of 0.3 μm. The solid particles are introduced into the air stream through a settling chamber by means of an aerosol generator; the generator consists of a fluidized bed connected to a cyclone for the efficient removal of large particles or particle agglomerates.

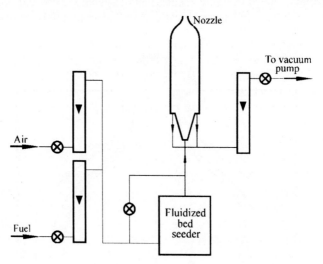

Figure 2: Experimental Facility

3 Measurement Techniques and Instrumentation

The main feature of the Particle Image Velocimeter used in this experiment is its capability to record two images in quick succession, from which the velocity field is derived using a cross-correlation algorithm. This is possible by integrating the PIV system's two main components: the Kodak ES 1.0 digital video camera and the dual Nd-Yag laser illumination system, each with a repetition rate of 15Hz. At the heart of camera is the CCD interline transfer sensor, KAI-1001 with a resolution of 1008(H) x 1018(V) pixels. Each square pixel measures 9 μm on the side with 60 percent fill ratio with microlens, and a center to center spacing of 9μm. The camera is also equipped with a fast electronic shutter and outputs eight bit digital images, via a progressive scan readout system, at a rate of 30 frames per second.

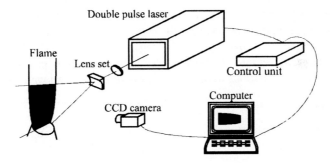

Figure 3: Particle Image Velocimeter

The unique feature of the Kodak ES 1.0 camera, implemented by Kodak in collaboration with the FMRL, is its ability to be operated in the "triggered dual exposure mode". Operation in this mode is possible due to the CCD sensor architecture, which incorporates both a light sensitive photodiode array and a masked register array. During the exposure cycle, light is converted to charge in the photodiode area of the array; after exposure, the charge on the photodiode is transferred to the masked area of the array. The maximum time for complete transfer of the charge is 5 μsec; using a programming feature of the camera's control electronics, this time setting can be made as small as 1 μsec; however image quality may not be preserved, as the times less than 5 μsec may be too short to ensure that all charge is transferred, especially in the case of very high intensity images. The image acquisition sequence in the "triggered dual exposure mode" is as follows: the image acquisition is initiated by an external trigger signal; the first image is illuminated by the first laser of the dual laser system, which is triggered in advance, to account for the usual delay between flash lamp trigger and Pockells cell trigger, i.e. by 165 μsec. The first image is then transferred to the read-out section of the sensor, within 1-5 μsec, and a second laser pulse illuminates the again the photodiode section of the sensor, which has been depleted of the charge. To achieve total separation between the two images, the trigger pulse to the flash lamp of the second laser is delayed with respect to the first flash lamp pulse by an amount that exceeds 5μsec.

The above described arrangement makes it possible to acquire up to 15 image pairs per second. The fact that the image pairs are recorded in separate frames, and that the image pair separation is variable from 1 microsecond up to several hundreds of microseconds, makes this instrument appropriate for general and simple use in flows with velocity reversals as well as very wide dynamic velocity range from very low speed (mm's/sec) up to very high speed (100's m/sec) flows.

The image data acquisition is done using an Imaging Technologies ICPCI board, which resides on a single slot of the PCI bus of a personal computer. The computer's CPU is an Intel 150MHz Pentium with 64 Mbytes of RAM.

An image matching approach is used for the digital processing of the image pairs to produce the displacement field. In this approach one sets up a cost function, C, to be maximized (or minimized), which models the match between two corresponding regions of the images. Typically, if I_1 and the I_2 are the image intensity distributions of the first and the second image, we write,

$$C(\vec{s}) = C\left\{I_1(\vec{x}), I_2(\vec{x})\right\}$$

where it is assumed that the second image is an exact translated copy of the first image, we may write,

$$I_2(\vec{x}) = I_1(\vec{x} - \Delta \vec{s})$$

or

$$I_2(\vec{x}) = I_1(\vec{x}) * \delta(\vec{x} + \Delta \vec{s})$$

where $\Delta \vec{s}$ is the average image translation and the function I_2 usually represents a small block (interrogation window) in a larger image, I_1. The match is obtained for the value $\vec{s}$ that maximizes (or minimizes) C. The cost function we choose to maximize is the cross-correlation, G, defined as:

$$G(\vec{x}) = I_1(\vec{x}) * I_2(\vec{x}) = \int_{-\infty}^{\infty} I_1(\vec{x}) I_2(\vec{x} - \vec{u}) d\vec{u}$$

The cross-correlation is effectively computed using Fourier transforms. Consider the Fourier transform of the first exposure image:

$$\Im\left\{I_1(\vec{x})\right\} = \int_{-\infty}^{\infty} I_1(\vec{x}) \cdot e^{-2\pi j(\vec{x} \cdot \vec{\omega})} d\vec{x} = \tilde{I}_1(\vec{\omega})$$

where $\Im$ is the Fourier transform operator, $j = \sqrt{-1}$, and $\vec{\omega}$ is the spatial frequency coordinate. Similarly, the Fourier transform of the second exposure image is:

$$\Im\left\{I_2(\vec{x})\right\} = \int_{-\infty}^{\infty} I_1(\vec{x}) \cdot \delta(\vec{x} + \Delta\,\vec{s}) \cdot e^{-2\pi j(\vec{x}\cdot\vec{\omega})} d\,\vec{x}$$

$$= \tilde{I}_1(\vec{x}) \cdot e^{-2\pi j(\Delta\,\vec{s}\cdot\vec{\omega})}$$

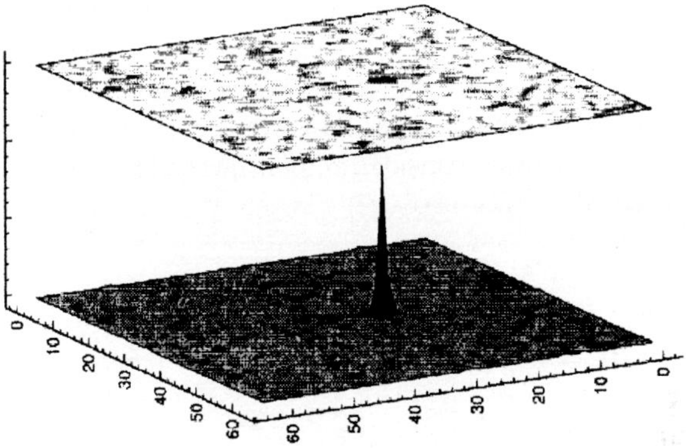

Figure 4: Typical Cross-Correlation Map

The cross-correlation function, G, is obtained computing the inverse Fourier transform, $\Im^{-1}$, of the product of the multiplication of the transform of the first image by the complex conjugate of the transform of the second image,

$$G(\vec{x}) = \Im^{-1}\left\{\tilde{I}_1(\vec{\omega})^2 \cdot e^{2\pi j(\Delta\,\vec{s}\cdot\vec{\omega})}\right\} =$$

$$= \Im^{-1}\left\{\tilde{I}_1(\vec{\omega})^2\right\} \cdot \delta(\vec{x} + \Delta\,\vec{s})$$

The cross-correlation function G, is thus the transformed image intensity pattern displaced with respect to the origin by the average displacement coordinates. A typical result for a square interrogation of 64 pixels is depicted in figure 4. The peak position is found with sub-pixel resolution by means of a Gaussian interpolator as described by Lourenco and Krothapalli (1995).

4 Global observations
In the following experiments the nozzle and suction collar geometry characterized by $L/D=1.0$ and $W/D=0.23$ were kept constant. The choice of the collar dimensions and shape were based on previous observations made in non-reacting flows by Strykowski and Niccum (1991).

Initial observations of the global behavior of the jet show a definite effect of the counter-flow on the flame, namely its blow-off characteristics. It is well known that the blow-off of the flame will occur if the mixture flow rate is increased beyond a certain limit. Figure 5 shows the relationship between the blow-off limit of the flame, and the jet velocity, with and without counter-flow.

It can be seen that the blow-off limit is dramatically extended when the counter-flow is applied to the flame. For example, with the counter-flow velocity at 0.01 m/s, a 280% increase in the jet velocity is observed at equivalence ratio of unity. Even for the quite lean mixture with $\phi = 0.59$, this amount of counter-flow still gives an increase in jet velocity, U_1, of about 230%. As shown in figure 5, the increase in the strength of the counter-flow will further extend the flame blow-off limit, showing that counter-flow can be a very effective method to improve flame stabilization, considering that these observations are carried out in a non-optimized burner geometry.

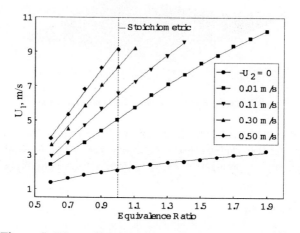

Figure 5: Blow-off limit with and without counterflow
(L/D=1.0, W/D=0.23)

The global effect of the excitation is also illustrated by the instantaneous, laser sheet, visualization images of the flame field with and without counter-flow in figure 6 (a-c). The images are obtained by illuminating the seeded flow with the laser sheet as it is used in the PIV configuration. The seeding consists of Aluminum Oxide particles with a mean diameter of about 0.3 μm. The distinct particle density difference in the images, due to the volume expansion generated by the large temperature gradients, is an excellent discriminator between the pre- and post-flame regions. These large particle density gradients which are accompanied by thin vorticity layers also create a set of unique conditions that make the application of PIV to reacting flows a challenging task (Mungal et al., 1995). However, the successful application of PIV to the study of these flows provides the means for proper identification of the mechanism responsible for the flow control as described in section 5.

Figure 6a is an image of a pre-mixed, laminar flame at Re=1790 and ϕ=1.5. It should be noted that under above condition the flame is quite near the blow-off limit if no counter-flow is applied, and a small increase in U_1 will cause the flame to be blown out. By means of a very small counterflow , with U_2=0.01 m/sec, the blow-out limit is extended to U_1=7.7 m/sec, at constant equivalence ratio ϕ=1.5, as evidenced by the image of figure 6b. This image clearly shows that the counter-current shear layer excites the jet flame, leading to the early breakdown of the inner cone of the flame (high particle density region) and the creation of large structures inside the flame. Further evidence of the creation of a flow instability is presented by the image of figure 6c which corresponds to an increased suction velocity, U_2 =1.1 m/sec.

Figure 6a: Instantaneous Visualization
$U_1 = 2.7$ m/s; $U_2 = 0$ m/sec

Figure 6 b: $U_1 = 7.7$ m/s; $U_2 = 0.01$m/sec

Figure 6c: $U_1 = 7.7$ m/s; $U_2 = 1.1$ m/sec

One important question that arises when the counter-flow is applied to the jet, is to what extent the gas mixture in the primary jet is drawn into the counter-flow. In the worst conditions, all the gases in the counter-flow may originate from the jet. To evaluate the relative contribution of the primary gas mixture to the ambient air in the counter-flow, the hydrocarbon (HC) concentrations both in the jet and in the counter-flow were measured using propane as the tracing element. In these measurements the air jet was running in the cold condition with a minute amount of propane added to it. A stainless steel probe was also used to sample the gases in the jet at the nozzle exit (HC_1) as well as in the counter-flow (HC$_2$); the sampled gases were then analyzed by a Horiba model FIA-510 total hydrocarbon analyzer and the results are summarized in figure 7. These results demonstrate that for velocity ratios less than 0.1 only a very small portion of the primary jet is ingested. This conclusion is reached taking in consideration that the relative concentration is about 30% for a velocity ratio of 0.1, equivalent to a volume flow ratio of 21%.

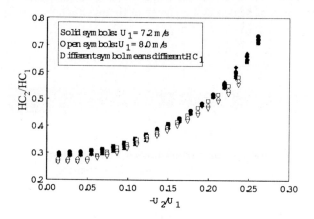

Figure 7: Hydrocarbon concentration vs. velocity ratio

5 Detailed Observations

A detailed characterization of the flow, with and without control is carried out with PIV. Initially the mechanisms responsible for the delay of flame blow-out by the suction flow are investigated. Figure 8 shows the velocity and the strain fields, at one instant in time, in the plane containing the streamwise axis of the jet, without suction, for the near blow-out condition depicted in figure 6a (laminar flow, U_1=2.7m/sec, Re=1790). The velocity field was obtained by the method described in section 3, with interrogation regions of 56x56 pixels corresponding to a physical dimension of 1.5x1.5 mm. This interrogation region size was determined mostly by the requirement for a minimum of 10 particle pairs in the post-flame gases. In the case of laminar flow, it has been shown that the condition for flame stabilization is that there is no region in the field where the flow velocity exceeds that of the combustion wave. Furthermore, Lewis and von Elbe (1961) show that there is a very good correlation between the boundary velocity gradients at blow-out for various equivalence ratios. In the near blow-out limit conditions corresponding to figure 8 the maximum measured strain is about 2000 sec^{-1} which is in general agreement with previously reported values of extinction rates of about 1750 sec^{-1}, by Law et al. (1988), for ϕ=1.35.

In the case of the flow with suction, the average velocity and streamlines from 60 samples, corresponding the flow conditions of the images in figure 6b and figure 6c are shown in figures 9a,b and 10a,b, respectively.

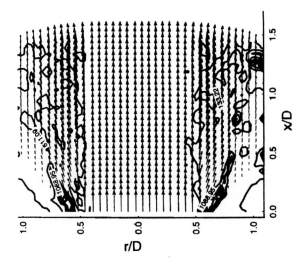

Figure 8: Velocity and Strain fields
$U_1 = 2.7$ m/sec

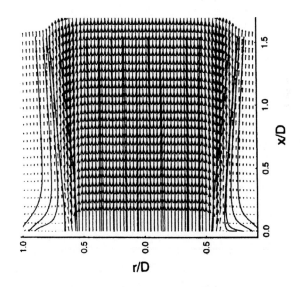

Figure 9a: Velocity and streamline pattern
U_1=7.7 m/sec; U_2=0.01 m/sec

434

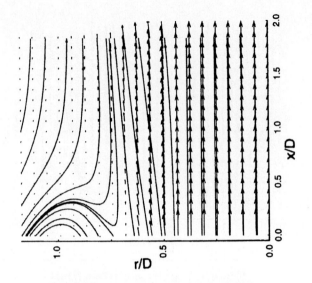

Figure 9b: Near exit region flow detail

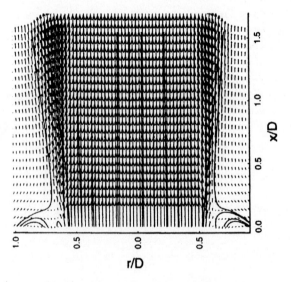

Figure 10a: Velocity and streamline patter
U_1=7.7 m/sec; U_2=1.1 m/sec

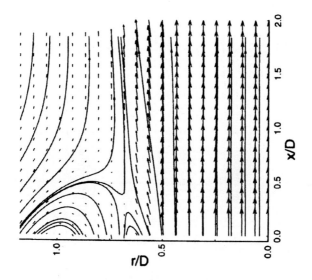

Figure 10b: Near exit region flow detail

These figures clearly show that the flow in the vicinity of the nozzle exit is significantly altered by the presence of the suction flow. The velocity field and streamlines in figures 9a,b and 10a,b, help establish several facts:

- A countercurrent shear layer is established, as regions of reversed flow are observed. It is also clear that most of the reverse flow is entrained ambient air; this observation is consistent with the measurements of HC concentration.
- The extend of the countercurrent shear layer is not affected significantly by the amount of suction, i.e., the velocity ratio, U_1/U_2, but rather by the length and shape suction collar. This conforms with a previous report (ref. 4).

Therefore, a very small mass of reverse flow can greatly extend the blow-off limit. Another significant observation, when the counter-flow is applied to the flame jet, is that a flow instability is introduced in the near exit region while in the case without counter-flow such vortices cannot be observed. We believe that such vortex structures is evidence of the self-excitation due to the counter-current shear layer and have positive effects on the local mixing process, also leading to the decrease of the boundary velocity gradient and the extension of the flame stabilization limit. The presence and strength of the vortical structures are characterized by the increased levels of turbulent shear stress as depicted in figures 11a,b.

436

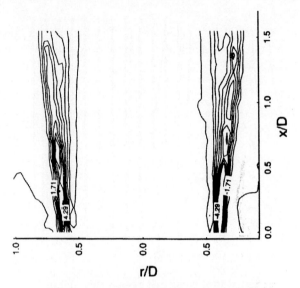

Figure 11a: Reynolds shear stress field
U_1=7.7 m/sec; U_2=0.01 m/sec

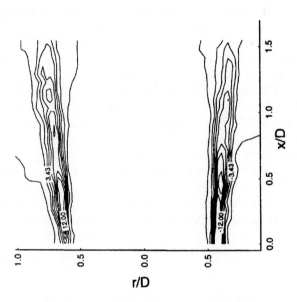

Figure 11b: Reynolds shear stress field
U_1=7.7 m/sec; U_2=1.1 m/sec

6 Conclusions
The novel results clearly illustrate both the usefulness of the on-line PIV system to the study of reacting flows, and merits of the shear layer control approach as a flame blow-out delay and excitation method.

The main advantages of the PIV set-up with the Kodak ES 1.0 camera are:

- Because the image pairs are stored in separate frames it does not require a velocity bias device to resolve the flow direction. In addition the method is quite tolerant of the variation in seeding densities.
- High-resolution (1008x1018 pixels) combined with high framing rate (30Hz), to produce up to 15 image field pairs.
- Asynchronous triggering image acquisition which allows for phase locking with the flow structures.

The PIV set-up is proven to be an essential tool in the characterization of the mechanisms for extension of the blow-off limit when suction flow is applied to the jet flame. In addition, it provides global instantaneous and mean measurements of other relevant mean and turbulent quantities.

Acknowledgments
This research was carried out under the sponsorship of the Office of Naval Research (N00014-92-J-1406) with Dr. Gabriel Roy as the technical monitor.

References

Law, C. K. (1988). Dynamics of stretched flames. *Twenty Second Symp. (Int.) on Comb.*, The Combustion Institute, 1381.

Lewis, B. and von Elbe, G. (1961). Combustion, Flames and Explosions of Gases. Academic Press, New York.

Lourenco, L. and Krothapalli A., (1995). On the accuracy of velocity and vorticity measurements with PIV. *Exp. in Fluids* **18**, 421-428.

Mungal, M. G., Lourenco, L. and Krothapalli, A. (1995). Instantaneous velocity measurements in laminar and turbulent premixed flames using on-line PIV. *Comb. Sci. and Tech.*, **106**, 239-265.

Strykowski, P. J., Krothapalli, A. and Wishart D., (1993). Enhancement of mixing in high speed heated jets using a counterflowing nozzle. *AAIA Journal* **31**, 2033-2038.

Strykowski, P. J., Krothapalli, A. and Jendoubi, S., (1996). The effect of counterflow on the development of compressible shear layers. *J. Fluid Mech.* **308**, 63-96.

Strykowski, P. J., Niccum, D. L., (1991). The stability of countercurrent mixing layers in circular jets. *J. Fluid Mech.* **227**, 309-343.

IV.5 On the Use of Laser Rayleigh Scattering to Study the Aerothermochemistry of Recirculating Premixed Flames

F. Caldas[1], D. Duarte[1], P. Ferrão[1], M. V. Heitor[1] and C. Poppe[2]

[1] Instituto Superior Técnico, Dept. of Mechanical Engineering, Av. Rovisco Pais, 1096 Lisboa Codex, Portugal

[2] Imperial College of Science, Technology and Medicine, Dept. of Mechanical Engineering, Exhibition Road, London SW7 2BX, United Kingdom

Abstract. The use of laser Rayleigh scattering, LRS, to study the thermochemistry of propane-air flames is shown to be accurate in the range of equivalence ratios $0.53 < \phi < 0.70$, making use of proper calibration procedures. The results have been analysed against thermocouple data and quantify the effect of flame luminosity on Rayleigh thermometry.

The measurements have been obtained in baffle-stabilised flames for $Re = 150000$ and the results are used to assess the effect of swirl, in the range $0 < S < 0.33$, on the aerothermochemistry of strongly sheared flames. To achieve these objectives, a previously reported LRS/LDV system was conveniently optimised and the results show that swirl attenuates the rate of turbulent heat transfer across the reacting shear layer, although it does not alter the existence of a large flame zone characterised by non-gradient scalar fluxes. The results have been evaluated against those obtained with the combination of LDV with digitally compensated thermocouples, which are shown to be attenuated by up to 50% mainly due to the lack of spatial resolution.

1 Introduction

Laser Rayleigh scattering, LRS, has been extensively used in laboratory flames to study the thermochemistry of non-luminous systems (e.g. Rajan et al., 1984; Namer and Schefer, 1985). Also, the results of Ferrão and Heitor (1996-a;-b) show that it can be conveniently combined with laser velocimetry and used, with adequate spatial and temporal resolution, to quantify the distribution of turbulent heat fluxes in recirculating flames, at least for lean flames (i.e., $\phi \leq 0.6$), see also the review of Ferrão and Heitor (1992) for details. Further, Almeida et al. (1995) have used a combined LRS/LDV system to analyse the effect of swirl on the structure of strongly-sheared baffle-stabilised flames. Nevertheless, the extent to which the technique can be used to study the details of the aerothermochemistry of flames with practical interest remains to be shown, mainly due to its limited application to luminous flames.

The work reported in this paper is intended to contribute to this discussion by providing experimental results on strongly-sheared recirculating propane-air flames as a function of equivalence ratio, in the range $0.53 < \phi < 1$.

Turbulent recirculating premixed flames stabilised downstream of baffles have been shown to be characterised by non-gradient scalar fluxes (e.g. Takagi et al., 1984; Takagi and Okamoto, 1987; Fernandes et al., 1994; Duarte et al., 1995), the extent of which appears to be particularly influenced by the magnitude of the mean pressure gradients associated with the streamline curvature or because of acceleration of gases across the flame front, Heitor et al.(1987), Ferrão and Heitor (1995), Duarte et al.(1995).

This paper provide further evidence of the process of turbulent mixing in recirculating flames and extends the results of Almeida et al. (1995) to improve the understanding of the interaction between gradients of mean pressure and density fluctuations. The effect of swirl on the aerothermochemistry of propane-air recirculating flames is particularly addressed, which has been possible due to the combination of laser Rayleigh thermometry and laser Doppler velocimetry.

The next section describes the experimental method and give details of the extent to which the LRS can be used in propane-air flames. Section 3 presents and discusses sample results and the last section provides the main conclusions of the work.

2 The Experimental Method and Accuracy

2.1 The Flames Studied

The experiments reported in this paper were conducted in unconfined swirling and non-swirling premixed flames of air and propane, stabilised on a disk with $D = 56$ mm in diameter, which is located at the exit section of a contraction with 80 mm in diameter. The annular bulk velocity is equal to Uo = 42.4 m/s, resulting in a Reynolds number, based on the disk diameter, of 1.5 x 105. Swirl could be imparted to the premixed reactants by a set of curved blades, located upstream of the contraction, resulting in a swirl number of S = 0.33. The equivalence ratio was varied between 0.53 and 1, although most of the flames characterised in this paper correspond to lean flames, with $\phi = 0.55$.

2.2 The Experimental Techniques

The instrumentation used throughout this work consists on a combined LDV/LRS system, which was based on a single laser light source (5 W argon-ion laser) as represented in figure 1. The system has derived from that described by Duarte et al. (1995) and Almeida et al. (1995), as the main data acquisition system includes a 16 bits analogue/digital converter, in place of the 12 bits data acquisition board previously used.

The velocimeter was based on the green light (514.5 nm) of the laser and was operated in the dual-beam, forward-scatter mode with sensitivity to the flow direction provided by a rotating diffraction grating. The calculated dimensions of the measuring volume at the e^{-2} intensity locations were 606 μm in length and 44 μm in diameter.

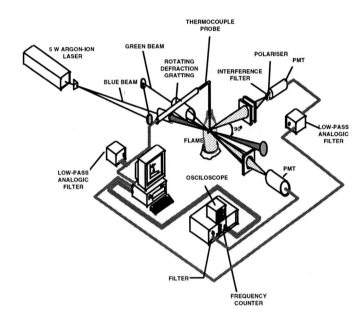

Fig. 1. Schematic diagram of the combined LDV/LRS system, together with thermocouple probe.

The Rayleigh scattering system was operated from the blue line of the same laser source, which was made to pass through a 5:1 beam expander. The light converged in a beam waist of 50 μm diameter was collected at 90° from the laser beam direction with a magnification of 1, and passed through a slit of 1 mm length. The collected light was filtered by a 1nm bandwidth interference filter and a polariser, in order to optimise the signal-to-noise ratio.

The signal was amplified and low pass filtered at 10 KHz before digitalisation. The temporal resolution of the system depends on the integration time associated with this filter which is quantified to be 50 μs. This value, associated with the typical flow velocities, give rise to path lengths up to 2 mm and, therefore, smaller than the integral length scales in the reaction shear layer. The resolution of the system was confirmed by the measured temperature distributions, which include instantaneous values close to either adiabatic or room temperature, confirming that the system is capable of resolving the temperature fluctuations associated with the premixed flames analysed in this work.

2.3 The Accuracy of LRS Thermometry

The amplitude of the LRS signal collected from a flame in the wavelength of the laser source, results from three main sources (e.g., Eckbreth, 1988):

i) The Rayleigh scattering process, as the electromagnetic radiation resulting from the elastic interaction between the incident electric field and the electric field of the molecules in the gas (with $d/\lambda \ll 1$, where d is the particle or molecule diameter and l is the wavelength of the laser beam);

ii) The Mie scattering process, as the dispersion of the incident light caused by particles (i.e., $d/\lambda > 1$) present in the flame, including soot;

iii) The radiation emitted by different molecular species formed during the combustion reactions.

As the collecting optics cannot discriminate the different light sources, but just maximise the overall signal to noise ratio, dedicated signal processing and calibration techniques were used to extend the utilisation of the LRS technique. This is achieved by compensating for the so-called "non-Rayleigh" contributions (namely, ii and iii above) possible as long as they are not dominant and are uncorrelated with the LRS signal. The related assumptions essential for the calibration procedure are discussed bellow.

The Mie scattering signals (i.e., ii above) are detected by derivative and amplitude criteria as, at least for large particles with $d/\lambda \gg 1$, the light dispersed by a particle causes a significative distortion in the collected signal (see Ferrão and Heitor, 1996-a, for details). The molecular radiation (i.e., iii above), V_{Lum}, is considered to be a function of the equivalence ratio and is evaluated by measuring the amplitude of the signal in the absence of the laser source.

A calibration procedure based on these premises was implemented and sample results obtained in the propane-air flame considered in this paper were compared with measurements made using digitally-compensated fine-wire thermocouples (Ferrão and Heitor, 1996), in order to assess the limits of the utilisation of Rayleigh thermometry as a function of equivalence ratio.

The procedure was implemented making use of measurements made at the centre of the present recirculation zone, where gas analysis has shown complete combustion. In addition, reference conditions were established for a lean flame, namely close to extinction, for $\phi = 0.53$. The related Rayleigh signal, V_H, at the thermocouple-based temperature, T_H, and the corrected Rayleigh signal for the reactants at ambient temperature were then used to calibrate the technique. Further corrections for any equivalence ratio, ϕ, were derived from the following equations:

$$T = \frac{1}{a(V - V_{Lum}) + b} \qquad [1]$$

$$a = \frac{T_a - T_H}{T_a T_H (V_H - V_a)} \qquad [2] \qquad\qquad b = \frac{T_H V_H - T_a V_a}{T_a T_H (V_H - V_a)} \qquad [3]$$

As the equivalence ratio is increased from $\phi = 0.53$, the Rayleigh signals V_a and V_H vary due to the modification in the composition of reactants and products, leading to different Rayleigh cross sections (e.g., Namer and Schefer, 1985; Ferrão and Heitor, 1996-a). Additionally, it should be noted that for different equivalence ratios, T_H do not represent the temperature of the products of combustion. Thus, two assumptions were considered as follows: i) The dependence of the Rayleigh cross section, s, on temperature is taken as linear, as suggested by Rajan et al (1984) and Shepherd and Daily (1984); and ii) The ratio between the mean temperature in the centre of the recirculation zone (i.e., for complete combustion) and the adiabatic flame temperature for each value of ϕ is considered as constant. As a result, the values of the parameters used for the calculation of equations [2] and [3] for different equivalence ratios, are corrected according to table 1.

Table 1 - Correction parameters for the calibration of the Rayleigh signal.

Values at $\phi = 0.53$	Corrected parameters for an arbitrary ϕ
V_a	$V_a \times \dfrac{\sigma_{react,\,\phi}}{\sigma_{react,\,\phi=0.53}}$
V_H	$V_H \times \dfrac{\sigma_{react,\,\phi} + \dfrac{(\sigma_{prod,\,\phi} - \sigma_{react,\,\phi})\,T_{ad,\,\phi=0.53}}{T_{H,\,\phi=0.53}\,T_{ad,\,\phi} - T_a\,T_{ad,\,\phi=0.53}}(T_{H,\,\phi=0.53} - T_a)}{\sigma_{prod,\,\phi=0.53}}$

Figure 2 quantifies the relative contribution of the flame luminosity for the overall time-averaged signal collected along the centre line of the flame, as a function of equivalence ratio in the range $0.53 \leq \phi \leq 0.92$. The results clearly show that the "non-Rayleigh" contribution to the overall signal, as mentioned before, is important for $\phi > 0.70$ and precludes the utilisation of Rayleigh Thermometry. Figure 3 shows the related implications in terms of temperature data, by comparing the LRS results against thermocouple measurements at different locations within the flame. It is clear that for $\phi > 0.70$ the error associated with the Rayleigh measurements, namely due to contamination from flame luminosity, is larger than that expected to affect the thermocouple from radiation losses. Further, for $\phi > 0.80$, the error associated with Rayleigh thermometry is too large, that precludes the use of this technique in luminous flames.

It should be noted that the Rayleigh scattering measurements may be affected by the photomultiplier shot-noise, which results in an increase of the signal rms, according to the Poisson statistics (e.g., Ferrão and Heitor, 1996-a).

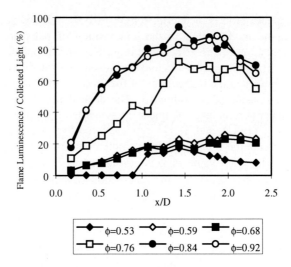

Fig. 2 - Contribution of flame luminosity for the overall time-averaged signal collected along the centre line of the flame, as a function of equivalence ratio.

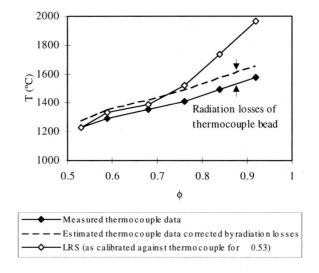

Fig 3 - Mean flame temperatures in the centre of the recirculation zone obtained by fine-wire thermocouples and calibrated LRS for different equivalence ratios.

The contribution of the photomultiplier shot-noise is, for a given experimental set-up, quantified by:

$$\frac{rms_{shot}}{V} = \frac{k}{\sqrt{V}} \qquad [4]$$

where k was evaluated according to equation [5] for the voltage obtained by the PMT at ambient conditions, V_a, and the corresponding rate of photons at the cathode, R_p, considering a cut-off frequency of the system of $f_c = 5KHz$.

$$k = \sqrt{\frac{2 f_c V_a}{R_p}} \qquad [5]$$

The velocity-temperature correlations reported in this paper were not compensated for the shot-noise as it influences the temperature fluctuations, but it is not correlated with the velocity fluctuations in the flame.

3 Results and Discussion

The analysis above has validate the use of Rayleigh thermometry to study non-luminous propane-air reacting mixtures for $\phi < 0.70$. We now turn to exemplify the use of the technique study two different recirculating flames with practical interest, namely with and without swirl. The results include those obtained by combining the Rayleigh signal with laser velocimetry in order to quantify the turbulent heat fluxes in the flames considered.

The most salient features of the mean flow characteristics of the two flames studied can be inferred from the streamlines represented in figure 4, together with the distribution of turbulent kinetic energy. For the non-swirling flame (figure 4a) the results are similar to those found in other baffle-stabilised recirculating flames (e.g. Heitor et al, 1987; Ferrão and Heitor, 1995), in that they exhibit a recirculation region extending up to $x/D = 2.21$, where the fluid has a large and fairly uniform mean temperature, surrounded by annular region of highly sheared fluid where gradients of mean temperature are large.

The single recirculation zone of the unswirled flame is to be contrasted to that of the swirling flame, figure 4 b), which is shorter, wider and annular in shape, because it includes an inner annular vortex with positive velocities along the centreline (Almeida et al., 1995). The inner recirculation zone is associated with positive mean velocities along the centreline up to the first stagnation point and rotates in the opposite sense to the outer recirculation zone. This nature of the swirling flame is characterised by a comparatively large inclination of the mean velocity vectors at the exit which, together with the aspects mentioned before, represents a direct consequence of the centrifugal forces associated with the swirl motion.

446

In general, the results quantify highly strained flames with maxima velocity fluctuations along the shear layer surrounding the recirculation zone. Turbulence is mainly generated by the interaction between shear strain and shear stress (Ferrão and Heitor, 1994), giving rise to a strongly anisotropic turbulent field with comparatively large axial velocity fluctuations. As the stagnation point is approached, the cross-stream turbulent components increase as a result of the augmented importance of the interaction between normal strains and normal stress in the conservation of turbulent kinetic energy, as in other recirculation flows with stagnation points.

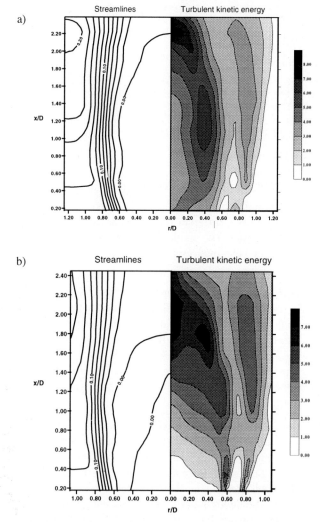

Fig. 4 Streamlines and distribution of turbulent kinetic energy along a vertical plane of symmetry:
   a) Non-swirling flame.
   b) Swirling flame.

Figure 5 shows profiles of the measured turbulent heat transfer rate of the two flames analysed, together with the corresponding mean velocity and the progress reaction variable.

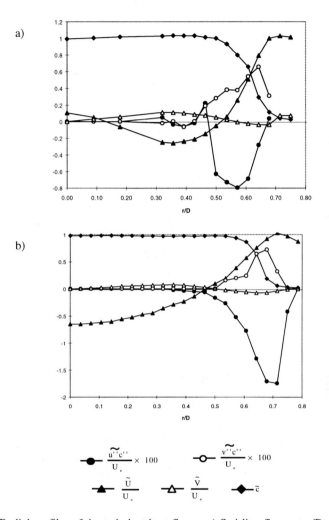

Fig. 5 - Radial profiles of the turbulent heat fluxes: a) Swirling flame at x/D = 1.04. b) Non-swirling flame at x/D = 1.09.

The results show that turbulent heat transfer is restricted to the reacting shear layers, where the temperature gradients are significative. Additionally, while in the non-swirling flame (figure 5 b), $\widetilde{u''c''}$ and $\widetilde{v''c''}$ do not change in sign, in the swirling flame (figure 5 a) the sign of both components of the turbulent heat fluxes change along the radius, as a result of the complex interaction between the pressure and temperature (or density) fields.

448

Prior to the analysis of the aerothermochemistry of the flames considered, it is convenient to discuss the accuracy of the results of figure 5, namely against those obtained by combining LDV with bare-wire thermocouples. Similar comparison in non-swirling flames (Ferrão and Heitor, 1996-b) has shown that the use of the thermocouple probes attenuates the measured velocity-temperature correlations mainly due to the lack of spatial resolution. The results of figure 6 extends this conclusion to swirling flames and shows that the results obtained from the two techniques are qualitatively in agreement, but the lack of spatial coincidence between the thermocouple bead and the LDV measuring volume leads to the underestimation of the values of the correlations obtained.

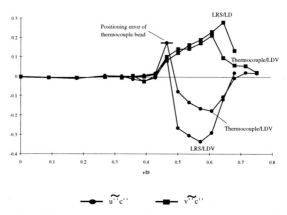

Fig. 6 - Radial profile of the turbulent heat flux components, measured by the combined LDV/LRS and LDV/thermocouple systems, for the swirling flame at x/D = 1.04.

It is clear that the accuracy of the measured velocity-temperature correlations is also dependent upon the absolute accuracy of the time resolved measurements of the progress variable, which is analysed in figure 7 making use of sample results obtained with LRS and digitally compensated fine-wire thermocouples. The results show that the compensation of the LRS signal for the pmt shot- noise (as in Ferrão and Heitor, 1996-a) gives values in agreement with those obtained by digitally-compensated thermocouples and confirms the ability of the procedures used for temperature fluctuation measurements. In general, analysis has shown that the errors induced in the results due to the photomultiplier shot-noise depend upon the experimental conditions used, but the net effect is to affect the r.m.s. of the temperature fluctuations up to 10% of the measured values.

We now turn to the analysis of the turbulent heat flux in the flames considered and figure 8 shows that a large component of the related vectors is directed along the isotherms, rather than normal to them, as would be expected form gradient transport models (Ferrão and Heitor, 1995).

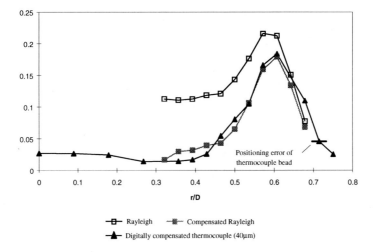

Fig. 7 - Radial profile of the rms of the fluctuations of progress reaction variable, measured by LRS and digitally compensated thermocouples, for the swirling flame at x/D = 1.04.

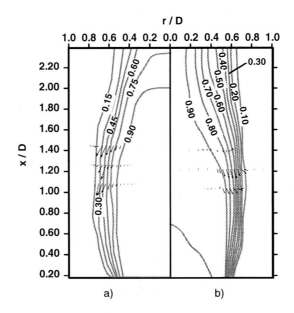

Fig. 8 - Turbulent heat fluxes superimposed on the isotherms: a) Non-swirling flame. b) Swirling flame.

The new features provided by these results is that swirl decreases the magnitude of the turbulent heat fluxes due to the attenuation of the mean temperature gradients across the reacting shear layer.

The results for the non-swirling flame have been explained before in terms of the interaction between the mean pressure field and the density fluctuations, which are important in the process of turbulent transport typical of reacting flows. The present results confirm the evidence first given by Almeida et al. (1995) that this interaction is affected by the degree of swirl imposed on the flows. In general, the results confirm that the prediction of these kind of flames must be based on second moment, rather than on effective viscosity, turbulent model closures so as to capture the effects of the mean pressure field in the conservation of turbulent heat fluxes.

It has been shown in the literature that the process of "counter gradient" heat transport can be explained by the preferential deceleration of the products of combustion, relatively to the cold reactants (e.g., Heitor et al., 1987; Hardalupas et al., 1996), and here this is clearly shown throughout the joint probability density function (pdf) of axial velocity and temperature of figure 9.

The change in sign of the axial turbulent heat flux across the reacting zone at x/D = 1.04 is explained in terms of the mean pressure distribution across the double recirculation zone associated with the swirling flame (see figure 4b), which results in the change of the patterns of the distributions of figures 9 a) and 9 b).

a)

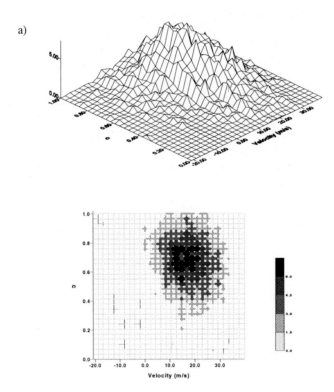

b)

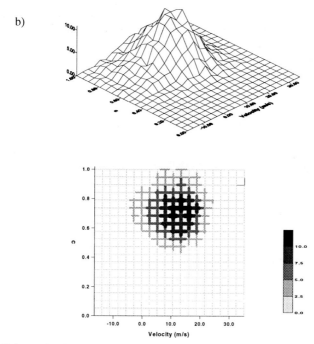

Fig. 9 - Joint probability density functions of axial velocity and temperature fluctuations, for the swirling flame at $x/D = 1.04$.

  a) $r/D = 0.57$ b) $r/D = 0.46$

In the zone of its curved flame front, where the flow is characterised by an adverse pressure gradient (as at $x/D = 0.57$, figure 9 a), the hottest "pockets" of fluid are associated with the lowest velocities. However, the radial heat flux is always positive, as previously shown for non-swirling flames. This is the expected direction of the heat flux in baffle-stabilised flames, which is not altered by the swirl motion. In summary, the evidence is that the net turbulent scalar flux is particularly dependent on the mean pressure distribution in the flow and occurs along directions in which the mean scalar gradient is not large.

4 Conclusions

The use of laser Rayleigh scattering, LRS, to study the thermochemistry of turbulent premixed propane-air flames is shown to be accurate for equivalence ratios smaller than 0.70, if adequate calibration procedures are used. The results quantify the effect of flame luminosity on the accuracy of Rayleigh thermometry and are extended to those of combined velocity-temperature correlations.

452

The shot noise associated with the photomultiplier used to collect the Rayleigh signal do not affect the turbulent velocity-temperature correlations obtained by the combined LDV-LRS system, which also avoids the limitation previously found associated with the lack of spatial resolution of combined LDV-thermocouple system.

Swirl alters the main aerodynamic features of the baffle-stabilised flames, in that it generates two annular counter rotating recirculation zones. The results give evidence of zones of non-gradient scalar fluxes which are associated with preferential deceleration of the combustion products relatively to the cold reactants. In general, the net turbulent scalar flux occurs along directions in which the mean scalar gradient is not large and is mainly determined by the mean pressure field associated with the flames studied.

5 References

Almeida, P., Ferrão, P. and Heitor, M.V., 1995, The effect of swirl on the interaction between pressure gradients and density fluctuations in baffle-stabilised premixed flames, Proc. of 10th Symposium on Turbulent Shear Flows, Vol. 2, pp. 16-7 to 16-12.

Chandran, S.B.S., Komerath, N.M., Grissom, W.M., Jogoda, J.I. and Strahle, W.C., 1985, Time resolved thermometry by simultaneous thermocouple and Rayleigh scattering measurements in a turbulent flame, Combust. Sci. and Tech., Vol. 44, pp. 47-60.

Dibble, R.W. and Hollenbach, R.E., 1981, Laser Rayleigh thermometry in turbulent flames, Proc. of 18th Symposium (Intl.) on Combustion, The Combustion Institute.

Duarte, D., Ferrão, P. and Heitor, M.V., 1996, Flame structure characterisation based on Rayleigh thermometry and two-point laser Doppler measurements, in: Developments in Laser techniques and application to Fluid Mechanics, Eds. Adrian et al, Springer Verlag, pp. 185-249.

Eckbreth, A.C., 1988, Laser Diagnostics for Combustion Temperature and Species, Abacus Press.

Fernandes, E.C., Ferrão, P., Heitor, M.V. and Moreira, A.L.N., 1994, Velocity-temperature correlations in recirculating flames with and without swirl, Experimental Thermal and Fluid science, Vol. 9, pp. 241-249.

Ferrão, P. and Heitor, M.V., 1992, Probe and optical techniques for simultaneous scalar-velocity measurements, in: Combustion Flow Diagnostics, Durão et al., pp. 169-231.

Ferrão, P. and Heitor, M.V., 1995, Turbulent mixing and non-gradient diffusion in baffle-stabilised flames, in: Turbulent Shear Flows - 9, Eds. Durst et al, Springer Verlag, pp. 427-437.

Ferrão, P. and Heitor, M.V., 1996-a, Measurements of velocity and scalar characteristics in premixed recirculating flames, part 1: probe and optical diagnostics, Submitted to Experiments in Fluids.

Ferrão, P. and Heitor, M.V., 1996-b, Measurements of velocity and scalar characteristics in premixed recirculating flames, part 2: simultaneous measurements, Submitted to Experiments in Fluids.

Hardalupas, Y., Tagawa, M: and Taylor, A.M.K.P., 1996, Characteristics of counter-gradient heat transfer in a non-premixed swirling flame, in: Developments in Laser Techniques and Applications to Fluid Mechanics, ed. Durst et al.., Springer Verlag, pp. 159-184.

Heitor, M.V., Taylor, A.M.K and Whitelaw, J.H., 1987, The interaction of turbulence and pressure gradients in baffle-stabilised premixed flames, J. Fluid Mechanics, Vol. 181, pp. 387-413.

Namer, I. and Schefer, R.W., 1985, Error estimates for Rayleigh scattering density and temperature measurements in premixed flames, Experiments in Fluids, Vol. 3, pp. 1-9.

Rajan, S., Smith, J.R. and Rambach, G.D., 1984, Internal structure of a turbulent premixed flame using Rayleigh scattering, Combustion and Flame, Vol. 57, pp. 95-107.

Shepherd, I.G. and Daily, J.W., 1984, Rayleigh scattering measurements in a two-stream free mixing layer, Proc. of The Combustion institute, Western States (USA), W98/CI, paper #84-15.

Takagi, T. , Okamoto, T. , 1987, Direct measurement of the turbulent transport of momentum and heat in the swirling flame, in: Laser Diagnostics and Modelling of Combustion, Eds.Iinuma, K. et al., pp 273-280, Springer Verlag, Berlin.

Takagi, T. , Okamoto, T. , Taji, M. and Nakasuji, Y. ,1984, Retardation of mixing and counter-gradient diffusion in a swirling flame, Proc. 20th Symposium (Intl.) on Combustion, pp. 251-258. The Combustion institute, Pittsburgh.

IV.6 Convenient Laser Diagnostics for Aerodynamic and Chemical Study of Axisymetric Non Premixed Bluff-Body Burner Flames

A. SUSSET[1], K. MOKADDEM[2], D.W. KENDRICK[2], J.C. ROLON[2], D. JAFFRE[3], D. HONORE[3], M. PERRIN[3], C. GRAY[4] and J.B. RICHON[4]

[1] Institut National des Sciences Appliquées de Rouen, F-76130 Mont Saint Aignan Cedex, France
[2] Laboratoire EM2C - CNRS - Ecole Centrale de Paris, F-92295 Chatenay Malabry Cedex, France
[3] Gaz de France, Research and Development Division, CERSTA, F-93211 La Plaine St Denis, France
[4] Optical Flow System Ltd., Mayfield Road, Edinburgh, Scotland

ABSTRACT

This paper presents an experimental investigation of a turbulent, non premixed, methane/air flame, produced by an axisymetric Bluff-Body burner. Only a few experimental data exists on time and spatially resolved fuel concentration and velocity fields on such reacting flows, despite their interest for a better understanding of combustion processes, modelling and numerical calculations in turbulent combustion.

The aim of this work is to provide these experimental data by using instantaneous, multiple point and spatially resolved measurements of methane concentration images by planar Mie scattering technique, and instantaneous velocity fields by cross-correlated Particle Image Velocimetry (PIV).

Results of methane concentration imaging using several tracer particles are compared to those obtained elsewhere from probe measurements. The Teflon marker is revealed to be the more efficient methane tracer in flow as it is capable to follow the aerodynamic flow fluctuations and to characterize the zones corresponding to CH_4 consumption in the flame by identifying the mean identical decomposition temperature of CH_4 molecules and Teflon particles. In this work we show that Mie scattering on Teflon particles is an accurate tracer in both reactive and non reactive flows as it gives information on methane concentration and flame structure.

The PIV vector fields are related to that previously measured using Laser Doppler Velocimetry. Analysis of the fluctuating velocity components in the PIV measurements reveals differences with LDV results which are mainly caused by the limited spatial resolution of the PIV method. The exploitation of PIV specific information, i.e. the spatial correlation of measurements, makes it possible to quantitatively study the coherent structures present in the flow.

1 - INTRODUCTION

The bluff-body model burner is a simplified geometry version of an industrial burner exploiting the wake effect of an obstacle to stabilise a turbulent non-premixed flame. It replicates the operating conditions of a full-scale burner while simplifying its study (axisymetric geometry). It is a valuable tool for the investigation of the influence of the conditions of the methane/air mixture on the characteristics of the flame (stability, length, temperature, polluting emissions). Until now the imaging techniques applied to bluff-body studies have been essentially flow visualization by Lorenz Mie scattering in non reactive case [Scheffer et al., 89], and flame species concentration measurements by Raman scattering and PLIF [Namazian et al., 88].

In the current study, two different laser diagnostics have been used for aerodynamic and chemical study of reactive and non reactive flows generated by a Bluff-Body burner. Turbulent non premixed methane-air flames are investigated using planar Mie scattering to render 2D map of the methane concentration. Different tracers are compared and new markers studied. The convenience of Teflon to be an accurate marker of the fuel has been demonstrated by comparison with previous measurements of CH_4 concentration by probe sampling. For the last few years, the development of PIV [Gray, 92] has been heralding genuine advances in the quantitative study of vortical structures and the complex flows they generate. Here, we present PIV measurements made on the bluff-body burner for one given flow case. The effect of each of the components of the acquisition system on the correlation calculation is evaluated, in order to underline the weakness of the Signal-to-Noise Ratio as a vector validation criterion. The comparison of average velocity fields obtained by PIV and LDV makes it possible to validate experimental choices and confirm the efficiency of the PIV method. A last approach based on the spatial correlation of the vectors shows the quality of the information provided by PIV.

2 - EXPERIMENTAL SET-UP

2.1 Description of the burner

Several Bluff-Body burner geometries have been already the subject of different studies [Perrin et al., 90 - Namazian et al., 95]. A schematic view of the axisymetric, unconfined burner is shown in figure 1.

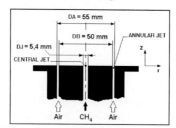

Fig. 1 Sketch of the Bluff-Body burner.

Methane is injected through the center of the Bluff-Body at a velocity of 21 m/s. Air is supplied through a coaxial jet surrounding the Bluff-Body at a velocity of 7.5 m/s. The Reynolds numbers of the methane and air jets are respectively : 7000 and 3300. In the present study, both reactive and non-reactive flows have been investigated with the two different techniques.

2.2 Experimental set-up and synchronisation

A multi-pulsed Nd-YAG laser (Quantel YG585 3V) is used for Mie scattering imaging of particles seeded in the flow. This system is able to deliver one or two laser pulses with an energy of 150 mJ/pulse at 532 nm, and a frequency rate of 25 Hz. The parallel laser sheet (0.2 mm thick- 150 mm height) is formed with a set of cylindrical an spherical lenses. Mean CH_4 concentration images are obtained on a Proxitronic HF1 ICCD camera which is triggered by the laser pulses. The synchronisation timing diagram for PIV experiment is presented on figure 2. The CCD camera (LHESA Electronique LH510), operating in field mode, drives the laser via a pulse generator (Stanford Research DG535) triggered by the odd/even signal of the VD/2 field [Lecordier et al., 94].

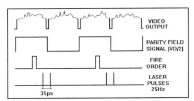

Fig. 2 Synchronisation timing diagram for PIV experiment.

In each experiment, the camera is equipped with a 532±10nm interferential filter in order to suppress spontaneous emission signal from the flame. Sequences of 250 images are directly acquired and digitised over 8 bits at video rate in a computer RAM via a Matrox Magic Color frame grabber, with a resolution of 512x512 pixels. Imaging processings are performed over an ensemble of 1000 instantaneous images.

For methane imaging, the ICCD camera provides a measurement volume of $0.15x0.15x0.2$ mm^3. This resolution is determined by the field of view dimensions and the need to have sufficient particles in the measure volume so as not to invalidate the assumption of marker continuum and to reduce the marker shot noise. In PIV experiment, the velocity is deduced from the calculation of the cross-correlation in square areas. Therefore, the result corresponds to the averaged velocity in this interrogation area. In order to reduce the spatial resolution, the measurement volume has been set to $0.11x0.11x0.2$ mm^3.

The Nyquist sampling theorem states that the size of the smallest detectable structures is two pixels, i.e. 0.3 mm in our cases, which should be compared to the Kolmogoroff length. For the studied flame, the microscale has the approximate value of 0.03 mm [Perrin, 93]. The experimental resolution is a factor 10 larger than the Kolmogoroff scale and well below the larger flow scales which vary between 5 and 25 cm. This results in flow scales being filtered below 0.3 mm. This is not critical since the Kolmogoroff scale is characterised by molecular diffusion processes. In this microscale, Mie scattering lead to important inaccuracies induced by the fact that particles effective diffusivity is completely different from that of gases.

3 - CALIBRATION OF THE TECHNIQUES

3.1 Calibration of the methane imaging technique

3.1.1 Data reduction and error analysis

The CH4 images are obtained by correcting each image for CCD and laser sheet spatial non uniformities, as well as aberrations and vignetting of the optics (for more details, see [Mokaddem et al., 94]).

Each image is rescaled so that the maximum intensity measured at the base of the jet potential core corresponds to 100% methane concentration and the minimum signal intensity out of the flow corresponds to 0 % of methane concentration. This correction takes into account the shot to shot variations in laser power. The random errors, taking into account seeding density fluctuations, shot noise, electronic noise of the camera and frame grabber have been evaluated to 2% of the maximum dynamic range. These effects are negligibles in the calculations of average methane concentration. The bias errors which depend on the spatial coordinates are corrected by using a reference image taking into account the spatial non uniformities and non linearity of the CCD response, non uniformity of the laser sheet and the optics' aberrations. The errors were found to be 2% of the maximum dynamic range. Moreover, statistical errors are introduced on the calculation of the statistical quantities over a finite ensemble of images. If we combine bias and statistical errors, the uncertainties obtained in the mean concentration data are about 5%.

3.1.2 Study of the different markers

A conceptual model for the dispersion of particles is the entrainment of particles by the vortex structure and ensuing centrifugal effects as the particle rotate within the organised structures [Crowe et al., 88]. This model is accurate in Bluff Body flames since the role of large scale structures widely determines the mixing and mass transport in these flows. The Stokes number St is an important parameter which affects the particles spreading in organised structures. It is defined as the ratio between the aerodynamic response of the particle τ_p and the time τ_f associated with the motion of large scale structures. τ_p is function of the density and diameter of the particles and the dynamic viscosity of the carrier fluid [Neveu et al., 94]. The flow time response is modelled as the ratio between the size of the structures and the mean velocity of the flow. If St<<1, the particle will faithfully follow the fluid path lines. If St~1, the particles will trend to be centrifuged by the structure. For St>>1, the particle will have insufficient time to respond to changes in the fluid and will continue along a near rectilinear path. Figure 3 provides a summary of these effects.

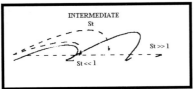

Fig. 3 Effect of St on the aerodynamic response of a particle

The particles used in this work are silicone oil, zirconium oxide and new markers such as Teflon and calcium hydroxide. Figure 4 gives the time response of these tracers in comparison the flow time macroscale response (non reactive case). It shows that the particles will follow the fluctuations of the large scale motions with an accuracy depending on the Stokes number.

Particle	diameter (μm)	Stokes number
Silicone Oil	5	0.02
Teflon	3	0.003
Calcium Hydroxide	10	0.2
Zirconium Oxyd	$30 . 10^{-3}$	$1.3 . 10^{-4}$

Fig. 4 Time response and diameter of the particles used

For methane imaging, the particles are injected with the fuel jet flow. In the cold flow experiment, the presence of the seeded particles is correlated with the methane presence within the volume domain. In order to extract the methane concentration from the images of flames, the particle must yield information on regions corresponding to the methane reaction zones and the seeding quantities must be sufficiently low to have little effect on the chemistry of the combustion process. It is known that CH_4 molecules dissociates at 400°C, two possibilities have been therefore explored :
- the use of particles disappearing before or at 400°C such as silicone oil or a new particle such as Teflon,
- the use of particles that generate a noticeable decrease in signal luminance above 400°C by a decrease in their molecular dimensions. A new particle such as calcium hydroxide which losses a H_2O molecule at this temperature satisfies this criterion. These particles used herein are compared to an inert reference : zirconium oxide.

3.2 Calibration of the PIV rig

The variation of velocities between the central methane jet (35 m/s maximum) and the recirculation zone (zero and near-zero velocities) imposes to resolve a large dynamic range. Such a resolution is only possible with cross-correlation processing, where zero and reversing displacements can theoretically be measured. Moreover, in our case, presence of zones with large seeding concentrations in the vicinity of the burner exits adds another limitation to the use of an auto-correlation technique

460

[Trinité, 93]. We therefore selected the cross-correlation image analysis technique, available with the use of the field mode of the camera (§ 2-2).

Figure 5 shows how the information flows across the elements of the acquisition system. In the following, the effect of each of the components of the acquisition system on the correlation calculation is evaluated, in order to underline the weakness of the Signal-to-Noise Ratio (SNR) as a vector validation criterion.

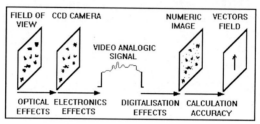

Fig. 5 Information flow across the acquisition system.

The aberrations of the recording optics which noticeably affect the images are mainly geometric distortion and vignetting. These two effects, commonly encountered in imaging, are corrected for during an image pre-processing phase.

The response of each pixel on the CCD sensor to uniform lighting is not in itself uniform, therefore it is necessary to construct a correction file to counter this effect, by exposing the chip to different light levels cast on a uniform plane and performing a linear regression on the whole of the averaged images. The displacement changes resulting from the non-linearity corrections are not significant. If during a batch analysis we apply vector filtering using the Signal to Noise Ratio of the correlation computation, the number of vectors left intact by the filter at one given point of the analysis grid can vary by as much as 6% for a 1.18 SNR threshold. The SNR is a parameter far too sensitive to image definition to be adopted as main vector validation criterion.

Image defects induced by the synchronisation and the digitisation of the video signal are mainly caused by line jitter and pixel jitter. The direct effect of the line jitter, i.e. a horizontal shift of the digitised lines, can be observed on the fluctuations of the grey levels in an image column. The effect of this shift on a vector field has been shown elsewhere [Rouland, 94]. Images of a target with randomly set white spots were taken to assess the accuracy of zero velocity measurements. An accuracy better than 0.04m/s was found with our set-up. The incidence of the pixel jitter on the accuracy of the correlation computations is quantified using synthetic digital images. Accuracy of the displacement measurements is not significantly modified (about 5%). This fluctuation of the particle edge definition mainly degrades the SNR.

We seek to determine the accuracy on the localisation of the maximum of the correlation peak. The interpolation of the correlation peak by a gaussian function leads, in ideal conditions, to a maximum theoretical accuracy near 1/100th of a pixel [Willert et al., 91]. The accuracy of the calculation is calibrated with synthetic images. In the range of measured displacements, i.e. from 0 to 8 pixels, the accuracy on the determination of the correlation peak position is evaluated to 0.08 pixel.

4 - METHANE IMAGING RESULTS

Mie scattering is demonstrated using several particles such as silicone oil, Teflon, calcium hydroxide and zirconium oxide. Figure 6 shows the average methane concentration in a non reactive, Bluff Body flow using Teflon particles. This figure reveals the symmetry of the flow and the weak penetration of the entrained air within the central jet. These results qualitatively agree with LDV velocity data discussed in Neveu and al.

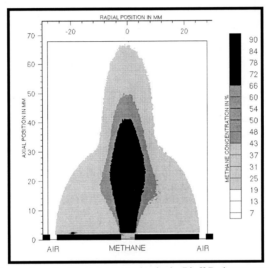

Fig. 6 Mean value of methane concentration in the Bluff-Body non reactive flow
by Mie scattering of Teflon particles

Figure 7 presents the mean value of methane concentration profile along the axial direction of the fuel jet. The data from Mie scattering are compared with those obtained from probe measurements (horizontal probe, 4 mm in diameter) [Prwzeswa et al., 91]. In this experimental method, the absolute uncertainties in the mean CH4 concentration are about ±1%. The results show good agreement between the two experimental methods for non reactive flows. The differences noticed in the particles' behaviour are due to their time response. The corresponding Stokes number for each particle indicates that centrifugal effects will be important for hydroxide calcium. This effect, coupled the inhomogeneities in the seeding particle diameter, also explains the fluctuations in the mean axial profile for hydroxide calcium.

462

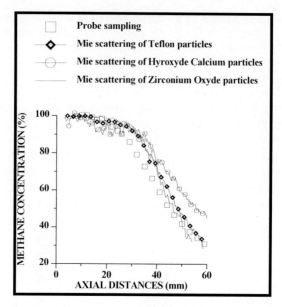

Fig. 7 Mean axial methane concentration in the Bluff-Body non reactive flow
by Mie scattering of different particles

Figure 8 presents the mean methane concentration, 20 mm from the burner and shows good agreement between both the probe and Mie measurements. The mixing of the fuel into the recirculating zone is evident, indicating that this region consists of a nearly uniform mixture of fuel and air.

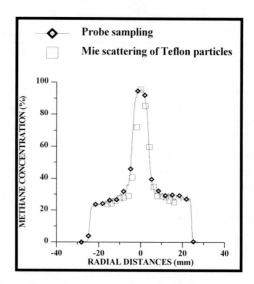

Fig. 8 Mean radial methane concentration in the Bluff-Body non reactive flow
(20 mm from the base of the burner) by Mie scattering of different tracers

Figure 9 reveals the mean CH4 concentration for the combusting case using Teflon particles.

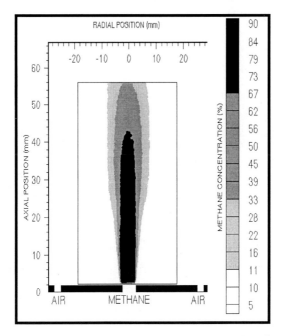

Fig. 9 Mean methane concentration in the Bluff-Body by Mie scattering of Teflon particles

One sees that the methane concentration, in the recirculating zone, is beyond 0.5% (theoretical limit of detectability by Mie scattering). This result shows that this region corresponds to a reacting zone where the fuel is being consumed. It is also apparent that there is no penetration of the ambient recirculating air into the potential core of the fuel jet.

Figure 10 depicts the axial methane concentration by Mie scattering using different particles. Together with these results, the mean temperature profile at the same locations from fine wire thermocouple measurements is also presented [Neveu et al., 94]. This graph shows that Teflon particles are able to depict the mean CH4 concentration until a height of 50 mm from the basis of the burner. Oxide zirconium particles which act as an inert, provide methane concentrations measurements below the values obtained for a jet seeded with Teflon particles jet. This effect, which was not observed in the non reactive case, is most likely due to changes in the optical properties with temperature.

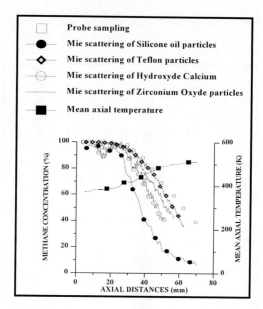

Fig. 10 Mean axial methane concentration and temperature in the Bluff-Body flame
by Mie scattering of several particles

Figure 11 shows that above 50 mm from the base of the burner, Mie scattering of Teflon particles yield methane concentration data below the probe measurements. In this low velocity region, the effects of molecular diffusion are more pronounced and thereby explain the difference in behaviour between the particle and the CH_4 molecules.

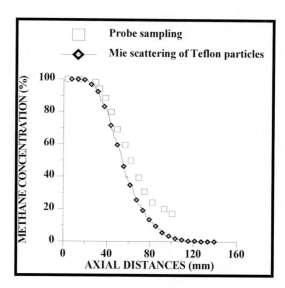

Fig. 11 Mean axial CH_4 concentration in the Bluff-Body flame by Mie scattering of Teflon particles

The radial concentration profile at 20 mm from the burner's base (fig. 12) shows the accuracy of the Teflon particles in comparison with the probe measurements. The difference observed at the edge of the fuel jet between the Teflon measurements and the probe data is probably due to the intrusive effect of the latter method. During combustion, methane is consumed at the edge of the fuel jet. In this region, the horizontal probe generates a recirculation zone that increases the fuel air mixing and hence the combustion, thereby under estimating in the CH_4 concentration. This effect, however, does not exist with Mie scattering.

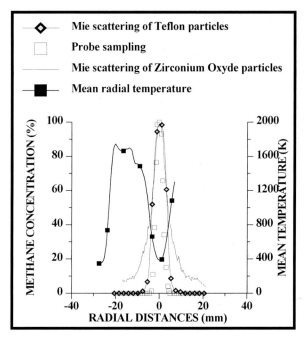

Fig. 12 Mean radial methane concentration and temperature in the Bluff-Body flame.

5 - PIV RESULTS

5.1 Comparison with LDV

A PC-based analysis software with batch processing facility (VidPIV 2.09, Optical Flow Systems) is used to obtain instantaneous velocity fields as shown figure 13. The size used for the interrogation areas is 32x32. Each instantaneous field is filtered by an SNR threshold of 1.18 and by thresholding of each velocity component. The calculation of a field composed of 854 vectors takes 70 seconds on a 130 MHz Pentium. The addition of a DSP card would at least double the processing rate.

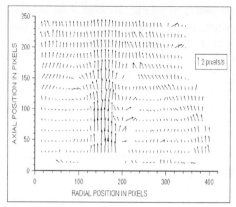

Fig. 13 Instantaneous velocity field for the non-reacting flow.

The convergence of the cross-correlation PIV measurements towards a field average over 480 vector files is presented in figure 14. We calculate the difference between partial averages and total averages, weighted by the value of the average velocity at the point of interest. In order to eliminate the coupling between the instant at which images are recorded and certain natural frequencies of the flow, the convergence calculations are repeated after shifting the analysis window.

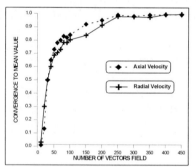

Fig. 14 Convergence of the PIV velocity components
towards the average velocity field (non-reactive flow).

The convergence towards the average dynamic field is very quick, as illustrated by figure 14. We consider that a set of 250 vector fields, i.e. 10 seconds worth of video imaging, is sufficient to obtain a good representation of the average vector fields in the flow. For the LDV measurements [Neveu et al., 94], the average velocity at one point is obtained in about 3 seconds, i.e. about 3000 particles crossing the measurement volume. For the PIV images, we count in average around 30 particles per grid cell in the recirculation zone, i.e. 7500 particles for the convergence towards the average dynamic field. Figure 15 shows the average fields resulting from LDV and PIV measurements. The original LDV measurement grid is of much finer step than that for PIV measurements, therefore an interpolation of the LDV measurements to the PIV grid is performed. The average velocity field in the recirculation zone shows the existence of two vortices : a first one rotating counterclockwise, which is driven by the air flow; the second one, smaller, is located at the root of the methane jet, and rotates clockwise.

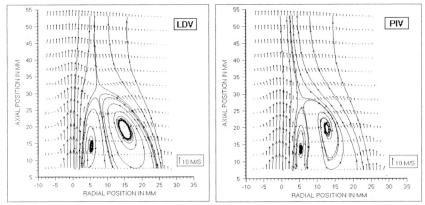

Fig. 15 Comparison of PIV and LDV average dynamic fields (non-reacting flow).

The overall structure of the average dynamic field, which consists in the different zones of interaction between the two jets, is similar for the two methods. We think that the main difference seen in the central fuel jet and in the recirculation zone may be due to a fuel flowrate control drift. Figure 16 shows radial velocity profiles at a height of 5.6 mm. The profiles from PIV and LDV have similar features. The amplitude difference at the location of the methane jet is due to a flowrate control drift. The profiles obtained by PIV are wider in the region of the jets. This effect is the consequence of the coarse spatial resolution of PIV and of the large overlap of the interrogation windows in the radial direction (75%). The lack of spatial resolution of PIV is mostly felt in the air jet, where the interrogation window (3.2 mm in the x - direction) is larger than the jet (2.5 mm).

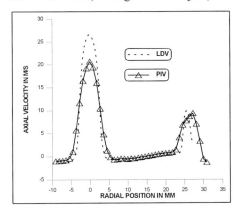

Fig. 16 Comparison of radial profiles of axial velocity (z=5.6mm).
(LDV: 95 measurement points. PIV : 45 measurement points).

The comparative fluctuation profiles are shown in figure 17. The same general trend is found in both axial RMS profiles. The amplitude differences and the absence of a double maximum are due to the lack of spatial resolution of PIV compared to LDV. Recently, algorithms have been devised to bring a solution to the loss of information in the analysis window [Okamoto et al., 95]. Such algorithms allow in theory the identification of rotation, expansion and shear in the analysis window by linking each particle by a fictitious spring and then analysing the deformation of these systems from one image to the next. This technique, which is based on the identification and tracking of particles, could be applicable in PIV in the recirculation zone.

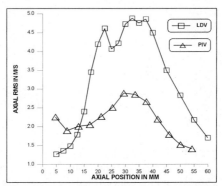

Fig. 17 Comparison of axial profiles of RMS axial velocities. Non-reacting flow.

The three first points of the RMS profile of the axial component, as obtained by PIV, show greater velocity values than LDV. For the first point, the position of the analysis window in the image is mainly responsible for the high RMS value. Indeed, the edge of the Laser sheet is located a few millimetres above the burner, causing half of the first interrogation area to be empty of particle images for the analysis grid chosen. As SNR filtering is used, some erroneous vectors computed in this interrogation window are not eliminated by the filtering operation. If the SNR threshold is increase to 1.4, the number of vectors drops from 350 to 250 out of a total of 480.

The position of the three first points corresponds to the plateau of high velocities (see fig. 16). For these velocities, the calculated displacement in the 32x32 pixel interrogation area is about 6 pixels, which reduces the useful area for the calculation. The SNR of these 3 first points is 25% less than the average SNR over the vector field. It is then difficult to appreciate the incidence of erroneous vectors on the value of the fluctuation.

In the first part, the comparative study of the two anenometric methods has mainly enabled us to validate the PIV measurements. The following section is based on an aspect which is specific to PIV, i.e. the spatial correlation of vectors, which makes it possible to identify in the flow coherent structures such as vortices.

5.2 Vortex Identification

Figure 18 is an instantaneous image of the reactive flow. Two different structures can be identified in the eddy driven by the air jet in the recirculation zone. Figure 19 presents the vector field calculated over the interrogation area shown in figure 18.

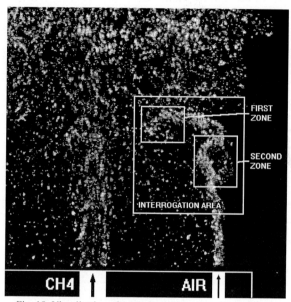

Fig. 18 Visualisation of two coherent structures in the flame.

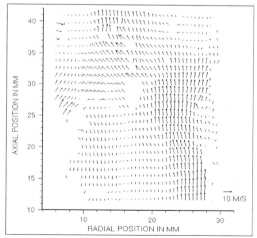

Fig. 19 Instantaneous vector field
in the interrogation area shown in figure 18.

470

The first structure identified in figure 18 is directly visible on the vector field, without post-processing. We see a set of vectors which have a group velocity with an orientation differing from that of the surrounding velocity vectors. The second structure in the image in figure 18 is not apparent in the raw vector field. Simultaneous measurements by fine wire thermocouple and LDV [Neveu et al., 94] have shown that hot gas "packets" were ejected from the recirculation zone for two distinct velocity values corresponding each to a specific gas temperature Tg. The following table shows the two velocity values at r = 17.5mm and z = 30 mm.

Tg (K)	axial velocity (m/s)	radial velocity (m/s)
900	6	5
1700	4.5	1.5

The bimodal nature of the joint velocity distributions at this point explains the intermittent behaviour of the flow in the recirculation zone area, this behaviour being driven by the radial velocity component [Scheffer et al., 87]. Figure 20 shows a grey colour-coded plot of the radial velocity component for the visualised field.

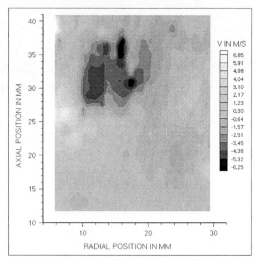

Fig. 20 Radial velocity component for the dynamic field visualised in figure 19.

One sees clearly a zone with a group velocity breaking the continuity of the surrounding values. This set may correspond to the hot gas "packet" ejected out of the recirculation zone. The measurements of velocity and position of this structure corroborate this assumption, as the results obtained with PIV are identical to that obtained with LDV (900K component). In the near future, experiments will be conducted in order to correlate the velocity measurements with the position of the flame front. We emphasise that the observed structure has mainly a translation component, since the subtraction of the average entrainment velocity does not produce a vortical velocity field.

The second structure seen on the image bears resemblance to the shedding of a vortex. A first processing pass is performed on the image, then an average velocity value corresponding to the entrainment velocity of the structure is subtracted from the vector field. The final result is presented in figure 21, where a vortical velocity distribution is clearly visible.

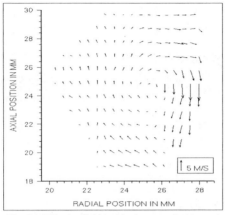

Fig. 21 Velocity field inside the second structure.

The difficulty in this type of processing is to locate the structure in the image to calculate its average entrainment velocity. A systematic study over the whole image could be undertaken if the characteristic dimensions of the structure were known, but the associated processing would be time-consuming and the smaller-scale structures would be missed. In our case we could take advantage of the presence of a strong concentration in particles in the vortex to define a skeleton the size and location of which would serve as starting point for the calculation of the vector field.

6 - CONCLUSIONS

The data obtained by Mie scattering indicate the accuracy of Teflon particles as a methane tracer in both reactive and non reactive flows. Due to the difficulties in using spectroscopic method such as Raman diffusion in reacting turbulent flows, Mie scattering on Teflon particles is a complementary diagnostic technique for studying turbulent flames. The images resulting from Mie scattering on Teflon particles provide information on the CH_4 concentration, the flame structure, the recirculating zone and the stagnation regions. This method is capable of producing future quantitative results of RMS methane concentration fluctuations, probability density functions and spatial correlations.

The analysis of PIV data has shown that the production of an average dynamic velocity field was much faster than with LDV, while the fluctuating component bandwidth and the spatial resolution are inferior. A convergence criterion cannot be the only guarantee of integrity of the PIV method. The knowledge of the accuracy on the correlation calculations for each point of the interrogation grid, as well as the vector validation methods still remain key parameters for the quantitative use of PIV measurements. While currently available spatial resolution and imaging rate limit the ability of PIV to produce temporal flow statistics, it is the only measurement method allowing the spatial correlation of velocity data across a two-dimensional region. The development of post-processing algorithms based on instantaneous velocity fields, such as the identification and tracking of coherent structures in the flow, is the next challenge to overcome in order to systematize the use of PIV as a velocity measurement technique.

Furthermore, the 25 Hz recording rate should also provide insight into the temporal development of large scale motions in similar flow fields. Finally, these two techniques based on Mie scattering are non intrusive methods which can be easily used at industrial scale in high power boilers or furnaces.

ACKNOWLEDGEMENTS

We would like to thank P. Mahé and J. Imbach (Gaz de France) for their advice on experimental matters, and F. Neveu and M. Trinité (CNRS - CORIA) who performed the LDV and fine thermocouple measurements at Gaz de France R&D.

REFERENCES

Crowe C. T., Chung J. N., Troutt T. R., "Particle Mixing in Free Shear Layers", Prog. Energy Combust. Sci., Vol 14, pp 171-194, 1988.

Gray C, "The evolution of Particle Image Velocimetry", IMechE Optical Methods and Data Processing in Heat and Fluid flow, Londres, pp 19 - 35, 1992.

Lecordier B., Mouqallid M., Vottier S., Rouland E., Allano D., Trinité M., "CCD recording method for cross-correlation PIV development in unstationnary high speed flow", 7th International Symposium on Applications of Laser Techniques to Fluid Mechanics, Lisbon, 1994.

Mokaddem K, Perrin M. Y., Rolon J. C., Perrin M., Levinsky H., "Flame Front Visualization by C2 Spontaneous Emission and OH LIF in Axisymmetric Laminar Methane Air Premixed Flames", 7th International Symposium on Application of Laser Techniques to Fluid Mechanics, Lisbon, 1994.

Namazian M., Kelly J. T., Scheffer R. W., "Near Field Instantaneous Flame and Fuel Concentration Structure", 22th (International) Symposium on Combustion / The Combustion Institute, pp 627-634, 1988.

Namazian M., Kelly J. T., Scheffer R. W., Perrin M., "Effects of Confinement and Bluff-body Burner Recirculation Zone and Flame Stability", IGRC, Cannes, 1995.

Neveu F., Corbin F., Perrin M., Trinité M., "Simultaneous velocity and temperature measurements in turbulent flames obtained by coupling LDV and numerically compensated finewire thermocouple signals", 7th International Symposium on Application of Laser Techniques to Fluid Mechanics, Lisbon, 1994.

Okamoto K., Hassan Y.A., Schmidl W. D., "New tracking algorithm for particle image velocimetry", Experiments in Fluids, Vol 19, pp 342-347, 1995.

Perrin M., Namazian M., Kelly J., Scheffer R.W., "Effect of confinement and blocage ratio on non-premixed turbulent bluff-body burner flames", 23th International) Symposium on Combustion / The Combustion Institute, Orléans, 1990.

Perrin M., "Mesures de Conditions Aerodynamiques Initiales dans un Brûleur Bluff Body ", GDF Internal Report, Ref: M.CERSTA-Bcn N° 931060, 1993.

Prwzeswa M., Albert S., "Mesures de Concentration de méthane dans un Brûleur Bluff Body ", GDF Internal Report, Ref: M.CERSTA-Bcn N° 931000, 1991.

Rouland E., "Etude et developpement de la technique de vélocimétrie par intercorrélation d'images de particules et application aux écoulements en tunnel hydrodynamique", Thèse Rouen, 1994.

Trinité M., "La vélocimétrie par Intercorrélation d'Images de Particules, Séminaire Européen, Le laser outil de Diagnostic en Milieu Industriel, GDF, La Plaine Saint Denis, 1993.

Scheffer R.W., Namazian M., Kelly J., "Velocity measurements in a turbulent nonpremixed bluff-body stabilized flame", Combustion Science and Technology, Vol. 56, pp 101-138, 1987.

Scheffer R.W., Namazian M., Kelly J., "Comparison of Turbulent Jet and Bluff Body Stabilized Flames", Combustion Science and Technology, Vol 67, pp 123-146, 1989.

Willert C.E., Gharib M., "Digital particle image velocimetry", Experiments in Fluids, vol 10, pp 181-193, 1991.

Chapter V

Engines

V.1   Combined Application of Particle Image Velocimetry (PIV) and Laser Doppler Anemometry (LDA) to Swirling Flows Under Compression

J. Volkert, C. Tropea, R. Domann, W. Hübner

Lehrstuhl für Strömungsmechanik, University of Erlangen-Nürnberg, Cauerstr.4, 91058 Erlangen, Germany

Abstract

The velocity field of a strongly swirling flow has been examined using an LDA and PIV in combination. The PIV revealed an asymmetric component of the flowfield, in particular a movement of the swirl center away from the geometric center of the rotating cylinder. A statistical description of this asymmetric component has been developed, from which the *apparent* turbulence measured with an LDA has been estimated. These results are compared to the actual LDA measurements. The conclusion is that a large portion of the measured turbulence using a one-point technique (LDA) can be attributed to this effect and can also explain the previously observed discrepancy between experiment and numerical simulation.

1   Introduction

This paper investigates the use of PIV and LDA in a strongly swirling flow, however the methodology introduced to combine the results of the two measurement techniques for a better interpretation of the measured data has a much wider range of applications. It is the methodology which is considered to be the main contribution of this paper.

The problem will first be stated, using some of the experimental data for illustration. In the present case we are dealing with a swirling flow generated by a rotating cylinder, which after a period of constant rotational speed, is abruptly stopped and the resulting swirling flow is left to decay (spin-down). Optionally, the flow can also be subjected to compression during the spin-down phase, through movement of a piston. Figure 1 illustrates the velocity field in a radial plane at three time steps during the spin-down phase. Despite the exact axisymmetry of the cylinder, the center of flow rotation does not remain in the geometric center of the cylinder. A similar behaviour is noticed for the case of constant rotation of the cylinder.

A point velocity measurement (eg. LDA) will register not only turbulent fluctuations with time but also changes of the mean velocity (through movement of the swirl center) in the measured standard deviation.

478

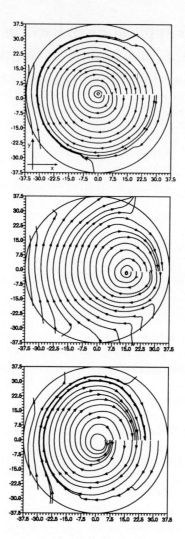

Figure 1: Streamline plots measured by PIV at different times during spin down showing the precession of the vortex core

A two-dimensional numerical simulation (axisymmetric) will necessarily compute turbulence without this added component attributable to the precession of the rotation center. This problem is well known for measurements in internal combustion engines, usually referred to as the problem of cycle-to-cycle variations. Without defining too precisely what is meant by turbulence, efforts have been made to separate components of velocity fluctuations of low and high wavenumber, on the premise that the higher wave number fluctuations were turbulence, whereas the lower wave number fluctuations may not all be irreversibly lost energy. Previous approaches to this problem involved the use of filters in frequency domain [1,2], more elaborate decomposition schemes [3-5], scanning LDA measurements [6] and two-point velocity measurements [7].

Generally the problem arises because with LDA the velocity is only known at one point and any global flow movement cannot be deduced from such limited data. In this respect the PIV, being a whole-field technique, can be helpful. On the other hand, the PIV does not yet offer the required time and spatial resolution necessary for a statistical description of the turbulence field. Therefore a complementary combination of the two techniques is sought.

The approach used in the present experiment can be summarized briefly as follows. From PIV data similar to that illustrated in Fig. 1, statistics of the movement of the rotation center are derived by examining an ensemble of 50-100 repetitions of the one-shot experiment. These statistics are accumulated either for the constantly rotating cylinder or for each time delay during the spin-down, and take the form of a joint probability density distribution in an x and y plane with its origin at the geometric center of the cylinder. Mean and RMS velocity profiles of the tangential velocity component on the same axial plane as the PIV measurements are then obtained with the LDA. The movement of the center of rotation, derived from the PIV data, is now used to postulate movement of the measured mean velocity profile (LDA), at least in a statistical sense. Knowing the spatial gradient of the mean velocity, the artifact (or apparent) turbulence appearing in the one-point LDA measurements, in addition to actual turbulent fluctuations, can be estimated. Of course the global movement of the flow will be a maximum in the center of the cylinder and reduce to zero at the cylinder wall. This must be taken into account by some spatial decay relation with respect to r of the joint probability function describing the center of rotation.

The remainder of the paper is organized as follows. The experimental apparatus is described in section 2 and the measuring equipment in section 3. The data processing of the PIV measurements is described in detail in section 4 and applied to the LDA measurements in section 5. Concluding remarks are given in section 6.

2 Experimental Apparatus

A Rapid-Compression-Machine (RCM) allows observation of the time-dependant behavior of an in-cylinder flow with a specified swirl strength under the influence of a one dimensional compression. Fig. 2 and Table 1 show the design and the dimensions of the RCM. With the current apparatus, the available range of operational parameters corresponds to those typically found in combustion engines, but also encompasses conditions in adherence with Rapid Distortion Theory.

The experimental operation consists primarily of two phases. The first phase is that related to the generation of swirl inside the cylinder. The cylinder consists of a rotating part (section $x_h + x_c$) and a non-rotating part (section x_s), see Fig. 2, where the rotating section is driven by an external motor. Both sections are separated by a small gap ($\Delta x = 1mm$).

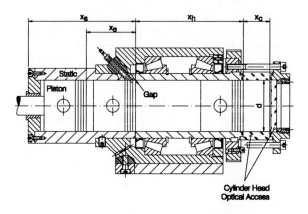

Figure 2 : Design of the Rapid Compression Machine

After stopping the cylinder rotation, a compression stroke is performed by a hydraulically driven piston. For optical access, the cylinder head (module section x_c) is made out of glass. Its length may easily be changed to enable experiments at different compression ratios $\epsilon_c = \frac{x_h + x_c}{x_c}$. Note that section x_e serves as acceleration length of the piston to its desired initial compression velocity when passing the gap.

	SIZE (MM)	DESCRIPTION
d	75.0	Cylinder diameter
x_e	74.0	Piston acceleration length
x_h	160.0	Stroke
x_c	40.0	Head clearance

Table 1 : Apparatus dimensions

Although the major goal of the overall project is to study the influence of compression upon the in-cylinder flow, experiments are also carried out without compression as a reference. The behaviour of the decaying swirl during the spin-down is discussed here, since its features, namely the precessing vortex core phenomenon, complicate the experimental evaluation of measurement data as mentioned in section 1 and described in the following.

The range of operating parameters for the swirl mode is characterized by the rotation rate of the cylinder. The motor used to drive the cylinder is capable of rotation rates up to 3000 rpm, with experiments being performed at 300 and 1200 rpm.

3 Measurement Equipment

3.1 Laser Doppler Anemometer

The two-component backscatter LDA system employed in the present study is shown schematically in Fig. 3.

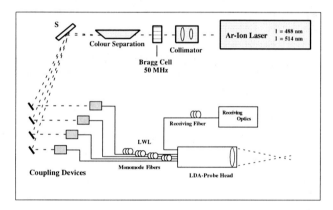

Figure 3: LDA Sending Optics

Transmitting Optics: The laser beam emitted by an Ar-Ion Laser is passed through the Bragg Cell, which operates at a nominal frequency of 50 MHz . The Bragg Cell acts as a beam splitter and frequency shifter. A colour separation prism follows. The power supply of the Bragg Cell driving unit is adjusted to yield nearly equal intensities for all beams. The four beams (one shifted and one unshifted beam of each color) are spatially separated by the use of mirrors before they are coupled into the monomode, polarization preserving fibers. The collimation before the Bragg Cell is used to condition the beam for the fiber launch.

2-COMPONENT-LDA PROBE		
LASER POWER/BEAM:	$P_{beam} \approx 40mW$ (GREEN)	
	$P_{beam} \approx 30mW$ (BLUE)	
BEAM SEPARATION:	D = 22 mm	
NUMBER OF FRINGES :	$N_f = 9.96$	
FOCAL LENGTH f	120 mm	
	$\lambda = 488nm$	$\lambda = 514nm$
INTERS. HALF ANGLE ϕ	5.237 deg	5.237 deg
MCV DIAMETER d_m	25.82 μm	27.22 μm
LENGTH OF MCV l_m	281.65 μm	296.9 μm
FRINGE SPACING Δx	2.67 μm	2.82 μm
MCV : MEASUREMENT CONTROL VOLUME		

Table 2 : Specifications of the LDA probe

The fiber based system allows a relatively fast change of the probe head for application of different techniques (two-component / two-point measurements). In the present state a 36mm two-component probe is used, its specifications are given in Tab. 2.

Receiving Optics/Signal Processing: The light scattered from tracer particles is collected through the receiving aperture in backscatter mode and focused onto a graded index fiber of $50\mu m$ core diameter. Oil-water droplets (mixture 1:1) with diameters of $2\mu m \leq d_P \leq 5\mu m$ produced in a pressured air atomizer are used for seeding. The scattered light, which contains light coming from both measurement volumes, is directed to a colour separation optic based on a combination of three prisms. Two photomultipliers are mounted on the optics to detect the signals. Counter processors (TSI 1980) are used after passing the signals through a two-channel downmixer and bandpass filters. The data is transferred to a PC computer for further processing (Dostek 1400A LDA Interface).

3.2 Particle Image Velocimeter

Measurement System: For the PIV measurements presented in this work the *FlowMap* PIV System of Dantec MT was used, combined with a double pulsed frequency-doubled Nd:Yag laser (*Quanta Ray*, Spectra Physics). The data analysis procedure was based on vector processing by computing the auto-correlation function of double exposed interrogation areas. The details of the PIV measurements are as follows.

The in-cylinder flowfield was seeded by tracers as mentioned in section 3.1. A plane perpendicular to the cylinder axis was illuminated through the transparent cylinder head by a double pulsed light sheet ($\lambda = 532nm$). The light sheet was produced by a cylindrical lens, which was mounted at the end of a light transmitting system designed by tubes and "joint-mirrors" (*light-guiding arm*). Due to the high laser power (peak energy $E_{Puls} = 300mJ$) a fiber based system was not feasible.

The CCD array of the *FlowSense* Camera comprises 1024 x 1024 pixels with a pixel size of $19\mu m$. The object/image magnification when mapping the whole flow field was $m = \frac{O}{I} = \frac{75mm}{19x1024\mu m} = 3.85$. The size of the interrogation areas was varied between 32x32 pixels and 128x128 pixels, giving a maximum spatial resolution of 1.172 mm. A velocity bias is incorporated by electronic pixel shift (displacing the first image a selected number of pixel lines) which allows the determination of the flow direction.

Synchronization with RCM Experiment: An important feature of the system is its capacity for controlling the timing of all hardware components. The experiments were synchronized as shown in Fig. 4.

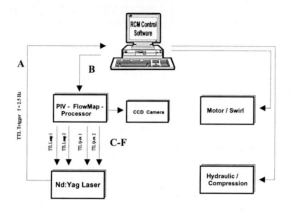

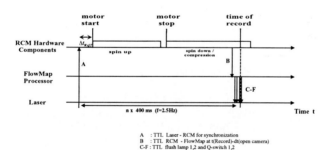

A : TTL Laser - RCM for synchronization
B : TTL RCM - FlowMap at t(Record)-dt(open camera)
C-F : TTL flush lamp 1,2 and Q-switch 1,2

Figure 4: PIV Timing

Corresponding to the repetition rate of the laser and the readout time of the camera, measurements could be taken every 400 ms. Therefore, the program had to be synchronized with the periodically flashing pre-lamp of the laser. A TTL signal from the pre-lamp (**A**) was used to start the system: rotation - stopping the cylinder - start of compression. The PIV Processor was then triggered (**B**) to control the timing of the camera, the laser flash lamps and the Q-switches (**C-F**). By adjusting the time interval Δt_{WAIT}, measurements could be taken at every point in time after stopping the cylinder rotation.

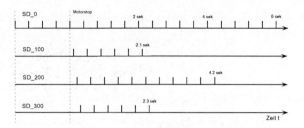

Figure 5: PIV timing pulses, showing 4 overlapping runs

As shown in Fig. 5, a series of time interleaved measurements have been carried out, beginning simultaneously with the motor stop and being delayed for 100ms, 200ms, and 300ms, respectively. In this way, a sampling of $\Delta t = 100ms$ until 2.4s after motor stop, $\Delta t = 200ms$ until 4.4s after motor stop and $\Delta t = 400ms$ for $t > 4.4s$ after motor stop has been achieved.

4 Data Processing

The processiong of the PIV data and their subsequent use to interpret the turbulence profiles measured with LDA can be divided into the following steps,

- determination of the swirl center location for a single PIV frame

- formulation of a joint probability function describing the local movement of the mean velocity gradient in the radial plane of the cylinder

- application of this function using measured mean velocity gradients to estimate artifact turbulence of LDA measurements

which are described in sequence below.

4.1 Swirl Center Location

Any attempt to localize the momentary position of the swirl center will necessary be an approximation, firstly because the flow by definition is no longer exactly axisymmetric and thus "center" must be newly defined and secondly because of the finite spatial resolution of the PIV velocity data.

Nevertheless a good approximation can be achieved by restricting the search to a more local area. Working on a $x - y$ coordinate system, the lines of zero $U, V, U + V$ and $U - V$ velocity are interpolated (Fig. 6a) leading to a first estimation of the swirl center location and to a reduced set of interrogation spots for further analysis (Fig. 6b,c).
Again the lines of zero velocity are interpolated and this reduced set results in a zone where all lines cross. The center of a circle encompassing all crossings is specified as the swirl center.

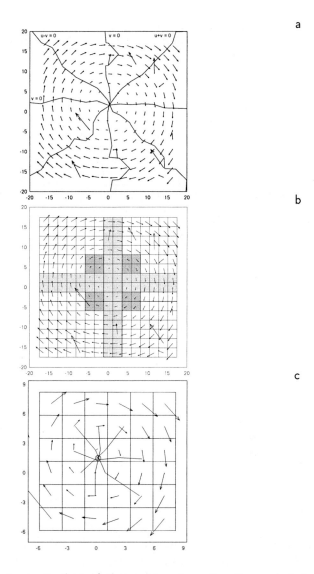

Figure 6: a) measured velocity field, b) first swirl center estimation and choice of reduced pixel set c) final interpolation of swirl center location

The movement of the swirl center, either in time for the constant rotation case or over an ensemble at equal time delays for the spin-down case, can now be expressed as a joint probability function in x, y.

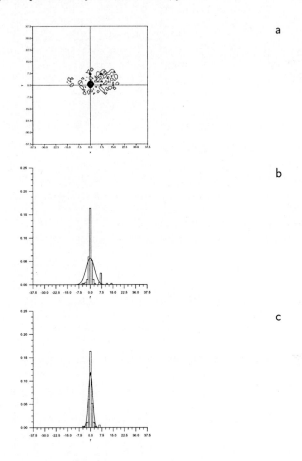

Figure 7: a) measured probability of swirl center movement, b) fitted Gaussian distribution for movement along the x axis, c) Gaussian fit using clipped data ($x < \pm 6.5$)

The choice of this function and the estimation of its parameters is a rather sensitive step, since only a relatively small sample set is generally available. For the present work a Gaussian distribution is postulated

$$P_{sc}(x, y) = \frac{1}{2\pi\sigma_x\sigma_y} e^{-\frac{1}{2}\left[\frac{x^2}{\sigma_x^2} + \frac{y^2}{\sigma_y^2}\right]} \tag{1}$$

where σ_x, σ_y are to be determined from the PIV measurements.

An example measured distribution is illustrated in Fig. 7a. Assuming that for $x = y = 0$ the variances σ_x and σ_y are equal for a sufficiently large sample size, the value of σ_x is taken as the variance of the measured distribution along the $y = 0$ diameter.

The subsequent fit is illustrated in Fig. 7b for the case of a constant rotation at 300 rpm. This particular case shows outlying values in the measured data. Some experimentation with data clipping (e.g. $x < \pm6.5$) before the estimation of σ_x was performed, as discussed below and illustrated in Fig. 7c.

4.2 Movement of Mean Velocity Gradient

At the swirl center, the probability given in Equation (1) can be applied directly to describe movement of the velocity gradient off center, for instance at the position x_0, y_0. However, the values of σ_x and σ_y are expected to decrease, ultimately becoming zero at the wall.

Two approaches were used to quantify this radial damping of σ_x, σ_y, a linear decay and a third order polynomial

$$\sigma_x = \sigma_x(0)f(x) \tag{2}$$

$$f(x) = 1 - \frac{x}{R} \qquad \text{linear}$$

$$f(x) = a_0 + a_1 x + a_2 x^2 + a_3 x^3 \qquad \text{3rd order polynomial} \qquad \text{where } R \text{ is the cylinder}$$

radius. An expression for the movement of the local mean velocity gradient can therefore be given as

$$P_U(x', y', x_0, y_0) = \frac{1}{2\pi\sigma_x(x_0)\sigma_y(x_0)} \tag{3}$$
$$e^{-\frac{1}{2}\left[\frac{x'^2}{\sigma_x(x_0)^2} + \frac{y'^2}{\sigma_y(x_0)^2}\right]}$$

where σ_x and σ_y are given as functions of x_0. The coordinates x', y' are now local to x_0, y_0 ($x' = x - x_0, y' = y - y_0$).

4.3 Application to Measured Mean Velocity Gradient

The strategy to estimate the apparent turbulent velocity measured by LDA is now to first describe the mean velocity field around the measurement point and then to use the above probability distribution to "sample" this field. The mean velocity field is known from LDA measurements and is assumed not to be influenced by the swirl center movement, at least to first order. Such a mean velocity profile is shown in Fig. 8, together with a numerical prediction using a full Reynolds Stress Model and low Reynolds number wall treatment,

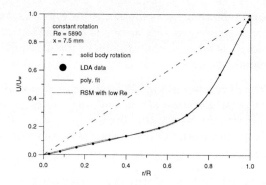

Figure 8: Radial profile of tangential velocity at 300 rpm

as described in detail in [8].

The mean tangential velocity is therefore available as an interpolation curve along the $y = 0$ diameter. A linear approximation can be used locally at the position (x_0, y_0) in the form

$$U_\Phi(x, y) = n(x_0) + m(x_0) \cdot r \tag{4}$$

where $n(x_0)$ and $m(x_0)$ is a local intercept and slope and r ist the radius from the geometric center. Out of the $y = 0$ diameter the y component of the velocity can be approximated by

$$U_y(x, y) = \frac{n(x_0)x}{\sqrt{x^2 + y^2}} + m(x_0)x \tag{5}$$

The measured variance of the fluctuating tangential velocity component can now be evaluated as

$$\overline{u^2}(x_0, y_0) = \int \int u_y^2(x, y) P_U(x', y', x_0, y_0) dx dy \tag{6}$$

Assuming a linear mean velocity gradient (Eq.4) this integral can be analytically solved :

$$
\begin{aligned}
n(x_0) &= 0 \\
\sigma_x &= \sigma_y = \sigma(x_0)
\end{aligned}
$$

$$
\begin{aligned}
\overline{u^2}(x_0, y_0) &= m^2 \sigma_x^2(0) \left(1 - \frac{x_0}{R}\right)^2 + m^2 x_0^2 - \\
&\quad 2m x_0 \bar{U}(x_0, y_0) + \bar{U}^2(x_0, y_0) \\
&= m^2 \sigma_x^2(x_0)
\end{aligned}
\tag{7}
$$

Alternative to Eq.(4), the actual measured mean velocity gradient can be used in the integral of Eq.(6), in which case a numerical integration using a 9 point quadrature method is suitable [9]. For this purpose the measured mean velocity profile is first approximated by a 7th order polynomial, as shown in Fig. 8.

5 Results and Discussion

The first results to be presented are taken from the case of constant cylinder rotation at 300 rpm and serve to illustrate the sensitivity of various processing parameters, as introduced above.

In Fig. 9 the choice of σ_x, σ_y, the integration limits and the mean velocity gradient is investigated. To begin, σ_x and σ_y were computed using Gaussian fit, as illustrated in Fig. 7b and applied with either a linear or polynomial decay function to a locally velocity gradient of the form Eq. (4).

Fig. 9a compares the computed apparent turbulence with the actual turbulence (RMS) measured with the LDA. Two discrepancies are immediately obvious. In the center ($r/R = 0$) the apparent turbulence is larger than the measured, which is physically impossible. Furthermore, the trend of the apparent turbulence towards the wall is to decrease, whereas the measured turbulence remains approximately constant.

Physically the integration of a Gaussian distribution to $\pm\infty$ is unrealistic. Therefore the integration was repeated using limits of $\pm 5\sigma, \pm 4\sigma, \pm 3\sigma$ and $\pm 2\sigma$. In fact little difference was observed so that subsequent results were computed using $\pm 2\sigma$, which represents about 96% of the probability mass. The next step was to use the actual measured mean velocity gradient, as shown in Fig. 8, instead of the linear approximation, Eq. (4). The results are shown in Fig. 9b, which reflect also the increased gradient towards the wall. Still however, the apparent turbulence is larger than the measured turbulence in the center of the cylinder. Finally data clipping was applied before the Gaussian fitting of the measured probability of the swirl center movement, as illustrated in Fig. 7c. This lead to the results shown in Fig. 9c, which shows also the unclipped result for comparison. Clearly the initial estimates of σ_x, σ_y are very important and should be based on a larger number of single PIV frames to obtain the necessary certainty. Further results have therefore been computed using a polynomial decay of σ_x, a linear decay of σ_y, an integration limit of $\pm 2\sigma$ on the measured mean gradient and data clipping for the initial estimate of σ_x, σ_y.

A final result for the constant rotation at 300 rpm is given in Fig. 10, where the LDA turbulence data have now been corrected by the estimated apparent turbulence.

490

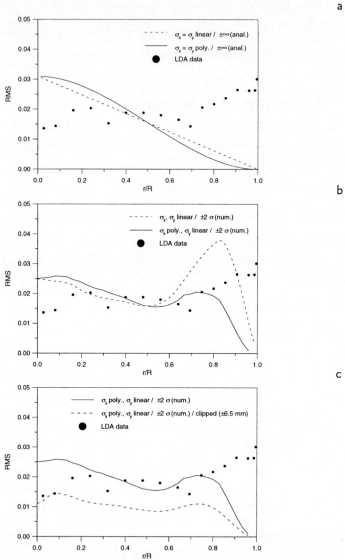

Figure 9: Apparent and measured RMS velocity : a) Gaussian fit for σ and linear mean velocity profile; b) restricted ($\pm 2\sigma$) numerical integration of actual mean velocity profile; c) as in b) with data clipping ($\pm 6.5 mm$) for Gaussian fit

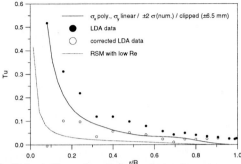

Figure 10: Radial profiles of turbulence intensity - comparison of numerical simulated data, estimated apparent turbulence, actual measured LDA data and "corrected" LDA data

The agreement with the results from numerical simulation is much improved, except near the wall. Numerically laminar flow conditions are predicted. Therefore the remaining variance in the LDA data is presumbly attributed to noise.

In Fig. 11 the measured mean velocity profile for one spin-down case is shown, including the polynomial fit for subsequent integrations.

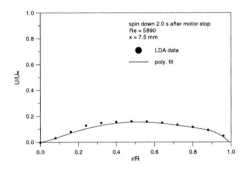

Figure 11 : Mean velocity profile - spin-down case

The result of the analysis is given in Fig. 12, again showing the measured turbulence, the computed apparent turbulence and the corrected measurements. In this case, despite the clipping, the apparent turbulence is larger than the measured turbulence over large regions of the profile. This indicates clearly that the actual turbulence level is very low, in fact lower than the uncertainty in the corrected estimates.

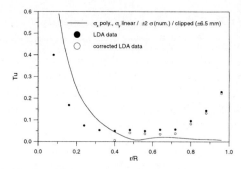

Figure 12: Radial profiles of turbulence intensity - comparison of estimated apparent turbulence, actual measured LDA data and "corrected" LDA data

6 Summary and Conclusions

In the present paper, a combined application of the LDA and the PIV techniques to investigate the contribution of large scale fluctuations to the turbulence intensities measured by LDA has been proposed and tested. A statistical description of the vortex motion based on PIV measurements allows the determination of *apparent* turbulence, which obviously is related to the precession of the vortex core. A better interpretation and even "correction" of the measured turbulence data is then possible. The presented methodology offers a better comparison of experimental and numerical results, even in the presence of three dimensional flow motion, which cannot be reproduced by numerical simulations. The results can be summarized as follows :

I: A significant part of the actual measured turbulence (LDA) can be attributed to the effect of vortex motion. "Correction" of the LDA data by estimated apparent turbulence leads to an improved agreement with results from numerical simulations.

II: For most of the presented results the apparent turbulence is larger than the measured turbulence. Since the number of single frames measured by PIV was limited to a maximum of 100, the certainty of the estimated probability distribution of the vortex core location, and therefore the first estimation of the σ_x, σ_y represents a very critical point.

III: Parametric studies showed the sensitivity of various processing parameters, such as the integration limits, the damping of σ_x and σ_y towards the wall and data clipping limits. Further work will be focused on a more reliable data analysis, based on an increased number of PIV frames, to increase the certainty of σ_x, σ_y- estimation. The data reduction procedure will be refined to show its potential for a wide range of applications as already mentioned before.

Acknowledgements

The authors gratefully acknowledge the support of DANTEC Measurement Technology, Skovlunde and Dantec/invent MT, Erlangen, who made the PIV equipment available. Also Spectra-Physics is acknowledged for the loan of the Nd-YAG laser. The work has been supported by the Deutsche Forschungsgemeinschaft (DFG Tr194/ 7-1). Thanks are also due to S. Jakirlic for providing the numerical predictions shown.

References

[1] FANSLER,T.D. AND FRENCH, D.T. (1988) *Cycle-Resolved Laser-Velocimetry Measurements in a Reentrant-Bowl-in-Piston Engine*, SAE 880377

[2] BRERETON, G.J. AND KODAL,A. (1991) *A Frequency-Domain Filtering Technique for Triple Decomposition of Unsteady Turbulent Flow*

[3] RASK, R.B. (1981) *Comparison of Window, Smoothed-Ensemble and Cycle-to-Cycle Data Reduction Techniques for Laser Doppler Anemometer Measurements of in-Cylinder Velocity*, **Fluid Mechanics of Combustion Systems**, ed. Morel, T., Lohmann, R.P. and Rackley, J.M., 11-20, ASME

[4] HILTON, A.D.M., ROBERTS, J.B. AND HADDED, O. (1991) *Autocorrelation Based Analysis of Ensemble Averaged LDA Engine Data for bias-free Turbulence Estimates: A Unified Approach*, SAE 910479

[5] DAW, C.S. AND KAHL, W.K. (1990) *Interpretation of Engine Cycle-to-Cycle Variation by Chaotic Tine Series Analysis*, SAE 902103

[6] GLOVER, A.R., HUNDLEBY, G.E. AND HADDED, O. (1988) *An Investigation into Turbulence in Engines using Scanning LDA*, SAE 880378

[7] CORCIONE, F.E. AND VALENTINO, G. (1990) *Turbulence Length Scale Measurements by Two-Probe-Volume LDA Technique in a Diesel Engine*, SAE 902080

[8] JAKIRLIĆ, S. AND HANJALIĆ, K. (1995) *A Second-Moment Closure for Non-Equilibrium and Separating High- and Low-Re-Number Flows*, Proc. 10th Symposium on Turbulent Shear Flows, The Pennsylvania State University, USA, August 14-16

[9] ABRAMOWITZ, M. AND STEGUN, I.A. (1970) **Handbook of Mathematical Functions**, New York, S. 892f.

[10] HEYWOOD, J.B. (1987) *Fluid Motion within the Cylinder of Internal Combustion Engines - The 1986 Freeman Scholar Lecture*, Journal of Fluids Engineering, Vol. 109, 3-35

V. 2 Characterization of Spray Flows Under High Fuel Temperature Using Phase Doppler Anemometer

Tomio Obokata[1], Tsuneaki Ishima[1], Tetsuji Koyama[2], Kouichi Uehara[1], Kazumitsu Kobayashi[3] and Masayoshi Tsukagoshi[3]

[1] Department of Mechanical System Engineering, Gunma University, 1-5-1, Tenjin-cho, Kiryu, 376 Japan
[2] Tsukasa Sokken Co., LTD,1-19-4, Tamazutsumi, Setagaya, Tokyo, 158 Japan
[3] Unisia Jecs Co. 1671-1 Kasukawa, Isesaki, 372 Japan

Abstract. Effects of fuel temperature on the characteristics of the intermittent-cyclic spray have been experimentally clarified using the phase Doppler anemometer. The spray was initiated from the air assisted fuel injector. The injection frequency was 50Hz, and the pressures of the liquid fuel (n-heptane) and air were 250kPa and 147kPa, respectively. The fuel temperature was changed from 293K (20°C, room temperature) to 413K (140°C).

Experimental results showed that the mean velocity of the droplets increased with the fuel temperature. The Sauter mean diameter increased in the fuel temperature region from 293K to 343K for intermittent spray and from 293K to 323K for continuous steady spray. After the regions, it decreased. The droplets with higher fuel temperature and larger diameter have smaller decaying ratio of the velocity. These features are well explained by the analyses using grouped data classified by the droplet diameter or temporally divided velocity history of the intermittent spray. Evaluation of the droplet size distribution has also been performed by a log-hyperbolic (LH3) type fitting function.

Keywords. Air-Assisted injector, Spray Flow, PDA, Fuel Temperature

1 Introduction

Spray characteristics are one of the dominant phenomena to affect the performance of internal combustion engines. In this field, a phase-Doppler anemometer (PDA) allows one to perform local and instantaneous measurement of both particle size and velocity. Many investigators have studied the subject with the PDA [1-5]. They indicated successfully the experimental data of particle size and velocity not only for steady condition but also for intermittent spray. Differences between continuous and intermittent sprays were also shown in them [5]. Since the fuel spray is injected intermittently in the actual engine, establishment of evaluating methods of it is needed for spray characterization.

The authors' group have investigated an intermittent gasoline type spray initiated from an air assisted spray injector [6, 7]. The injector was proposed for obtaining better quality of air/fuel mixture. In these experiments, a time dividing analysis proposed by Obokata and Long [5], a size distribution using 3-parameters log-hyperbolic function (LH3) reconsidered by Xu et al [8], and a size classification method were used for evaluating the spray. They indicated that the spray cloud of air assisted injector had smaller droplet diameter than that of existing system (pintle injector). An effect of the fuel temperature was also reported in the study [7]. The high fuel temperature resulted in an increase in the droplet diameter and velocity. In those studies, however, it was difficult to have detailed discussion on the effect of the fuel temperature because alternative fuel of Laws which was used for spray instead of gasoline was a composite fuel.

The present work was undertaken to clarify the effect of the fuel temperature on the characteristics of the intermittent spray initiated from the air assisted injector by using PDA system. An pure fuel of n-heptane has been chosen for the present study. Its evaporation temperature is 371.4K and that value is lower than that of the Laws (423K). The characteristics of the intermittent spray are discussed by using the time dividing analysis and by using velocity information on droplets grouped by diameter. Whole and time divided size distributions of the spray droplets have also been discussed with 3-parameters log-hyperbolic function.

2 Experimental Apparatus

2.1 Fuel Injection System

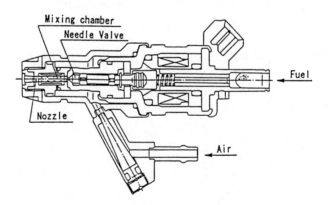

Fig. 1. Cross section of air assisted fuel injector.

Figure 1 shows the inner mixing type air assisted fuel injection nozzle. The injector had a mixing chamber which was mounted on a pintle type fuel injector. The fuel was intermitted by the needle valve and the assisted air continuously flowed in the nozzle. The nozzle outlet was 2mm in diameter. The frequency of injection was 50Hz, the pressure of fuel was 250kPa and that of air was 147kPa.

Table 1. Physical properties of the fuels.

		n-heptane	Laws	Gasoline
Density kg/m^3	(293K)	684	797	746
Kinematic viscosity m^2/s	(293K)	0.61×10^{-6}	1.12×10^{-6}	0.54×10^{-6}
Refractive index	(293K)	1.385	1.433	1.41
Surface tension N/m	(293K)	20.9×10^{-3}	26.4×10^{-3}	22.5×10^{-3}
Coefficient of volume expansion	K^{-1}	1.22×10^{-3}	0.66×10^{-3}	0.83×10^{-3}

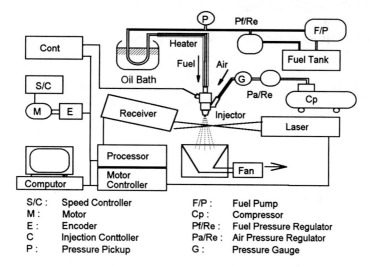

S/C : Speed Controller	F/P : Fuel Pump
M : Motor	Cp : Compressor
E : Encoder	Pf/Re : Fuel Pressure Regulator
C Injection Conttoller	Pa/Re : Air Pressure Regulator
P : Pressure Pickup	G : Pressure Gauge

Fig. 2. Experimental set up

The n-heptane was used for injection instead of gasoline. Typical fluid properties of n-heptane, Laws and gasoline are listed in Table 1. The fuel was injected into atmosphere which had normal room temperature and pressure. Flow rate of fuel was changed by valve opening duration, Δt=2.5ms, 4.5ms, 6.5ms (intermittent condition) and ∞ (steady condition). Figure 2 shows a schematic diagram of the injection tester used in this experiment. Spray flow and radial directions set Z-axis and r-axis, respectively, and the origin of the coordinates was set at the center of the nozzle outlet. Fuel temperature was varied from 293K (20°C) up to 413K (140°C). The fuel flowed through a copper pipe which was in a thermostatic oil bath and the temperature of fuel was set the certain value. The temperature of the assisted air was not controlled and was around 293K.

498

2.2 Phase Doppler Anemometer

A phase Doppler particle analyzer of Aerometrics (PDPA) was used to measure the droplet velocity and diameter. The diameter of the measuring volume was adjusted to more than five times of the expected maximum particle diameter, to remove measurement errors which would be affected by the Gaussian distribution of light intensity of the measuring volume [9]. The diameter of the measuring volume used in this experiment was about 0.3mm and the total intersection angle of incident beams was 1.5° and forward scattering mode was used with the off-axis angle of 30° from the incident beams. The measuring diameter range was chosen from 3.4~510.5 μm at 1.385 of refractive index of the fuel. High voltage of the detectors was set as 350V.

3 Results and Discussions

3.1 Distribution of Droplet Velocity and Diameter of the Spray

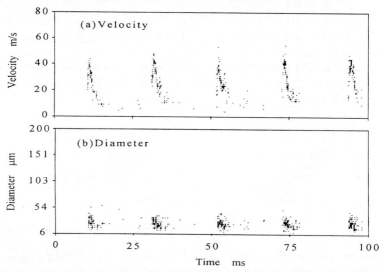

Fig. 3. Time series of droplets size and velocity (Z=50mm, r=0mm, Δt=2.5ms and fuel temperature 413K)

Time series of droplets velocity and diameter measured at Z=50mm on the spray center line (r=0mm) are plotted in Figure 3 where the fuel temperature is 413K, the valve opening duration Δt=2.5ms and the flow rate of the air Q=0.255Nl/s at the normal condition. In this figure, the instantaneous maximum data rate is about 20kHz and the mean data rate is 10kHz. Axial distributions of droplet mean velocity (V_m) and Sauter mean diameter (D_{32}) on the centerline (r=0mm) at the all period of intermittent spray with various temperatures are plotted in

Figures 4(a) and (b). The mean velocity decreases along downstream distance (Figure 4(a)). The Sauter mean diameter decreases slightly just after the nozzle exit and after then increases with downstream distance (Figure 4(b)). The value at the initial point (Z=5mm) is much smaller than that of the Laws used in the previous work [7]. The decreasing rate of them at the upstream is also smaller than that of the Laws. In the downstream, the range of the values are almost same as that of the Laws. In the upstream, atomization and breakdown at n-heptane is much more active than Laws and therefore the Sauter mean diameter is decreased. In the region, the Laws spray probably has a lot of non-spherical droplets due to their large kinematics viscosity. Then, the large value of the Sauter mean diameter is also explained by an incomplete atomization in n-heptane. In the downstream, the breakdown becomes inactive and small droplets disappear or their diameters become smaller than the measuring range of PDA due to evaporation. The tendency is also caused by that small droplets with slower velocity are overtaken and pushed away from the centerline by large droplets or merged into those [5]. These tendencies of the velocity and the diameter are little influenced by the fuel temperature. At the 413K of the fuel temperature, both of the mean velocity and the Sauter mean diameter have a different feature from those on another fuel temperature condition. Since the fuel temperature of 413K is higher than the evaporation temperature of the n-heptane, the evaporation seems to affect the results. The actual fuel temperatures would be lower than the given values because the fuel was cooled by the assisted air and the atmosphere. Then, the fuel droplets were able to exist in the measuring region and the PDA gives information on their velocity and diameter.

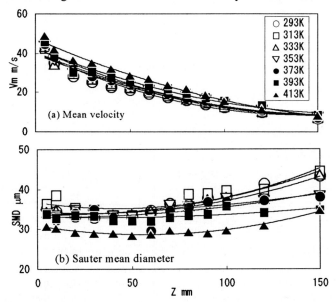

Fig. 4. Axial distributions of droplets mean velocity and Sauter mean diameter with various fuel temperatures (r=0mm, Δt=2.5ms).

Figures 5(a) and (b) show the radial distributions of mean velocity. In these figures, fuel temperatures are 293K (room temperature) and 373K, and the measurement axial positions are Z=25mm, 50mm, 75mm and 100mm. The mean velocity and radius are normalized by the maximum mean velocity (V_{max}) on the axis and by the half-value width of the velocity distribution curve ($r_{0.5}$). The velocity distributions are close to the Gaussian distribution which is indicated by the solid curve (Gauss). The velocity on the outskirts of the spray is slightly greater than the Gaussian distribution. This feature is caused by the detection of droplets spreading out from the spray center and velocity bias at the LDA measurement near the edge of jet. The distributions of the mean velocity have similar tendency on the both temperature conditions. Figures 6(a) and (b) show the Sauter mean diameter with non-dimensional radius. The distribution shape of the Sauter mean diameter has a minimum value around the center (r=0mm). The edge of the spray becomes asymmetric due to measurement uncertainty and decreasing data rate or absence of droplets. The Sauter mean diameter at the edge for the low fuel temperature of spray is larger than that for the high fuel temperature.

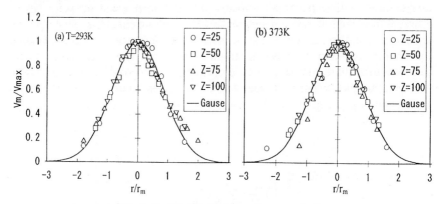

Fig. 5. Radial distributions of mean velocity.

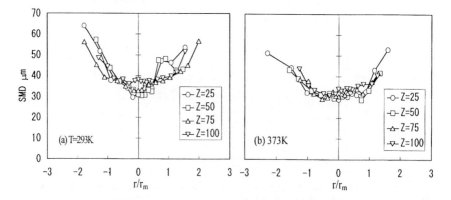

Fig. 6. Radial distributions of Sauter mean diameter.

To clarify the effect of the fuel temperature on spray characteristics, the mean velocity and the Sauter mean diameter, which were measured at Z=50mm on the spray center line (r=0mm) with various fuel flow rates, are illustrated in Figures 7(a) and (b), respectively. The fuel flow rate is changed by the valve opening duration Δt, and flow rates at fuel temperature of 293K are 5.1mg/shot, 10.1mg/shot, 15.2mg/shot and 1.8g/s at Δt=2.5ms, 4.5ms, 6.5ms and ∞ (steady continuous spray), respectively. The mean velocity increases with increasing fuel temperature and fuel flow rate. The Sauter mean diameter increases in the region from 293K to 343K for intermittent spray and from 293K to 323K for continuous spray, after the region, they decrease. The high fuel temperature causes fuel evaporation, and it also causes promotion of the atomization due to the decrease in the fuel kinematics viscosity and in the fuel surface tension. The increase in Sauter mean diameter can be explained that the small droplets tend to decrease or lose their diameter due to the evaporation because their total droplet surface area is greater than that of large droplets with same total fuel volume. In the region of increasing in the Sauter mean diameter, the evaporation affected the spray. After the region, the decrease in the fuel surface tension and the kinematics viscosity can mainly influence on the spray characteristics and therefore the Sauter mean diameter decreases. For increasing the fuel flow rate, duration of the injecting of the fuel becomes long where the large droplets with high velocity rush in the flow field. Then, the mean velocity and the Sauter mean diameter increase due to many fuel droplets with large diameter and high velocity passing at higher flow rate.

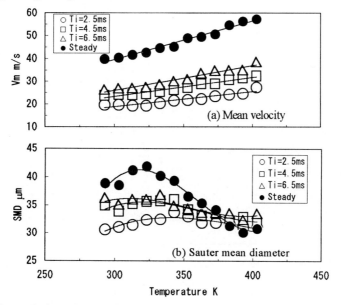

Fig. 7. Mean velocity and Sauter mean diameter with various fuel temperature and various fuel flow rate (Z=50mm, r=0mm)

3.2 Velocity Decay at Droplets Group Classified by Diameter

The PDA measurement can provide a diameter and a velocity of each droplet. Figures 8(a) and (b) show decays of the mean velocity (Vm) at each classified size where V_0 is the exit velocity at the nozzle estimated from measured data and d is diameter of the nozzle outlet (d=2mm). An interval of the classified size diameter was set as about 4μm by the PDA system and six classified sizes were chosen for the figures. In these figures, a straight line indicates a theoretical decay line of air circular jet [10]. The experimental results indicate non-liner change in the decay of the droplet velocity although the decay of the velocity for the air is linear. Horizontal origin seems to be shifted to reverse axial direction with increasing the diameter of the droplet. Each droplet has a similar velocity at the nozzle outlet. In the upstream of Z/d $\leqq$ 50, the small classified droplets have larger velocity decay compared with the large droplets. In contrast, the small particle has smaller velocity decay in the downstream. In the downstream of Z/d $\geqq$ 50 for 293K of fuel temperature, the velocity decay rate of the each droplet size can be observed to become almost same rate as air jet. The larger droplet keeps its velocity longer distance due to its large inertia depending on its diameter. Therefore, the large classified size droplets have smaller velocity decay. Comparison between the both conditions of the fuel temperature indicates small velocity decay for high fuel temperature in the whole experimental region. For high fuel temperature, the velocity decay rate maintains smaller than that for the low fuel temperature in the upstream shown by broken line.

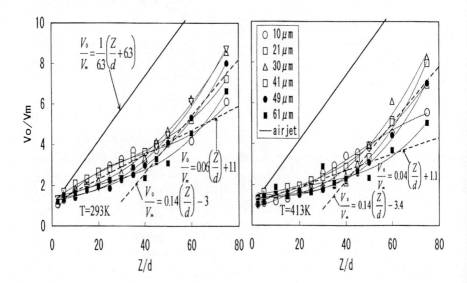

Fig. 8. Decay of mean velocity at classified droplets on the centerline (r=0mm, Δt=2.5ms)

3.3 Time Dividing Analysis of Intermittent Spray

In order to estimate the intermittent spray, the temporal velocity and diameter data were divided into some parts according to their phase angle [5-7]. This estimating method was named a time dividing analysis. The method is explained

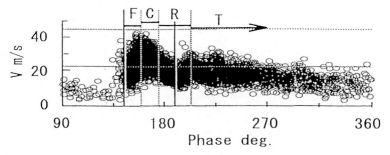

Fig. 9. Example of phase matched plot of droplets velocity.

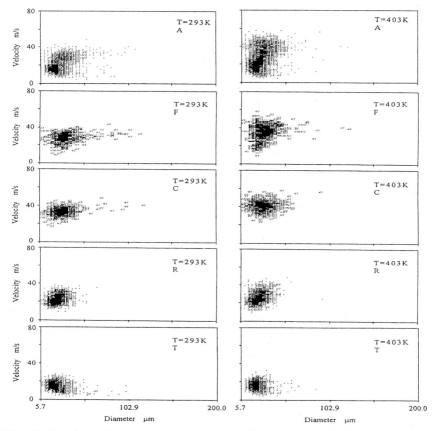

Fig. 10. Correlation between droplet velocity and diameter at each divided part of spray (Z=50mm, r=0mm, Δt=2.5ms)

as follows. Figure 9 shows an example of ensemble plots of velocity for intermittent spray, where the time axis of Figure 3 was converted into the injection phase angle. A duration ($t_{0.5}$) where the droplet velocity is larger than the half value of the maximum velocity is determined on the figure. The duration was divided by three parts. The first and the second $t_{0.5}/3$ durations are represented by fore part (F) and central part (C). Next duration of $2t_{0.5}/3$ is defined as rear part (R). The rest of the duration is represented by the tail part (T).

Figure 10 indicates correlation between droplets velocity and size for intermittent spray with time dividing analysis for fuel temperature of 293K and 413K at the Z=50mm and on the centerline (r=0mm). The data of entire spray (A) is also shown in the figure. A critical difference between the correlations of both fuel temperature conditions exists in F and C. The correlations at the parts of R and T are not sensitive to the fuel temperature. In the parts of F and C under the high fuel temperature condition, it can be observed an existence of a lot of droplets with large velocity. It is caused by that the droplets have a small velocity decay as indicated in Figure 8 and that they can reach a certain distance remaining a large velocity. In the parts of R and T, an instantaneous flow rate of the fuel is very low and the assisted air continuously flows in the nozzle. Since the assisted air cools the fuel, the fuel temperature would be lower than the given value. Therefore, the correlations at R and T are insensitive to the fuel temperature.

Figure 11 shows the axial distribution of the mean velocity of each temporal part. For the high fuel temperature, the droplet velocity remains large in the parts of F, C and R. This result is caused by the small velocity decay.

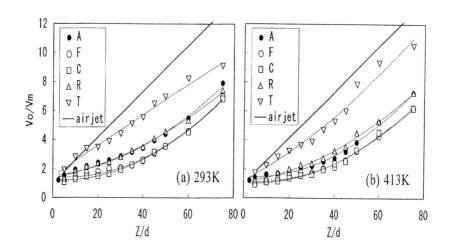

Fig. 11. Axial distribution of the mean velocity of each temporal part (r=0mm, Δt=2.5ms).

Figure 12 shows the axial distribution of the Sauter mean diameter. The Sauter mean diameter at the F and C is larger than that at the R and T. It will be supposed that the large droplets were accumulated in the F and C. At the high fuel temperature, the variance among the values of the parts becomes undistinguished. It is related with the decrease in the fuel kinematics viscosity and the fuel surface tension.

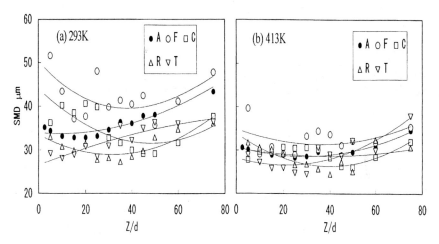

Fig. 12. Axial distribution of Sauter mean diameter of each temporal part on the centerline (r=0mm, Δt=2.5ms).

3.4 Evaluation of Droplet Size Distribution

Recently, the three parameter log-hyperbolic function (LH3) was reconsidered for evaluating the distribution of droplet diameters [8]. The function provided a

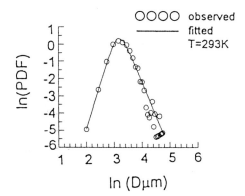

Fig. 13. Example of measured droplet size distribution with result of log-hyperbolic fitting (Z=50mm, r=0mm, T=293K, Δt=2.5ms)

good fitting curve for Diesel type spray [5] and gasoline type spray [6, 7]. This method plots the particle sizes and number density distribution on a diagram of logarithmic scales and fit a hyperbolic curve to the size distribution. The function represented by the hyperbolic curve has three parameters, α, θ and μ. The α defines the opening angle of the hyperbola, the θ is angle of the axis of the hyperbola relative to the coordinate system, and the μ is the location parameter defining the peak of the distribution. Since these parameters of α, θ and μ are related to the shape of the fitting curve, the droplet size distribution can be discussed by using them. Figure 13 shows an example of the measured size distribution of the droplets with the result of the LH3. In the figure, the abscissa is logarithmic scale of droplet diameter of the histogram and the ordinate is logarithmic scale of the probability density. The result of the LH3 is obtained by using maximum likeness estimation proposed by Barndorff & Nielsen [11]. The fitting curve of the LH3 provides a good approximation to the size distribution of the air assisted spray. Figure 14 indicates the variation of the three parameters. For high fuel temperature, the α have more extent than that for the low fuel temperature at just after the nozzle exit. These parameters of θ and μ have a similar tendency with each temperature condition. The LH3 seems to be useful for describing the characteristics of the spray.

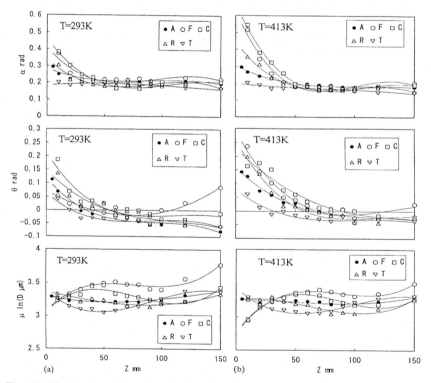

Fig. 14. Distribution of three parameters of LH3 (r=0mm, Δt=2.5ms)

4 Concluding Remarks

In order to clarify the effect of the fuel temperature on the characteristics of the intermittent fuel spray initiated from the air assisted injector, the experimental study has been performed under the conditions of the fuel temperature changing from 293K to 413K. N-heptane which is one of the pure fuel was used for detailed discussion. The concluding remarks are indicated as follows:

1. The mean velocity of the droplets in the fuel spray become large with increasing fuel temperature. These phenomena tend to become remarkable with increase in fuel flow rate. Axial flow velocity patterns are similar under each temperature condition.
2. The Sauter mean diameters increase in the region from 293K to 343K for intermittent spray and from 293K to 323K for steady continuous spray. After the region, they decrease.
3. For high fuel temperature, correlations between droplet velocity and diameter are different from that of room temperature of fuel. In the results of the correlation at high fuel temperature, the number of droplets with high velocity becomes larger than that for the low fuel temperature.
4. From the results of diameter classified velocity for various fuel temperatures, small decay of the droplet velocity is observed in high fuel temperature and in large diameter droplets.
5. The difference in the correlation at the different fuel temperature is observed at the fore and central parts of the spray cloud with time dividing method. In the result of the part of F and C under the high fuel temperature condition, it can be observed an existence of a lot of droplets with large velocity.
6. Log-hyperbolic function (LH3) well fits the droplets size distributions of the spray. The three parameters at the LH3 are useful for describing the characteristics of droplets size distribution of the spray.

References

1 Hosoya, H. and Obokata, T., 1993, Effect of Nozzle Configuration on Characteristics of Steady-State Diesel Spray, SAE Paper No. 930593.
2 Coghe, A. and Cossali, G. E., 1994, Phase Doppler Characterization of a Diesel Spray Injected into a High Density Gas under Vaporization Regimes, Proc. of 7th Int. Symp. on Appli. of Laser Techniques to Fluid Mech., Lisbon, Paper No. 36. 3.
3 Kobashi, K. et al., 1990, Measurement of Fuel Injector Spray Flow of I. C. Engine by FFT Based Phase Doppler Anemometer, Proc. of 5th. Int. Symp.on Appli. of Laser Techniques to Fluid Mech., Lisbon, Paper No. 21. 2.
4 Vannobel, F. et al. D., 1992, Phase Doppler Anemometry Measurements on a Gasoline Spray Inside the Inlet Port and Downstream of the Induction Valve: Steady and Unsteady Flow Conditions, Proc. of 6th Int. Symp. on Appli. of Laser Techniques to Fluid Mech., Lisbon, Paper No. 25.1.

5 Obokata, T. and Long, W. Q., 1994, LDA/PDA Characterization of Conical Spray for Diesel Engine, Proc. of 6th. Int. Conference on Liquid Atomization and Spray Systems., Rouen, Paper No. IC94/90.

6 Obokata, T. et al., 1994, PDA Analysis of Transient Spray Flows Initiated from Air Assisted Injector, Proc. of 7th Int. Symp. on Appli. of Laser Techniques to Fluid Mech., Lisbon, Paper No. 36.5.

7 Obokata, T. et al., 1995, Characterization of Spray Flow Initiated from Air Assisted Injector Using Phase Doppler Anemometer, FED-Vol. 229 Laser Anemometry, ASME

8 Xu, T.-H. et al., 1991, The Three-Parameter Log-Hyperbolic Distribution and its Applications to Particle Sizing, Proc. Int. Conf. on Liquid Atomization and Spray systems, ICLASS 91, Gaithersburg, Paper No.31, pp. 315-324.

9 Grèhan, G. et al., 1991, Evaluation of Phase Doppler System using Generalized Lorenz-Mie Theory, Proc. Int. Conf. Multiphase Flows. Tsukuba, pp. 291-294.

10 Rajaratnam, N., 1976 Turbulent Jets. Elsevier Scientific Publishing Company, Amsterdam.

11 Barndorff and Nielsen, O., 1977, Exponentially Decreasing Distributions for the Logarithm of Particle Size, Proc. R. Soc. of London, A. 353, pp. 401-419.

V.3 Velocity Profiles in Shear-Driven Liquid Films: LDV-Measurements

S. Wittig, A. Elsäßer, W. Samenfink, J. Ebner, K. Dullenkopf

Lehrstuhl und Institut für Thermische Strömungsmaschinen
Universität Karlsruhe (TH), D-76128 Karlsruhe, Germany

Abstract. In design and layout of fuel preparation systems for SI-engines or gas turbines as well as other technical systems working with shear-driven liquid films accurate information on the velocity distribution within the film as a function of film height is necessary for numerical modeling. To meet this requirements an improved LDV system was built which allows for the first time accurate velocity measurements in thin and wavy liquid films down to an average thickness of about 100 μm. Layout and the optical arrangement of the new system is presented in details together with some examplary applications.

1. Introduction

The prediction of internal fuel films plays an important role in the design of combustion engines. Typical examples are prefilming airblast atomizers (Wittig et al. (1992)), prevaporating combustors in gas turbines (Pfeiffer (1992)), oil films in aero engine bearing chambers (Glahn et al. (1995)), or intake manifolds of sparc ignition engines (Elsäßer et al. (1994)). Especially in sparc ignition engines liquid fuel stored in wall films introduces a characteristic delay time to engine reaction, also causing high pollutant emissions during load changes or in cold start conditions.

A major problem in numerical modeling of the film flow is the lack of information about the velocity distribution as a function of the film height. For an accurate prediction of the film transport and evaporation characteristic the knowledge of the precise velocity profile is required. If the assumed profile is incorrect, the liquid film thickness and as a consequence the interaction with the air flow cannot be calculated adequately. This leads to a poor prediction of the heat and mass transfer at the phase boundary.

In the literature different theories for the internal film velocity profile are proposed. Some investigators assume that the film is of turbulent nature, others suppose laminar conditions. However, most of the authors were not able to verify the velocity profiles in shear-driven films directly, due to the small average layer thickness of $50 \leq \bar{h}_F \leq 200\ \mu m$ of typical films for example in intake manifold flows.

2. Velocity Measurements in Liquid Layers

To our knowledge none of the instruments available is suitable to performe velocity profile measurements within thin shear-driven films. Due to the thin layer an optical, non intrusive technique has to be applied. Especially using the Laser-Doppler-Velocimeter (LDV) has some potential in this case. Some investigators have already applied the LDV technique to perform velocity profile measurements inside thicker liquid films or microscopic flows within small gaps. Some of the methods are discussed in the following, especially with respect to the applicablity to thin and wavy films under the influence of high air velocities.

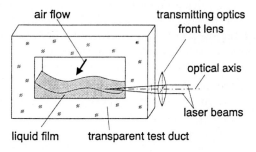

Fig. 1: Transmitting optics located horizontally at the side wall of the test duct

The methods described in the literature differ primarily in the orientation of the probe volume with respect to the film surface. Mudawar and Houpt (1993) used a transparent film splitter to cut a slice from a vertical falling film. This method allows lateral access to the film by focusing the beams of a LDV system in a certain film height, as shown in **Fig. 1**. The transmitting and receiving optics were aligned in the optical axis of a forward scattering setup. The probe volume had a length of 150 μm and a diameter of 29 μm. The main problem using this method in shear-driven films was, that measurements were possible only when the surface waves are two-dimensional line shaped. This is due to the fact that the convergent beams of the transmitting optics, the probe volume itself, and the path of the scattered light to the photomultiplier might not be disturbed by the air/liquid interface and therefore have to be completely inside the film. However, shear-driven liquid films in contrast to falling films have a distinct three-dimensional wavy structure. They contain large regions with a typical thickness of only $20 - 30\ \mu m$ embedded in wave fronts with maximum heights of $150 - 500\ \mu m$, depending on the liquid volume flux and the shear force conditions. Due to this conditions the path of the light will be interrupted most of the time and therefore a different optical arrangement is required.

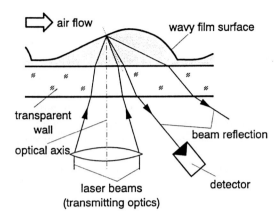

Fig. 2: Transmitting optics arrangement and beam reflections at the wavy
film surface

Another method to arrange the LDV optics with respect to the film is shown
in **Fig. 2**, where the transmitting optics is applicated vertically to the bottom
wall. This method is used by some investigators to determine velocity profiles
in thin liquid layers. Tieu et al. (1995) measured the single-phase flow profile
in a small gap using a diode-laser LDV system in 180° backscattering mode.
By means of a extremely short focal length in the transmitting optics the
resulting probe volume length was 10 μm and the diameter 5 μm. This very
small probe volume size offered an excellent spatial resolution in the liquid
layer. The major drawback of this arrangement to study shear-driven films
results from the wavy surface, as outlined in **Fig. 2**. If measurements are
performed near the wavy boundary for a wide range of surface angles strong
reflections are directed towards the detector. In shear driven films these conditions may be present quite often, therefore the applicability of this method
for the films under investigation in this study is questionable.

A similar method was used by Paras and Karabelas (1992) to analyze annular
shear-driven liquid layers. In their case a probe volume of 121 μm length and
23 μm in diameter was utilized. The air flow velocities had been low for their
study, and only films with a thickness in the range of $3.4 \leq \bar{h}_F \leq 7.6\ mm$
had been investigated. Due to these thick films the resolution along the optical axis was sufficiently high. However, in their measurements the effect of
beam reflections at the air/liquid interface could clearly be detected by the
photomultipliers. If these reflections were treated as Doppler signals they
would lead to wrong results. Furthermore Paras and Karabelas (1992) set a
limit for measurements with their system to $1 - 2\ mm$ below the film surface.

512

Measurements closer to the surface had not been possible.

Plimon (1991) used a similar arrangement of the transmitting optics. It was located vertically below a transparent horizontal bottom wall of a rectangular duct also. The difference to the methods described previously was the position of the receiving optics above the film surface at an off-axis angle of 45° to reach an optimum between the signal to noise ratio and scattering efficiency. The resolution along the optical axis was about 25 μm by using a short focal length in the transmitting optics. In his paper Plimon presented a velocity profile determined at low air velocity (20 m/s) and high liquid volume flux. These conditions were ideal for the detection of the scattered light through the air/liquid interface. At higher air velocities it can be assumed that the signal quality would become significantly worse due to the strongly wavy film surface.

Because of their destinct shortcomings, none of the previosly described systems was applicable in our study. Therfore a different setup had to be developed especially adapted to the conditions of thin wavy liquid films.

3. New Approach for Film Profile Measurements

3.1 Test Rig

The basic item of the test rig used for the experiments is a transparent duct with a cross section of 30 x 196 mm and a length of 1000 mm. It is connected with the suction side of a blower. To vary the free-stream turbulence intensity turbulence grids can be inserted in the inflow.

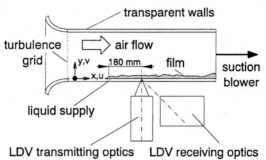

Fig. 3: Test section

The test liquid (demineralized water) is seeded with MgCa-particles and supplied to the channel floor wall through a row of small holes and forms a continuous film. Free stream and film conditions including temperature, pressure and volume flux are controlled continuously.

3.2 Transmitting Optics

The newly designed setup consists of a transmitting optics looking vertically up through a transparent bottom wall of the horizontal test section. This provides good optical access to the film flow. The probe volume can be exactly located without any disturbances due to the wavy film surface, see **Fig. 2.** As basic system a two-component fiber optic LDV system was used. The LDV system includes a 4 W Argon-ion Laser in multi mode operation, a standard optics with 40 MHz Bragg-cell, color and beam seperators, and a 2D-fiber probe. To achieve the resolution suitable for film flow under investigation a beam expander and a short focal length front lens ($f = 80$ mm) have been used in the transmitter. They reduce the probe volume size to a length of $\Delta y_F \simeq 200$ μm and a diameter of $\Delta x_F \simeq \Delta z_F \simeq 25$ μm.

However, the size Δy_F is still larger than the desired spatial resolution of about 30 μm in the direction of the optical axis. Therefore, the probe volume has to be reduced further by the receiver design.

3.3 Receiving Optics

As shown already mentioned the 180° backscattering mode is unfavorable for detailed film profile measurements. Firstly, the sensible area of the detector is still too large, and secondly the reflections at the film surface cause serious problems considering the signal to noise ratio in a wide range of surface angles. To solve these problems the detector was shifted to an off-axis angle with respect to optical axis of the transmitter, thus cutting a narrow section of the probe volume as shown in **Fig. 4** for one component. Simultaneously, this helps to solve the reflection problem at the film interface. An angle of 45° to the optical axis was chosen, because previous investigations have shown that reflections in this range occures relatively seldom for conditions used in the tests. To observe only a defined section of the probe volume in the y-axis direction a slit aperture is positioned in front of the color splitter in the receiver. The signals have been detected by a two-component photomultiplier setup in combination with counter processors and a self developed software for data aquisition and analysis.

As demonstrated in **Fig. 5**, the receiving optics is located below the test section in line with the transmitting optics. Its optical axis would cross the bottom wall under 45° in air. However, the arrangement with a simple flat bottom wall is unfavorable due to multiple refraction at the air/glass and glass/liquid interfaces, as pointed out in the top part of **Fig. 5**. In this case the beams have already different angles $\vartheta_{Li1,Glass} \neq \vartheta_{Li2,Glass}$ when passing the bottom wall and consequently a strong distortion of the focus occures. This distortion may be reduced by using thin walls, but this is not possible

514

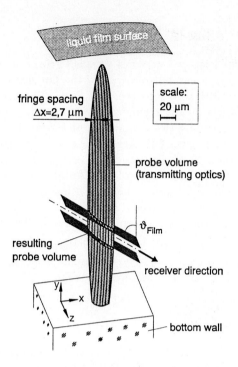

fringe spacing
Δx=2,7 μm

scale:
20 μm

probe volume
(transmitting optics)

resulting
probe volume

ϑFilm

receiver direction

bottom wall

Fig. 4: Reduction of the resulting probe volume length by off-axis location
of the receiver aperture

in our case. Therefore, as shown in **Fig. 5b** a prism is fixed perpendicular
to the optical axis in order to obtain equal refraction conditions for both
beams at the air/glass interface. Different angles of incidence appear only at
the glass/liquid interface. This interface has a small variation in the index
of refraction and the optical pathes are short (about 500 μm) to the intersection point $P_{Measure}$. By means of this measure a minimal abberation for
all beams parallel to the optical axis and intersecting in $P_{Measure}$ can be
achieved.

In **Fig. 6** a sketch of the optical receiver arrangement is given. Two lenses
with equal focal length give a 1:1 projection of the probe volume onto the
slit aperture. To determine the exact value of the section height caused by
different beam path and the distortion of the focus a numerical analysis of the
beam trajectories is provided. The analysis indicates that due to the different
angles of incidence at the glass/liquid interface the focal points of all parallel
beams are not located at an identical position inside the film. The results of
the calculation are plotted in **Fig. 7**. It shows the computed 1:1 projection
of the slit aperture focus inside the liquid film. The co-ordinates of the film

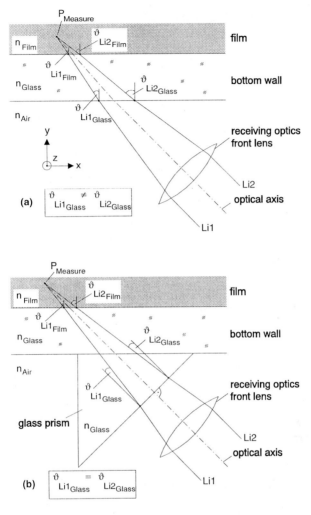

Fig. 5: Improvement of projection quality by a prism perpendicular to the optical axis

internal foci are calculated for the envelope in the drawing plane ($\parallel$-mode) and the plane perpendicular to this ($\perp$-mode). In the figure the locations of the minimum focus size in both directions as envelopes of the whole beam are plotted. The symbols characterize the points achieved by a receiving optics traverse of $\Delta y = 50$ μm and $\Delta x = 0$ μm. The corresponding points for discrete positions of the receiving optics are connected with lines in the figure. These lines are very close to the optical axis which is inclined under $\vartheta_{SF} = 53.9°$ in the fluid. To illustrate this effect, a schematic of the focus shape in the film is given in **Fig. 7** also. Between the two modes the distance

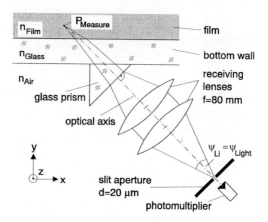

Fig. 6: Sketch of the receiving optics realization including the glass prism

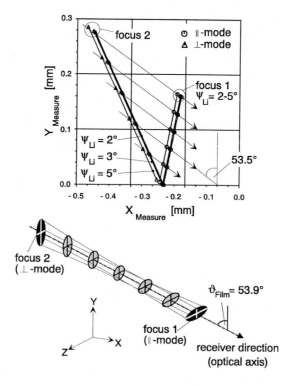

Fig. 7: Computed optical conditions in the focal point

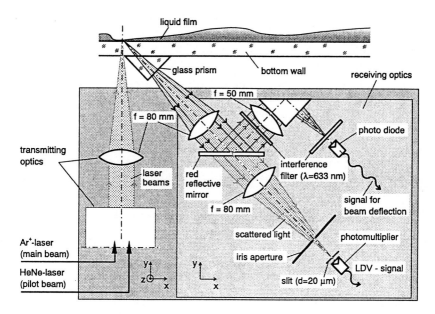

Fig. 8: Setup of the LDV system for velocity profile measurements in thin and wavy liquid films

of the foci is very short, so a useful projection is achieved. The difference between the focus location of both modes increases with larger height in the liquid layer. For a detailed construction of the optics the mode parallel to the drawing plane is of special interest, because it determines directly the size in the film height direction. The focus dimension of the perpendicular mode can be negelected when using a slit aperture. Beams with this orientation pass the slit along its long expansion and therefore will be detected anyway.

Additionally the maximum collection angle is varied in the calculation in the range of $2 \leq \Psi_L \leq 5°$. An angle of $\Psi_L = 5°$ is the maximum value in the realized optics, limited by an iris aperture. For the design it is important to keep in mind that the slit is projected nearly at the same position for different values of Ψ_L. The angle dependent deviation is therefore negligible. However, as mentioned above, between the foci of the $\|$-mode and the $\perp$-mode a noticable difference appears.

Furthermore, the focus point of the $\|$-mode moves not only vertical by a y-motion of the collecting optics. Therefore, the receiving optics has to be traversed in both, y-axis and x-axis direction, to obtain a soley vertical movement of the probe volume. The corresponding ratio between Δx and Δy for an exclusive vertical movement can be derived by this calculation, too.

Finally the solution of the reflection problem at the film surface will be discussed. Because it is necessary to protect the photomultipliers from reflected light to avoid overload and errorneous Doppler signals, the laser beams are switched off by a Bragg-cell for reflecting conditions. An additional laser beam of different wavelength ($\lambda_{Pilot} = 633\ nm$) is used as a indicator for reflecting conditions. The pilot beam is coupled into the receiving fiber of the standard fiber probe, thus forming a focus by the front lens, too. The time span necessary for the electronic and the Bragg-cell to switch off the main beams is gained by the specific optical arrangement in the receiver. The front lens of the receiver collects scattered light in an angle range of $\Psi_L = 20°$ and this complete field is used in the red light detection for the switch-off device. The angular range of the green and the blue photomultipliers is reduced by an iris aperture to $5°$. Therefore, light is reflected into the pilot beam detector first and the main beams can be switched off prior to full illumination of the photomultipliers by reflection.

The layout of the newly designed optics is shown in **Fig. 8**. The complete system is mounted on a high precision 3D-traversing device. The independant spatial movement of the receiving optics relative to the transmitting optics is achieved by a second extremely accurate 2D-traverse unit.

4. Results

4.1 Measurements in a Typical Shear-driven Water Film

An exemplary result of the time mean velocity profile in main stream direction is shown in **Fig. 9** for an air velocity of $30\ m/s$ and the liquid volume flux of $\dot{V}/b = 0.434\ cm^3/cm\ s$. This condition leads to a shear stress of $\tau = 3.9\ N/m^2$ at the air/liquid interface. The time mean film thickness for this layer is $\bar{h}_F = 177\ \mu m$. To determine the film thickness value a novel optical measurement system developed at the Institut für Thermische Strömungsmaschinen at Karlsruhe University and described by Samenfink et al. (1996) was used.

The step width of the profile measurement in y-direction was $\Delta y = 15\ \mu m$. In addition to the time mean velocity distribution the velocity fluctuations are given too, centered around the mean value for different film heights and marked by the dotted lines. On the right of **Fig. 9** three probablity density charts are given for three specific film heights. The histograms show, as expected, that the velocity fluctuations are weaker near the wall than in the wave crest region.

As expected by the two dimensional character of the flow the time mean velocity in the cross-flow direction is close to zero for all heights, as plotted in **Fig. 10**. The RMS value shows increasing values with increasing wall distance, too. This is also present in the histogram charts of the velocity distribution in different y-positions inside the film.

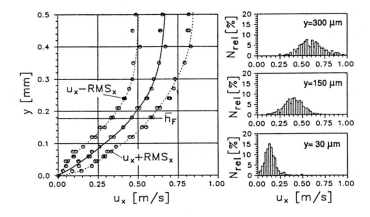

Fig. 9: Mean stream velocity measurements, $u_{Air} = 30 \ m/s$, $T_F = 20°\ C$

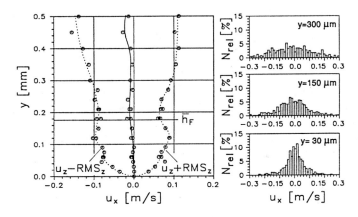

Fig. 10: Cross stream direction velocities, $u_{Air} = 30 \ m/s$, $T_F = 20°\ C$

4.2 Influence of Volume Flux on the Film Velocity Profiles

To determine the effect of liquid volume flux on the profile shape within shear-driven films additional investigations have been performed. In **Fig. 11** the results of a volume flux variation in the range of $0.175 \leq \dot{V}/b \leq 0.868 \ cm^3/cm \ s$ are shown. The time mean thickness of the film was $108 \leq \bar{h}_F \leq 245 \ \mu m$ for the conditions plotted in the figure. By varying the volume flux of the liquid the shear-stress induced on the film is changed too, because of different wave structure and roughness effects. **Fig. 11** indicates that the velocity profiles are nearly identical from the bottom wall up to a film height of about $150 \ \mu m$. Above this height the profiles split in the individual curves corresponding to the volume flux. This also points out, that the film differ only in the wave region under constant air flow conditions. The profiles have a nonlinear shape in height and therefore a turbulent character. This confirms the trend in

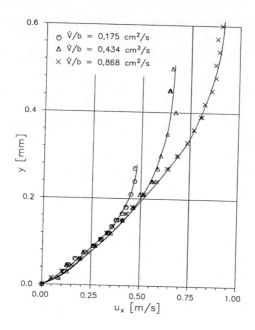

Fig. 11: Mean velocity profiles in water films, $u_{Air} = 30 \ m/s$, $T_F = 20°\ C$

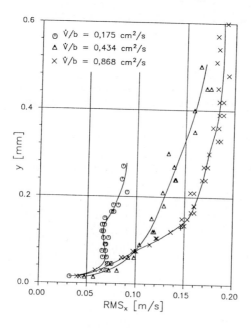

Fig. 12: RMS values in water films, $u_{Air} = 30 \ m/s$, $T_F = 20°\ C$

the results of Plimon (1991) where the parabolic shape of the film velocity profile is shown for a thicker shear-driven film. First comparisons with profile assumptions given by Deissler (1954) show a good agreement for water films, too.

In **Fig. 12** the RMS values of the measurements described earlier are presented. The RMS values deviate much earlier than the mean velocities. Considering the velocity fluctuations the distributions start to diverge significantly at a height of about 40 μm. For a detailed explanation of this effect corresponding film thickness measurements are necessary. The height of 40 μm is approximately the minimum film height being present for all investigated volume fluxes. This minimum film thickness represents a thin sublayer on top of which waves of different heights, depending on the volume flux, move driven by shear forces.

For an accurate prediction of liquid film flows the knowledge of the different transport mechanisms is crucial. More detailed information about these complex phenomena can only be obtained by simultaneous measurements of film thickness and film velocity. Comprehensive efforts have been directed at the Institut für Thermische Strömungsmaschinen at Karlsruhe University to combine the two measurement systems previously mentioned, and some first results will be given in Samenfink et al. (1996). By the simultaneous use of both systems a detailed characterization of mass transport profiles in wavy shear-driven films will be possible.

5. Conclusions

To improve the knowledge of the transport mechanisms in wavy shear-driven liquid films important in many technical applications a novel LDV system was designed and presented. For the first time it allows accurate velocity profile measurements in thin and wavy films with average thickness of about 100 μm from the bottom up to the wave crest region. The instrument combines a active small probe volume and an acousto-optical switching device to protect the photomultiplier-tubes from the intensive reflections from the film surface.

The results presented in the paper show that the time mean velocity profiles inside the film have a parabolic shape and are non-linear with the height. These first results can alrady be used to validate film velocity assumptions required to improve the prediction of mass and heat transfer in shear force driven film flows.

References

[1] Deissler, R.G.: Heat Transfer and Fluid Friction for Fully Developed Turbulent Flow of Air and Supercritical Water with Variable Fluid Properties. ASME–Journal of Fluids Engineering, 76, pp. 73 – 85, 1954.

522

[2] Elsäßer, A., Samenfink, W., Hallmann, M. and Wittig, S.: Mixing Phenomena of Fuel Sprays in Intake Manifolds. In: Proceedings on the Sixth International Conference on Liquid Atomization and Spray Systems (iclass94), Rouen, France, Paper VII-10, 1994.

[3] Glahn, A. and Wittig, S.: Two-Phase Air/Oil Flow in Aero Engine Bearing Chambers - Characterization of Oil Film Flows. ASME-Paper 95-GT-114, 1995.

[4] Mudawar, I. and Houpt, R.: Measurement of Mass and Momentum Transport in Wavy-Laminar Falling Liquid Films. International Journal of Heat and Mass Transfer, 36, No. 17, pp. 4151 – 4162, 1993.

[5] Paras, S.V. and Karabelas, A.J.: Measurements of Local Velocities Inside Thin Liquid Films in Horizontal Two-Phase Flow. Experiments in Fluids, 13, pp. 190 – 198, 1992.

[6] Pfeiffer, A.: Entwicklung einer keramischen Kleingasturbinen-Brennkammer: Neue Möglichkeiten zur schadstoffarmen Verbrennungsführung. Dissertation, Institut für Thermische Strömungsmaschinen, Universität Karlsruhe (T.H.), 1992.

[7] Plimon, A.: Velocity Profiles in Shear Force Driven Wall Films. Experiments in Fluids, 11, pp. 339 – 340, 1991.

[8] Samenfink, W., Elsäßer, A., Wittig, S. and Dullenkopf, K.: Internal Transport Mechanisms of Shear-driven Liquid Films. Eighth International Symposium on Applications of Laser Techniques to Fluid Mechanics, Lisbon, Portugal, 8-11 July, 1996.

[9] Tieu, A.K., Mackenzie, M.R. and Li, E.B.: Measurement in Microscopic Flow With a Solid-State LDA. Experiments in Fluids, 19, pp. 293 – 294, 1995.

[10] Wittig, S., Himmelsbach, J., Noll, B., Feld, H.J. and Samenfink, W.: Motion and Evaporation of Shear-Driven Liquid Films in Turbulent Gases. ASME–Journal of Engineering for Gas Turbines and Power, 114, pp. 395 – 400, 1992.

V. 4   Steady Intake Flow Characteristics Through a Diesel Four-Valve Cylinder Head

C. Arcoumanis, B. French* and J.M. Nouri

Department of Mechanical Engineering, Imperial College of Science, Technology
& Medicine, London, UK.
*Diesel Engineering,, Ford Motor Company Ltd, Essex, UK.

Abstract.   The annular flow around the two intake valves of a production four-valve diesel cylinder head has been investigated by laser Doppler velocimetry under steady flow conditions at a mass flowrate of 85 Kg/h corresponding to an engine speed of 1000 rpm. The results showed that the primary velve generates much stronger swirl than the secondary valve, which in turn generates more than 20% higher mass flow rate. When both valves are open, the generated swirl and mass flow levels fall between those obtained with the primary and secondary valves but much closer to the former than to the latter. The exit flow around the intake valves was found to be non-uniform, with more pronounced the case of the primary valve with the maximum exit flow at the planes towards the end of the helical ramp of the port. Similar non-uniform flows were observed when both valves were open except for the two planes of the primary valve which are affected by the jet flow exiting from the secondary valve. Turbulence, in general, was anisotropic with maximum levels of about $5V_p$(V_p=mean piston speed) near the cylinder head.

1 Introduction.   The importance of the in-cylinder air and fuel motions on the performance and emissions of direct-injection Diesel engines has long been recognised. The desirable conditions is to achieve rapid mixing between the swirling air and injected fuel in order to complete combustion within a short crank angel interval close to TDC. High fuel-air mixing rate, which in turn controls the fuel burning rate, is even more important in passenger car diesels where the mixing time scale is very short relative to engines at the large end of the size range. Improved understanding of the characteristics of the introduction air motion over a wide range of speeds is a prerequisite for the development of high speed direct-injection (HSDI) diesels, as the legislation on NO_x and particulate emissions becomes world-wide more stringent.

The present exprimental investigation aims to quantify the exit flow characteristics generated by the two intake helical ports of an advanced four-valve Diesel cylinder head in terms of the flow discharge and swirl coefficients and the mean/rms velocity profiles as a function of the valve lift. These results are expected to assist the development of multidimensional computer models, which require accurate experimental data as boundary conditions, and to allow the design of these ports to be improved by achieving higher volumetric efficiency especially at high engine speeds.

2 Flow Configuration and Instrumentation.   The four-valve cylinder head of a production Ford diesel engine was mounted on a steady flow rig, as

524

shown in Fig.1. Ambient air was drawn through the two inlet ports by a five-stage centrifugal suction pump wich was controlled by a bypass flow regulator. The mass flow rate was determined by a British Standard orifice plate, 1042, with an accuracy of ±1,3%. Digital and U-tube manometers were used to measure the upstream pressure of the orifice plate, the air pressure drop across the orifice plate and the pressure drop through the inlet manifold and the cylinder head assembly.

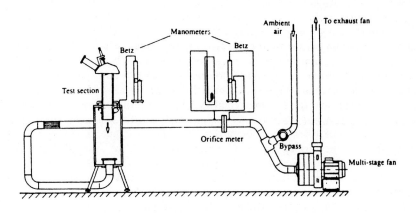

(a) Diagram of steady flow rig

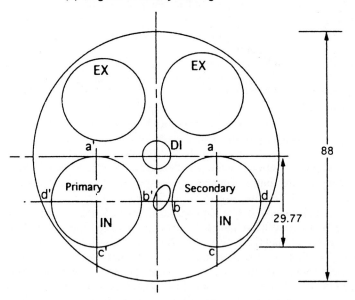

(b) Diagram of cylinder head

Fig.1. Schematic diagrams of flow circuit and cylinder head together with measuring locations around the intake valves (dimension in mm)

The ambient air pressure and temprature were measured with a mercury barometer and a thermometer , respectively. The two inlet ports were of helical type and produced dofferent levels of swirl; they are referred to here as primary (higher swirl) and secondary (lower swirl). An open-ended acrylic cylinder was fitted to the cylinder head with a bore diameter of 88mm, a wall thickness of 6mm and a length four times the bore diameter, wich provided access for the laser Doppler velocimeter. At the end of the acrylic cylinder four interconnected pressure tappings were made to monitor the presuure drop across the cylinder head prior to its connection to a settling chamber.

To obtain the swirl coefficient generated by the intake ports, either individually or jointly, a commercial torque meter (Quadrant Scientific) was placed four bore diameters downstream of the cylinder head. The meter consists of a honeycomb which converts the angular momentum to torque and a measuring device which gives the enduced torque with an accuracy of ±0,5%. The torque meter was calibrated over the measuring range with balance weights and the linearity of the response of the meter was found to be very satisfactory as shown in Fig.2.

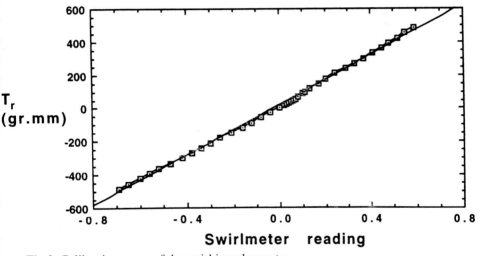

Fig.2. Calibration curve of the swirl impulse meter

The laser Doppler velocimeter comprised a water-cooled Argon-ion laser (Spectra Physics) operating with a wavelenght and power of 0.514 μm and 300 mW respectively, a diffraction-grating to divide the laser beam into two of equal intensity and provide frequency shifts up to 12 MHz, a collimating and a focusing lens to bring the two beams to an intersection volume, a photomultiplier (Dantec 55X34), and a frequency counter (TSI model 1990) interfaced to a microcomputer using a DMA board (Dostek 1400). The intersection volume was approximately 757 μm in length and 42 μm in diameter, with a fringe spacing of 4.6 μm. The signals from the photomultiplier were processed by the counter to form ensemble averages from 4000 individual velocities.

The measurement position and the magnitude of the velocity components were corrected , when required, for refraction effects of the laser beams through the cylindrical wall. Uncertainties due to velocity gradients were minimised by minimised by aligning the smaller dimension of the intersection volume along the velocity gradient, except for the swirl component and by collecting the light 30^0 off-axis. Statistical uncertainties were also minimised by considering a large number of samples and the maximum counting ambiguity of the TSI instrument, having a clock frequency of 1000 MHz, was 0,5% of the Doppler frequency shift. The overall maximum uncertainty in the measurements of mean and rms velocities is estimated to be 5% and 10%, respectively.

The flow was seeded with silicone oil droplets which were generated by an air-blast atomiser and introduced into the flow in the plenum chamber upstream of the inlet ports. This atomiser had been shown to produce droplets with Sauter mean diameter, d_p, of less than 2 μm and this corresponds to an effective Stokes number of less than 0.1, implying that droplets do follow the mean flow even with the occasional larger droplet and, thus, the uncertainty due to the wide size range of the droplets can be considered to be negligible.

3. Results and Discussion.

Measurements of the discharge, flow and swirl coefficients and the three components of the mean and rms velocities around the intake valves were obtained at constant mass flow rates as a function of the valve lift. For comparison purposes, three test cases were open. The initial tests reported here were made at a mass flow rate of 85 kg/h through the inlet ports, which corresponds to an engine mean piston speed of $V_p=3.155$ m/s for a stroke of 94.6mm and an engine speed of 1000 rpm. Subsequent tests were performed at a mass flow rate corresponding to 3000 rpm in order to examine the flow behaviour at higher speeds where the volumetric efficiency of the engine was found to deteriorate (Arcoumanis et al, 1995).

3.1. Discharge , Flow and Swirl Coefficients.

The discharge coefficiente is defined as the ratio of the measured mass flow rate, m, through the port and valve assembly for a given valve lift, L, and a pressure ratio, to the isentropic mass flow rate, m, through an ideal nozzle, i.e.

$$C_d = \frac{m}{m_i} \tag{1}$$

and

$$m_i = A_1 \sqrt{\left[2P_o \rho_o \frac{\gamma}{\gamma-1} \left(\frac{P}{P_o}\right)^{2/\gamma} \left\{ 1 - \left(\frac{P}{P_o}\right)^{(\gamma-1)/\gamma} \right\} \right]} \tag{2}$$

where P/P_0 i is the pressure across the port and valve assembly, P_0 and ρ_0 are ambient pressure and density, γ is equal to 1.4, and A_1 is the curtain area defined as $A_1 = \pi N D_h L$ where N is the number of intake valves and D_h is the valve head diameter (=29.77mm).

The flow coefficient is defined as the ratio of the measured mas flow rate to the theoretical mass flow rate through the valve inner sit area and is expressed as:

$$C_f = \frac{m}{\rho V_o A_i} \tag{3}$$

where A_i is the valve inner sit area, $A_i = \dfrac{\pi N D_i^2}{4}$ where D_i is the valve inner sit diameter (=26.2mm) and V_0 is the head velocity derived from the pressure drop across the valve using the compressible flow equation:

$$V_o = \sqrt{\left[\frac{2\gamma}{\gamma-1}\frac{P_o}{\rho_o}\left\{1-\left(\frac{P}{P_o}\right)^{(\gamma-1)/\gamma}\right\}\right]} \tag{4}$$

The swirl coefficient is essentially the ratio of the flow angular momentum to its axial momentum. With the impulse torque meter the simplest definition according to Heywood (1988) is

$$C_s = \frac{8T_r}{mV_0B} \tag{5}$$

where T_r is the induced torque and B is the bore diameter. The variation of the discharge, flow and swirl coefficients with valve lift is presented in Fig.3 for a constant mass flow rate of 85 Kg/h. The discharge coefficient, C_d, shows in all cases a sharp decrease up to a valve lift of L/D=0.083 (or 2.5mm) and a slightly smaller decrease thereafter; no measurements were obtained at valve lifts less than L/D=0.067 for the cases when only the primary or the secondary valve was open because the mass flow rate of 85 Kg/h could not be achieved. The results also show that the secondary valve produces a much higher discharge coefficient that the primary valve at all lifts (less than 0.083) where the difference becomes very small;

for example, at full lift (L/D=0.275 or 8.2mm) the C_d with the secondary valve is 20% higher than the primary valve. When both valves are open, C_d reduces from a value of 0.78 at the lowest lift to a value of 0.33 at full lift and the results follow those of the primary valve very closely in that C_d is higher with the primary valve up to L/D = 0.1 and consistently lower afterwards; the C_d with the primary valve open is higher by 5.6% at L/D = 0.067 while at full lift it is lower by 4%.

Unlike the C_d variation, the flow coefficient based on the inner valve sit cross-sectional area, C_f, shows an increase with increasing valve lift for all cases. Its value increase sharply and linearly up to a valve lift ratio of 0.125 after wich the rates of increase reduce considerably until L/D = 0.2 and then remain almost constant with a small and gradual increase until the full lift L/D of 0.275. The differences between the C_f values for diffrent flow cases are similar to those of C_d, so that at full lift the value of C_f when the secondary valve is open is higher by 14% and 20% than the corresponding values when both valves are open and when the primary valve is open, respectively.

The results of the swirl coefficient, C_s, show marked differences between the three flow cases but in a reverse order to that of the discharge coefficient, that is the highest swirl coefficient is generated by the primary valve followed by that when both valves are open while the lowest swirl coefficient is generated by the secondary valve wich produced the highest coefficient of discharge. With the secondary valve open the C_s increases mildly with valve lift to a maximum value of 0.16 at L/D = 0.15 and then reduces very slowly so that at full lift it is 0.125. The results of the primary valve and for the case when both valves are open exhibit similar trends in which the C_s, values increase sharply until L/D = 0.14 and then the rate of increase reduces till L/d = 0.2 after which it increases again. At full lift, the swirl coefficient generated by the primary valve is 0.675 compared to 0.125 when only the secondary valve is open. The value of C_s when both valves are open at full lift is 0.49 which is 27,5% lower and 390% higher then those of the primary and secondary valve, respectively. These results confirm that when degree of swirl is requierd, as for example at low speeds to improve mixing, the use of the primary valve is necessitated and when a higher mass flow rate is required, as for example at high speeds, both valves should be used.

3.2 Velocity Measurements. The three velocity components around the intake valves were measured for all three flow cases at a mass flow rate of 85 Kg/h and full valve lift, 8.2mm. Measurements were taken at different axial locations from the tip of the valves to a distance of 2mm from the cylinder head in cricumferential positions a, b, c and d around the secondary valve (SV) and a', b', c' and d' around the primary valve (PV) as shown in Fig.1(b), with a radial distance from the tip of the valve head of 0.5mm for the axial and radial velocity components and 1mm for the swirl component. The mean and rms velocities are normalised in all cases with the mean piston speed, V_p, equal to 3.155 m/s at 1000 rpm. The results are presented as velocity profiles and in vector forms and the three flow cases are examined below separately.

3.2.1 Primary Valve Open. The measurd velocity profiles and deduced vectors around the primary valve are shown in Fig.4 together with the position of the valve and the cylinder wall. The axial flow at the exit planes, Fig.4(a), shows a non-uniform distribution of the axial velocity around the valve so that the velocities are lowest with maximum of $5V_p$ at plane b', upstream of the end of the helical ramp of the port, and then increase considerably in the clockwise direction to the largest values in plane a' with a maximum value of $18V_p$; the maximum velocities in planes c' and d' are around $12V_p$ and $16V_p$, respectively. Surprisingly, the velocity profile across the valve gap in plane c' exhibits a double peak which may be due to the proximity of the measuring plane to the edge of the helical ramp of the port.

A similar non-uniform can be seen in the circumferential distribution of the radial mean velocity, Fig.4(b), with maximum values in plane a' of $24V_p$ which are more than 50% higher than the maximum values in planes b' and c'. Like the axial flow at plane c', a similar double peak profile can be seen here with a minimum value in the middle of the value gap . It is also interesting to note that despite the cylinder proximity at plane d', the axial and radial velocities in this plane are larger then those in plane b' where the cylinder wall is at a much larger distance. This is because plane d' is more aligned with the axis of the inlet pot, which is directed along the wall of the cylinder, than plane b' and the results emphaise the dominant influence of the port position on the flow around the intake valves.

The swirl velocity distribution around the valve, Fig. 4(c), is more uniform and reduced in magnitude than the other two velocity components, with the lowest values in plane b', similarly to the axial and radial velocity, and largest values in plane d' with magnitudes of up to $13V_p$; the velocity in all planes increases from the tip of the valve towards the cylinder head. The relative magnitude of the axial/radial and radial/tangential velocities indicates a variation of the discharge and angular flow directions, respectively, wich will be discussed again when the velocity vectors are presented later. The rms velocity variation of all three components follows the trends of the mean flow with the velocities not uniformly distributed around the valve periphery; the anisotropy of the flow is evident from the comparison of the three rms components. The level of turbulence is, however, relatively high in all three components varying from $1V_p$ to $5V_p$.

The non-uniformity of the discharging flow around the valve is more evident from the vector addition of the axial and radial velocity components shown in Fig.4(d), with largest and lowest values at planes a' and b', respectively. The results also reveal that the flow angle with respect to the valve axis varies slightly along the gap, so that in general the flow angle tends towards the valve axis as the flow approaches the valve tip; the reverse axial velocity at the tip of the valve and just below it is an indication of the reverse flow being part of the recirculation zone formed beneath the valve head. The flow angle variation from one plane to another is also evident so that the average flow angle at planes a', b', c' and d' is 43^0, 66^0, 38^0 and 50^0, respectively. The vector addition of the radial and tangential mean velocity components, shown in Fig.4(e), reveals more clearly the flow pattern around the valve indicates that the flow angle with respect to the valve radius varies only slightly within the valve gap except in plane d' where the flow angle close to the cylinder head wall, at z=2.2mm, is much larger than that close to the tip of the valve, at z=6.2mm. The variation of the flow angle around the valve periphery changes, however, more drastically with the average values at a', b', c' and d' being 25^0, 30^0, 34.5^0 and 45^0. Similar trends have been reported by Arcoumanis et al(1987) and Haghooie et al(1984) for helical ports in two-valve heads.

3.2.2 Secondary Valve Open. Similar measurements to those of the primary valve were made around the secondary valve and are shown in Fig.5. Similar trends to those observed with primary valve can be seen here but the distribution of the axial and radial velocities, Figs. 5(a) and (b), along the valve gap and around the periphery of the valve is more uniform in the case of the secondary valve; the maximum axial velocities in planes a, b, c and d are $12.5V_p$, $17V_p$, $13V_p$ and $7.5V_p$, and the corresponding maximum radial velocities are $15V_p$, $19V_p$, $16V_p$ and $7.5V_p$. It is also evident that the maximum axial and radial flow is in plane b which is expected since this plane is in-line with the inlet port axis; planes a and c seem also to be benefit from this proximity. The swirl velocity presented in Fig.5(c) shows a more uniform distribution around the valve than that of the primary valve but with slightly smaller magnitudes. All mean velocity components in plane d are smaller than in the other planes due to its proximity to the end of the ramp of the port and also due to the confinement by the cylinder wall. The values of normal shear stresses presented as rms velocities are high , like in case of the primary valve, and follow the mean flow variation; their levels vary from values as low as $1V_p$ just below the valve tip to a maximum of $5V_p$ close to the cylinder head and in the region of the strong shear layer above the tip of the valve.

The mean flow variation described above can be seen more clearly in Figs.5(d) and (e) where the vector addition of the velocities is presented. The axially and radially discharging flow around the valve, Figs.5(d), shows a good degree of uniformity except in plane d where the flow is significantly reduced for the reasons explained previously. The flow angle with respect to the valve axis varies from plane to plane so that at planes a, b, c and d the angles are 45^0, 59^0, 59^0 and 45^0, respectively, and tends to become more axially directed as the flow approaches the tip of the valve. The vector addiction of the radial and tangential mean velocity components, Fig.5(e), shows that there is a small variation in the flow angle within the valve gap and that the flow angle reduces as the flow approaches the valve head. The variation of the flow angle around the valve periphery is also small so that the average flow angles at planes a, b, c and d are 29^0, 9^0, 25^0 and 32^0, respectivelly, which are smaller than those of the primary valve suggesting lower overall swirl levels.

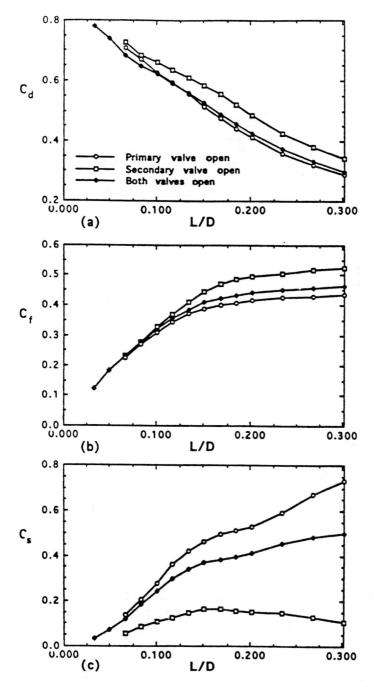

Fig. 3. Variation of discharge and swirl coefficients at a constant mass flow rate of 85kg/h and as a function of valve lift; (a) discharge coefficient, (b) flow coefficient, (c) swirl coefficient

532

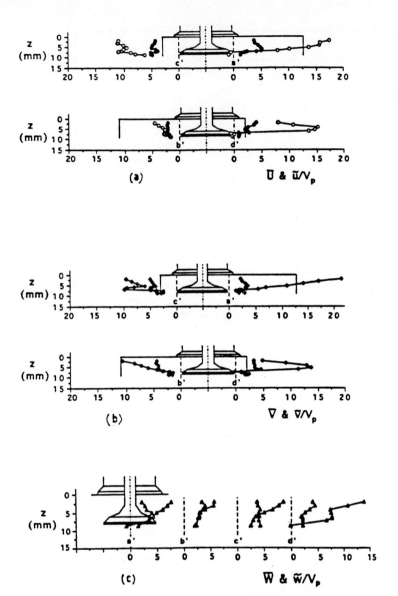

Fig. 4. Primary valve exit flow for full lift; (a) axial mean and rms velocity, (b) radial mean and rms velocity, (c) tangential mean and rms velocity, (d) Vector addition of the axial and radial mean velocity components, (e) Vector addition of the radial and tangential mean velocity components.

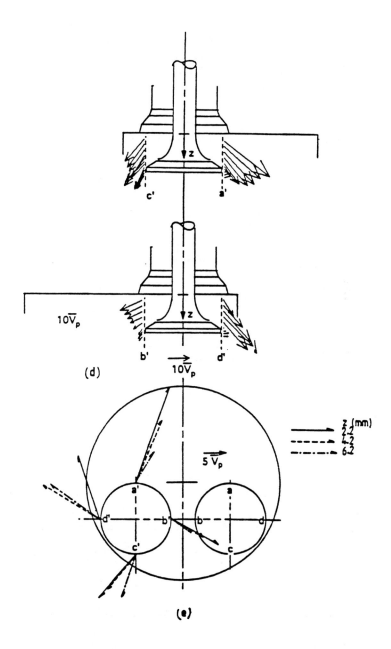

Fig. 4. Primary valve exit flow for full lift; (a) axial mean and rms velocity, (b) radial mean and rms velocity, (c) tangential mean and rms velocity, (d) Vector addition of the axial and radial mean velocity components, (e) Vector addition of the radial and tangential mean velocity components.

534

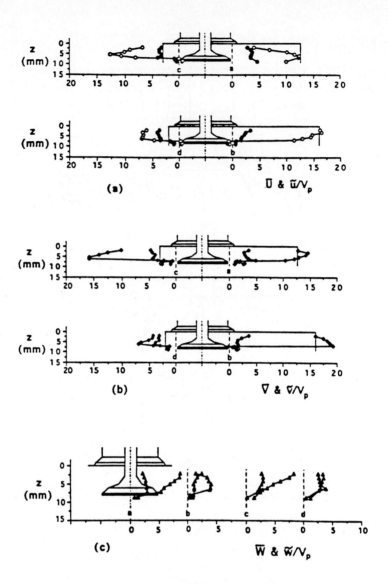

Fig.5. Secondary valve exit flow for full lift; (a) axial mean and rms velocity, (b) radial mean and rms velocity, (c) tangential mean and rms velocity, (d) Vector addition of the axial and radial mean velocity components, (e) Vector addition of the radial and tangential mean velocity components.

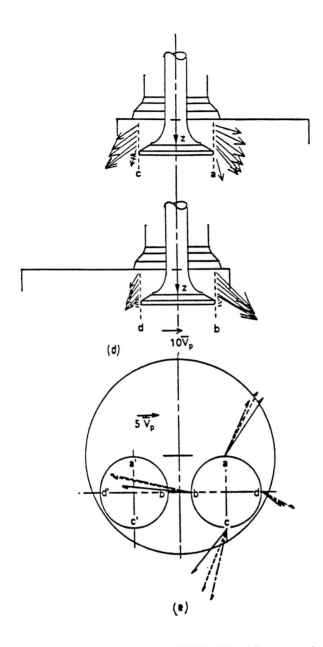

Fig.5. Secondary valve exit flow for full lift; (a) axial mean and rms velocity, (b) radial mean and rms velocity, (c) tangential mean and rms velocity, (d) Vector addition of the axial and radial mean velocity components, (e) Vector addition of the radial and tangential mean velocity components.

536

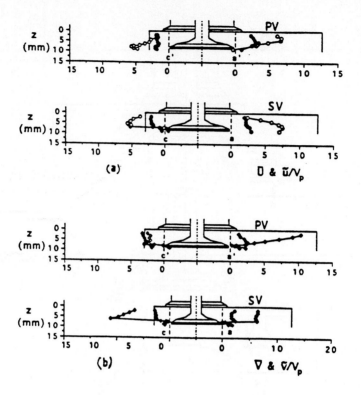

(a)

(b)

Fig 6. Primary and Secondary valve exit flow for full lift; (a) axial mean and rms velocity at planes a-c and a'-c', (b) radial mean and rms velocity at planes a-c and a'-c', (c) tangential mean and rms velocity, (c) axial mean and rms velocity at planes b-d and b'-d', (b) radial mean and rms velocity at planes b-d and b'-d', (e) tangential mean and rms velocity.

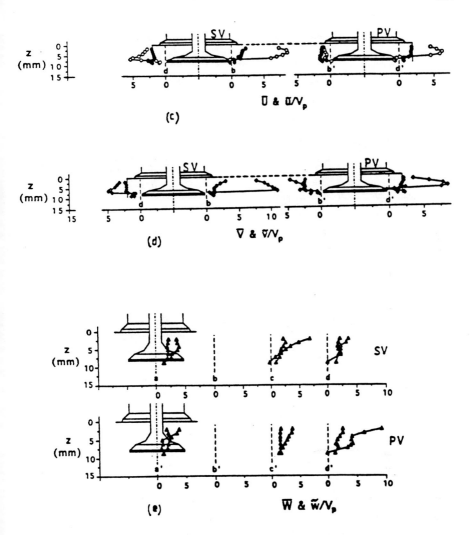

Fig 6. Primary and Secondary valve exit flow for full lift; (a) axial mean and rms velocity at planes a-c and a'-c', (b) radial mean and rms velocity at planes a-c and a'-c', (c) tangential mean and rms velocity, (c) axial mean and rms velocity at planes b-d and b'-d', (b) radial mean and rms velocity at planes b-d and b'-d', (e) tangential mean and rms velocity.

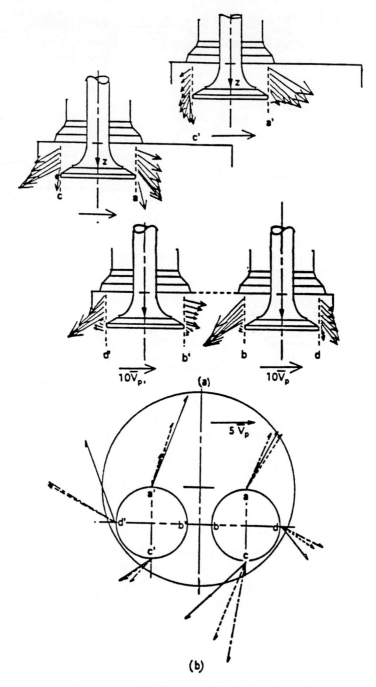

Fig 7. Primary and Secondary valve exit flow for full lift; (a) Vector addition of the axial and radial mean velocity components, (b) Vector addition of the radial and tangential mean velocity components.

3.2.3 Both Valves Open. Measurements were also obtained when both valves remained open and at the same lift and mass flow rate of 85 Kg/s to the two previous cases when either the PV or the SV was open; the results are presented in Figs. 6 and 7. Due to the presence of the valve head and stem in front of the laser beams, the measurement of the tangential velocity at planes b and b' was not possible. The variation of the mean and rms velocity profiles of all three components around the valve is presented in Fig.6 and shows similar trends to the similar trends to the corresponding profiles obtained when one of the valves was open, but with much smaller magnitude, around half, which is expected since the annular flow area here is twice as large as with one valve open; because of the different scaling between the flow cases, direct comparison is not straightforward. The distribution of the rms velocities indicates the high level of turbulence intensities in all three components with values of around $0.4V_p$ in plane b and c to a maximum of $3.4V_p$ in plane c'.

The interaction between the two flows discharging from the intake valves can be better seen in vector form in Fig.7(a). The flow angle of discharging flow around the SV is slightly affected, with the largest changes in plane b where the flow angle becomes smaller by up to 5^0. The discharging flow at plane b' of the PV is however affected by the strong jet coming from the opposite direction and is evident in plane b of SV so that the flow profile exhibits a bimodal distribution; this in turn is affecting the flow at plane b of SV too, although to a smaller degree, by reducing the flow angle as shown above. The flow in plane a' seems to be unaffected, but at c' the flow changes considerably compared to the flow when only the PV valve was open, and in such a way that the flow angle reduces from 55^0 close to the cylinder head to nearly zero at the tip of the valve head. This implies that the radial flow is impaired by the incoming strong jet from plane c of the SV which is directed along the wall by the surrounding liner and by the impingement of the two discharging jets at planes b and b'. This effect at plane c' can also be seen in Fig.7(b), where the addition of the radial and tangential mean velocities is presented, from the increase in the flow angle relative to the valve radius (50.5^0 compared to 34.5^0) and the smaller magnitude compared to the case when the PV was only open. This in turn resulted in lower flow angles in planes d' and a' by 3^0 and 7^0, respectively, compared to the case when only thePV valve was open. The average flow angles at the planes around the SV seem to be unaffected.

4 Conclusions. The steady flow characteristics around the two intake valves of a production four-valve diesel cylinder head were investigated by laser Doppler velocimetry for a mass flowrate of 85 Kg/h corresponding to 1000 rpm engine speed and the most important findings are summarised below.

The discharge and flow coefficients of the primary valve were lower than those of the secondary valve at all valve lifts, with a maximum difference of more than 20% around the maximum lift; when both valves were open the values of the coefficients were in between but much closer to those of the primary valve. The reverse trend was observed for the swirl coefficient in that the primary valve was generating much higher swirl than the secondary valve by more than sixfold; in the case when both valves were open, the value of the swirl coefficient was again found to fall between the values obtained with the two individual valves.

The discharging flow around the primary valve was found to be non-uniform with the maximum flow observed at planes which are in-line with the inlet port axis and the minimum flow at planes towards the end of the helical ramp of the port; this non-uniformity of the flow was found to be less pronounced around the secondary valve. When both valves were open the discharging flow from the secondary valve was little affected by the opening of the primary valve, while its affect on the exit flow of the primary valve was much more substantial, especially in two out of the four examined planes of the primary valve where the local flows were found to be greatly modified by the jet flow exiting from the secondary valve.

The distribution of the turbulence intensity was found not to be uniform around the valves, following the trend of the mean flow. The turbulence was, in general, anisotropic in all cases with maximum levels of up to 5VP near the cylinder head and in the strong shear layer just above the valve tip for the case of the individual valves being open; with both valves open the maximum turbulence intensity was about 3.4VP.

Overall, it is clear that certain improvements in the design of producing four-valve diesel cylinder heads are desirable and necessary in order to reduce the interaction between the flows discharging from the two intake valves. This effect is expected to be more pronounced at the higher air flowrates corresponding to the higher end of the speed range of passenger car diesel engines where volumetric efficiency deteriorates.

Acknowledgements. The authors would like to thank Mr. Roy Horrocks of the Ford Motor Company for his contribution to the researche programme, Dr. H.M. Xu for many discussions and useful suggestions during the course of this work and Mr. Paul Bruni for setting up the experimental rig.

References.

Arcoumanis, C., Vafidis, C. and Whitelaw, J.H. (1987) Valve and in-cylinder flow generated by a helical port in a production diesel engine. Trans. ASME J. Fluid Eng., 109, 368-375.

Arcoumanis, C., French, B. and Nouri, J.M. (1995) "Steady intake flow characteristics through a diesel four-valve cylinder head (Part 2: high flowrates), Report TF/95/25b, Mechanical Engineering Department, Imperial College.

Haghgooie, M., Kent, J. C. and Tabaczynski, R. J. (1984) Intake valve cylinder boundary flow characteristics in an internal combustion engine. Comb. Sci. Tech., 38, 49.

Heywood, B. J. (1988) Internal Combustion Engine Fundamentals. McGraw-Hill Internacional Edition, Chapter 8.

V.5   LIF Visualization of Liquid Fuel in the Cylinder of a Spark Ignition Engine

Peter O. Witze and Robert M. Green

Combustion Research Facility
Sandia National Laboratories
Livermore, California 94551 USA

Abstract. Laser-induced fluorescence (LIF) has been used to investigate the presence and location of liquid fuel films on the combustion chamber walls of a port fuel-injected engine. A frequency-tripled, Nd:YAG laser provided full-field illumination of the combustion chamber through a window in the piston crown. LIF from liquid-phase gasoline was imaged onto a gated, intensified solid-state camera, and recorded in real-time, synchronous with the engine, on video tape. The technique has been demonstrated on a production, four-valve head, comparing open and closed-valve injection timing. A cold start was simulated by motoring the engine at constant speed before enabling injection and ignition. For open-valve injection, liquid fuel was observed on and around the exhaust valves during the first injected cycle, and combustion occurred by the fourth cycle. For closed-valve injection, a detectable fuel film was seen near the intake valve seat by the fifth cycle, and combustion began by the eighth cycle.

Keywords. Spark-ignition engine, liquid-fuel visualization, laser-induced fluorescence

1 INTRODUCTION

In order to meet future, ultra low emission vehicle (ULEV) regulations in California, significant reductions will need to be achieved in unburned hydrocarbon (UHC) emissions during cold start. Typically, more than 60% of the UHC emissions measured during the Federal Test Procedure occur in the first two minutes when the catalyst is not yet hot enough to efficiently convert the UHC in the exhaust. Also during this period, the temperatures of the port walls, intake valve(s), and combustion chamber surfaces are too low to fully vaporize the liquid fuel. As a result, even for closed-valve, port fuel injection (PFI), liquid fuel can enter the cylinder during intake, and exist in the combustion chamber both as droplets and films on the walls and in crevices. While it is believed that all liquid that remains suspended as an aerosol during compression does vaporize, Shin et al. (1994) have shown that a significant portion of the liquid film on the walls survives compression heating, and exists there at the time of ignition. Perhaps even more surprising, the thicker of these liquid films appear to survive combustion. Thus, although it is well known that crevices are the main source of exhaust UHC under steady, operating conditions, it is likely that the diffusion-controlled burning of liquid-fuel films is a major contributor to UHC emissions during a cold start.

Three related, yet distinctly different areas for investigation are evident. The first is droplet sizing and dynamics, for which phase Doppler anemometry and direct imaging have been widely used in the past few years to study droplets in both the intake port and cylinder. The second is the measurement of film thickness. Based on earlier work on the measurement of cylinder-wall oil-film thickness by Smart and Ford, (1974) and Hoult et al. (1987), Le Coz et al. (1994) and Almkvist et al. (1995) used LIF to measure liquid-fuel film thickness in the intake port, both at a single point and along a line excited by a laser sheet. The third area of interest is the location of liquid-fuel films, which has been investigated using a number of di-rect-imaging techniques, and is the focus of this paper.

Shin et al. (1994) performed high-speed digital-video recordings (1000 fps) of liquid films on the flat head and quartz walls of a square-piston engine. They used indirect lighting to observe the evolution of the films from a 20 °C cold start until the cylinder wall temperature approached 100 °C (after approximately 2 minutes). The resulting visualizations led them to characterize the in-cylinder liquid film dis-tribution according to the three mechanisms by which they are believed to have formed: 1) A thick film on the valve face and around the seat, formed by a liquid film flowing from the port side of the valve; 2) A thin film on the combustion chamber surfaces from impingement of the droplet stream created by strip-atomization off the valve seat area; 3) Isolated puddles formed by the accumulation of splashed drops created by the intake-valve closing process (literally a squeezing of the liquid in the seat area as the valve closes). The interpretation of the results indicates that the latter two film-types disappear quickly (within one cycle) during the combustion process, even in the early cycles, while the first film-type persists for up to one minute. Because of the low compression ratio of this engine, 6.07, there is some question as to applicability of the results to a more realistic engine with a pent-roof geometry. The fuel used in these tests was indolene, the engine speed was 1000 rpm, and the manifold pressure was 50-60 kPa.

Fry et al. (1995) performed a similar study using a copper-vapor laser as the light source, and a high-speed movie camera to visualize liquid fuel in direct backscatter through a window in the piston crown. They found it necessary to polish the com-bustion chamber walls, so the liquid fuel would appear as dark shadows. During in-duction, they determined that wall films formed in the apex area of the wedge-shaped chamber and in the crevices around the intake and exhaust valves. These films remained in these locations during compression and combustion, and then were scavenged from the wall during the exhaust blowdown period, when the liquid was carried directly into the exhaust ports.

Both of these studies used direct, visible-wavelength illumination of the liquid fuel. This can be rather difficult, since the thin fuel films are essentially transparent, and thus their detection is dependent on scattering properties different from those of the surrounding walls of the combustion chamber. Saito et al. (1995) solved this discrimination problem by using fluorescence induced by full-field illumination with mercury lamps. They observed liquid fuel on the walls of a quartz cylinder liner. A high-sensitivity, conventional-speed video camera was used to record the images, implying a framing rate of either 30 or 60 Hz, depending on whether full-frame or single-field recording was used. They were only able to observe films on the cylinder wall; neither the head nor the piston crown were in the field-of view. Their results show considerably greater wall-wetting for open-valve injection (OVI) as compared to closed-valve injection (CVI), and greater wall-wetting for a conventional injector as compared to an air-assisted injector. During the compression stroke, the fuel that was deposited on the cylinder wall during induction was scraped off by the top ring and forced into the top-land crevice. As top-dead-center (TDC) was approached and combustion occurred, the increased pressure forced liquid fuel into the second land, where it may mix with oil. With the beginning of the expansion stroke, the liquid fuel in the top-land was laid-down on the cylinder wall as a thin film, with far greater area than it occupied during induction. The authors note that this thin film was not consumed by combustion, but rather that it vaporized as unburned HC when the pressure fell during the exhaust stroke.

Felton, et al. (1995) used a similar approach to observe liquid-fuel films on the walls of the intake port. They used a frequency-tripled Nd:YAG laser as the excitation light source, which entered the intake port through a quartz periscope. A small lens expanded the laser beam to illuminate the back walls of the port. The LIF signal was collected with a fiberscope and imaged onto an intensified CCD camera. Because of the slow framing rate of the camera, a single image was obtained each engine cycle, recorded either directly onto video tape or into computer memory. By integrating the intensity of the LIF signal over a selected region of each image, a semi-quantitative curve of signal intensity versus time was created.

The significance of the results obtained with this technique, together with its simplicity and ability to clearly discriminate liquid-fuel films from dry surfaces, inspired us to apply it to image liquid films on the combustion chamber walls, as observed through a window in the piston crown. Of particular interest are differences in the amount and location of liquid-fuel films during a simulated cold start for CVI versus OVI.

2 ENGINE FACILITY

The engine configuration, shown in Fig. 1, consists of an extended cylinder with transparent piston for optical access from below. The piston is unlubricated and uses a bronze-loaded Teflon® rider ring to prevent it from directly contacting the cylinder wall. The top ring is a sealing ring made of Vespel®, a graphite-filled polyimide material. The sapphire window in the piston provides optical access to 80 percent of the cylinder. The head is a complete, unmodified 1994 General Motors Quad-4, with the #3 combustion chamber aligned with the single cylinder block, and the compression ratio is 9.5. Both the coolant and oil systems are active along the entire length of the head, but all fluid passages that would normally mate with the block are plugged. A closed-loop system maintains the coolant temperature to a preset value, within a few degrees. Finally, an uncooled piezoelectric pressure transducer is installed, flush with the combustion chamber wall.

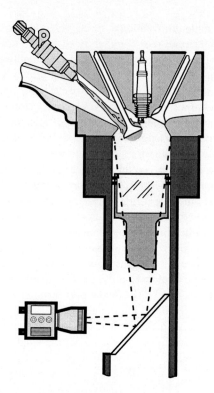

Fig. 1 Single cylinder engine with extended piston for optical access. The camera and laser illumination share the same field-of-view via a dichroic mirror.

The engine was fueled with commercial, 87 octane gasoline injected through a standard, single-spray fuel injector (with a nominal 12° cone angle). As opposed to the stock fuel rail, we used a 'once-through' system, where the fuel was supplied from a one-liter vessel pressurized with nitrogen to 270 kPa.

The cold start simulation consisted of motoring the engine at a constant speed, and then initiating fuel injection and ignition. A 300 rpm, wide-open-throttle condition was used in the current work for ease of implementation and demonstration of the new diagnostic techniques we are using. With the exception of engine speed, all engine and data acquisition events are controlled by a dedicated PC. An absolute-position, shaft encoder attached to the intake camshaft, with one crank-angle-degree (CAD) resolution, provides the crankshaft position for the engine controller. Start-of-fuel-injection, ignition, laser firing, and image-grab timing are all performed in the crank-angle domain, as determined from the shaft encoder. End-of-fuel-injection, however, and thus injection duration, is performed in the time-domain, and is controlled by a timing board in the computer. This is done to assure that the same amount of fuel is injected each engine cycle, irrespective of instantaneous changes in engine speed that occur when combustion first begins, and when combustion is unstable due to misfires and slow or partial burning cycles.

While it is straightforward to record LIF images directly into computer memory using a standard frame-grabber interface board, the large number of images involved in a typical cold-start test sequence make the use of real-time video-tape recording preferable. One image is obtained by the frame grabber each engine cycle, and is continually output to tape at normal video framing rates until the next image is grabbed. We have found that the playback of the video recording is much more useful if it is annotated with the cycle number of each image and the combustion chamber geometry, showing the location of valves, spark plug, and squish areas. We therefore have developed a software package that utilizes the overlay-plane capability of the frame-grabber board (Coreco Oculus TCX) that, in effect, creates a mask that overlays the LIF images. Only the portion of the mask that corresponds to the clear aperture of the window in the piston is transparent, permitting display of the LIF image. This blocks all irrelevant parts of the image outside the field-of-view through the window. The mask consists of a line drawing of the combustion chamber and is positioned and scaled to align with the actual head configuration.

3 OPTICAL SETUP

The optical system we used for the LIF imaging experiments is illustrated schematically in Fig. 2. The light source for excitation was a frequency-tripled, Nd:YAG laser operating at 10 pulses/sec and 30 mJ at 355 nm. The laser pulses were synchronized to the engine crank-angle via the shaft encoder and engine controller. The laser output beam was expanded into a light field by lens L_1, a spherical, plano-concave lens with a focal length of -100 mm. This expanding light field was then directed into the engine cylinder through the window in the piston by mir-

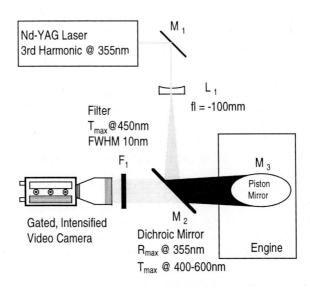

Fig. 2 Optical setup for full-field LIF imaging of the combustion chamber.

rors M_2 and M_3 (see Fig. 1), such that the laser light completely filled the window aperture to the combustion chamber. The fluorescence emission retraced the same optical path as the laser beam until it reached the dichroic mirror M_2, where it passed directly through to the video camera.

The dichroic mirror M_2 was coated for maximum reflectance at 355 nm to reflect the incident laser light field, and for maximum transmittance between 400 and 600 nm to transmit the fluorescence emission. The piston mirror M_3 was coated with UV-enhanced aluminum having a reflectivity of 90% at 355 nm. This coating is adequate because the energy density of the laser light is reduced by the expansion of the field to a level below the damage threshold. Between 400 and 600 nm it has a reflectance of 90 to 95%, which is quite acceptable for reflecting the fluorescence emission. The light entering the camera lens was passed through a narrow band-pass filter F_1 to reject extraneous background. The maximum transmission of the filter was at 450 nm, with a FWHM of 10 nm.

The intensified video camera captured the image on a 340 by 480 CCD array with a 105 mm, f/2.8 camera lens. The intensifier was gated synchronously with the laser Q-switch signal, while the Q-switch and frame-grabber control signals were synchronized to the engine by the engine controller.

4 FLUORESCENCE CHARACTERIZATION

A variety of compounds have been used to observe LIF from fuel films in engines, with the choice generally depending on the application. For quantitative measurements of fuel-film thickness, Almkvist et al. (1995) doped isooctane with 3-pentanone, selected for its close match with the boiling point of this single-component fuel (99°C). Le Coz et al. (1994) performed a comprehensive study of the liquid/vapor phase equilibrium of isooctane and six potential dopants, and showed that it is not adequate simply to match the boiling points, since when mixed with isooctane at low concentrations, the effective boiling point of the dopant is reduced. They obtained their best results using dopants with boiling points of about 130°C.

For multicomponent fuels, it is far more difficult to perform quantitative LIF because of the different boiling temperatures of the many components. However, if the fluorescing molecule has a boiling point near the high-temperature end of the fuel distillation curve, then it can be argued that LIF can be used to measure the shape and area of liquid films. In addition, if the area of the liquid film is always growing during the period of measurement, then a dopant with an even higher boiling point can be used. Saito, et al. (1995) used Xanthene dye added to an unspecified fuel to distinguish it from oil films. Xanthene has a melting point of 102 °C and a boiling point of 310 °C, which suggest that it will leave a residue if it is not consumed by combustion. Felton et al. (1995) used UNOCAL RF-A test fuel, which for visual identification purposes contains Oil Purple M Liquid dye. They found that the fuel fluoresced strongly when excited by UV light, and speculated that the dye was the source of the fluorescence. For the dye concentration used in RF-A, they observed no residual fluorescence signal from intake port walls after complete evaporation of the fuel. Because their experiment to observe liquid-fuel

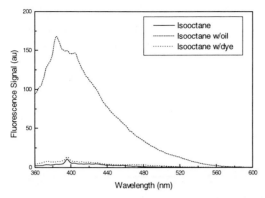

Fig. 3 Fluorescence signal of pure isooctane and with oil and dye added.

film was quite simple and yet successful, we chose to investigate the use of this dye as a fluorescing dopant for regular grade gasoline.

In order to understand better the fluorescence characteristics of the various liquids that might be encountered in the combustion chamber of our engine, we made use of a luminescence spectrometer. This device allows investigation of the characteristics of fluorescence emissions of liquid samples under tightly controlled conditions. Spectral excitation scans can be run at a fixed emission wavelength; or spectral emission scans can be run at a fixed excitation wavelength. For the purposes of the current work, we performed emission scans at an excitation wavelength of 355 nm to examine isooctane, a single-component, straight-chain, paraffinic fuel; 87 octane gasoline (direct from the pump), a multicomponent fuel containing various, unknown additives; 10w-30 motor oil, a potential, in-cylinder contaminant of liquid fuel films; and Oil Purple M Liquid dye, a fuel additive used to provide color for visual identification of special fuels, and a substance that has been suspected of yielding strong fluorescence emissions when excited with UV light.

We chose to examine isooctane because it is a pure, single component liquid known to have a weak fluorescence response. This is confirmed in Fig. 3, where we have plotted the emission scans for pure isooctane, isooctane with enough dye added to create an obvious purple tint, and isooctane with a small quantity (<5%) of motor oil dissolved in it. The pure isooctane exhibits only a weak fluorescence emission between 360 and 600 nm, with a small peak near 395 nm. The addition of the dye caused a small but insignificant increase in the amplitude of the fluorescence emission, and the spectral characteristics were unchanged. On the other hand, oil dissolved in the isooctane caused an order-of-magnitude increase in the amplitude of the signal, along with a change in the spectral characteristics. Note that the isooctane peak at 395 nm can still be observed along with the strong oil peak at 390 nm.

Having obtained the isooctane results described above as a reference, we next

550

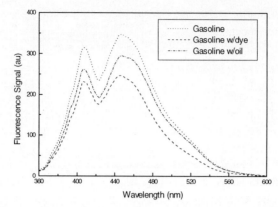

Fig. 4 Fluorescence signal of pure 87 octane gasoline and with oil and dye added.

performed the same series of tests with 87 octane, "pump" gasoline. These results are illustrated in Fig. 4, where the fluorescence from straight gasoline (without oil or dye) is seen to be many times stronger than that from pure isooctane. The spectrum exhibits a double-peaked behavior, with a minor peak at 410 nm and a major peak at 450 nm. Notice also that the peak fluorescence signal from gasoline is a factor-of-two stronger than that from isooctane with oil dissolved in it, and at 450 nm the ratio of their signals is about six. However, the most interesting aspect of these data is that the dissolution of small amounts of dye or oil (the same amounts as used with the isooctane) only has the effect of reducing the magnitude of the fluorescence emission. The fact that the characteristics of the fluorescence emission spectra are similar indicates that the oil or dye dissolved in the gasoline only absorbs the excitation light at 355 nm (reducing the excitation of the gasoline), and does not absorb or trap the LIF, since the latter would be wavelength dependent. Furthermore, while Fig. 3 showed fluorescence due to oil dissolved in isooctane in the spectral region from 360 to 600 nm, Fig. 4 indicates that it is weak at 450 nm compared to the fluorescence from gasoline alone.

The above observations led us to conclude that gasoline alone will provide a good fluorescence signal to locate the presence of liquid inside the engine intake port or combustion chamber, without the need to add dopants. Furthermore, if the fluorescence emission is high-pass filtered above 440 nm, the major peak of the gasoline fluorescence should be observed without serious conflict with fluorescence from engine oil that is either pure or dissolved in gasoline.

Finally, we performed two simple studies in controlled environments to examine the fluorescence from residue left behind after the evaporation of gasoline. In the first test, we placed several drops of liquid on thin quartz disks and allowed the liquid to evaporate by leaving the disks in an oven at 75 °C for 15 hours. We tested three liquids - gasoline, pure isooctane and isooctane with enough dye added to give the liquid a slight purple tint - with the objective of identifying how the residues respond to excitation from laser light at 355 nm. The results revealed that the residues from pure isooctane and isooctane with dye produced insignificant fluorescence, while the residue from gasoline fluoresced enough to suggest that it could confuse interpretation of images of liquid fuel films.

In the second test, we simulated the repeated wetting and evaporation of gasoline from a valve. We placed a production valve in a holding device with the valve face oriented upward. A liberal amount of gasoline was applied to the valve face, and then the valve was placed in an oven at 170 °C. As soon as the valve reached a temperature that evaporated the fuel, a second film of fuel was applied. This was repeated three times, and then the valve was baked for 20 minutes. We then exposed the valve to 355 nm laser light, and observed fluorescence from the polished surfaces of the valve face; a yellowish film was also visible to the unaided eye. Somewhat to our surprise, the unpolished, rough-casting region in the center of the valve did not fluoresce except in one small region. This result is consistent with the observation of Felton et al. (1995), who found that they did not need to clean the intake port walls between cold start tests using LIF.

5 CONTROLLED SIMULATION

Before beginning the LIF experiments in the operating engine, we performed a bench top simulation for evaluation and refinement of the technique under controlled conditions. To this end, we positioned a cylinder head on an optical bench and used the optical system described above to look at several different aspects of the problem under static conditions, without the complications of the transparent piston, periscope mirror and an operating engine.

The first issue we addressed was the primary background signal. This includes elastic scattering of laser light and broadband LIF from combustion chamber surfaces. The image illustrated in Fig. 5a is a typical example of these sources of background, where the two intake valves are at the top. The brightest areas in the image are the smooth, machined surfaces of the squish regions above the intake valves and below the exhaust valves, the apex of the pent-roof head, and the spark plug. The source of this light is most probably a combination of reflected laser light and LIF from deposits and soot. To create Fig. 5b, we placed a narrow bandpass filter (maximum transmission (T_{max}) at 450 nm, with 10 nm FWHM) in front of the camera, which effectively rejected most of the extraneous background signal. The remaining signal seen in Fig. 5b is due to oil that enters the ports through the valve guides and collects around the valve seats. Prior to performing these experiments, we carefully cleaned the intake valve seat in the upper left of the image to verify the cause of this effect. The total lack of signal from the clean valve seat indicates that oil and/or other contaminants do, indeed, cause background in the images. From these results, we concluded that the filter would remove most of the primary background signal, but we would still have to use background subtraction to remove the secondary background signal and further enhance the quality of the images.

In order to verify that background subtraction could reduce the secondary background signal without seriously affecting the desired data, we created a simulation of liquid films on two of the valves and recorded an LIF image. We placed several drops of oil on the lower left valve and several drops of gasoline on the lower right valve. Due to the high surface tension of the oil, it formed a thick, localized film on the face of the valve, while the lower surface tension of gasoline allowed it to form a thin film that spread over most of the lower half of the valve face. The LIF image of these films is shown in Fig. 5c, and illustrates a strong signal from the thick oil film along with a somewhat weaker signal from the thinner gasoline film. We removed the background from the liquid-film image by simply subtracting the image in Fig. 5b from the image in Fig. 5c. The result is shown in Fig. 5d. Clearly, the background subtraction procedure has enhanced the data without any detrimental effects.

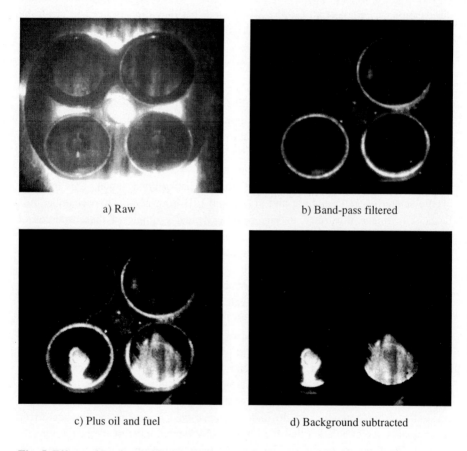

a) Raw

b) Band-pass filtered

c) Plus oil and fuel

d) Background subtracted

Fig. 5 Effects of band-pass filtering and background subtraction. The intake valves are at the top of each image.

6 ENGINE RESULTS

LIF images were obtained for two test sequences involving OVI and CVI of a stoichiometric mixture. End-of-injection was at 440 CAD and 700 CAD, respectively, where 720 CAD is TDC of compression. Ignition was at 705 CAD, and all LIF images were obtained at 700 CAD. We recorded the cylinder pressure and used it to compute the gross, indicated-mean-effective-pressure (IMEP). These results, which are summarized in Fig. 6, indicate first-measurable heat release in cycle #4 with OVI, while with CVI this did not occur until cycle #8 or 9. The large negative IMEP in cycle #11 for CVI is the result of flashback into the intake port that occurred when the intake valve first opened. Combustion in cycle #10 was probably complete, but somewhat slow, leading to residual gas that was severely underexpanded and unusually hot at the time of valve overlap. This resulted in ignition of the fresh charge in the intake port, leading to low-density products filling the chamber in cycle #11. This was followed by very good combustion in cycle #12, probably due to better fuel vaporization in the port caused by the flashback. Cycle #13 had a null IMEP, for unknown reasons, followed by another good burn because of a high level of fresh charge in the residual gas.

By cycle #20, combustion has stabilized and the IMEP's for OVI and CVI are nearly the same. We chose not to present results past this point, because after cycle #30 there is a considerable accumulation of soot clearly visible on the piston window. On first thought, one might expect soot to block the laser excitation and/or fluorescence, causing extinction of the LIF signal. However, it is also possible that the soot traps the heavy-ends of the fuel and/or fuel additives and oil, creating a fluorescing mixture of residual gunk that would no longer be representative of a pure liquid-fuel film.

LIF results for CVI and OVI are shown in Figs. 7 and 8, respectively. The numbers in the upper left-hand corner are the cycle number of the image; the particular frames shown were selected as a good representation of the evolution of the fuel

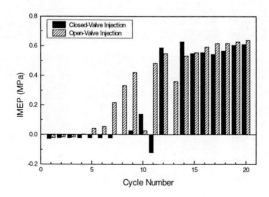

Fig. 6 Gross IMEP for the first 20 cycles of a simulated cold start, comparing OVI with CVI.

film for the two cases presented. The largest circle in the superimposed mask of cylinder features corresponds to the cylinder bore, and the smaller circle represents the clear aperture of the window in the piston. The triangular tip of the squish region between the two intake valves on the left side is visible in the field-of-view, whereas the exhaust side squish region cannot be seen.

For the CVI case shown in Fig. 7, liquid fuel can be observed in the intake-valve seat crevices by cycle #5, and perhaps even in cycle #2 as well. As the sequence progresses, this fuel film propagates away from the intake valve seat crevice and across the head surface; we believe it is on the head, and not the piston, because it flows around the spark plug and does not cover the faces of the exhaust valves.

A liquid-film first appears in the squish region on the intake-valve side in cycle #10. Because the film follows the contour of the squish area, it is definitely on the head, and not the piston. It would appear that liquid fuel stripped off the intake valve strikes the near-vertical wall leading to the squish area, distinctly marking its contour. By cycle #15, this squish area is totally covered by liquid, and the film has also grown to cover a significant portion of the intake-valve faces. We can only speculate as to the location of these later films, but because we can visually see fuel residue on the piston in these areas when we stop a test, and do not see similar amounts of residue on the valves and head, we are reasonably certain that they reside primarily on the piston.

For the OVI case shown in Fig. 8, liquid fuel is clearly evident in the first cycle. Because this image in not spatially resolved along the optical path, we can only speculate as to whether the fuel is on the piston, head or both. However, we believe

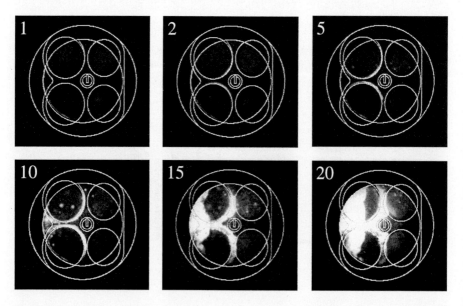

Fig. 7 Liquid-fuel fluorescence for closed-valve injection (CVI). The intake valves are on the left side in these images.

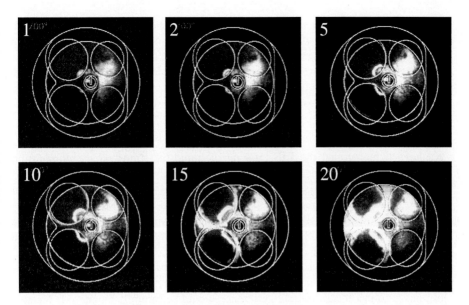

Fig. 8 Liquid-fuel fluorescence for open-valve injection (OVI).

that the large spot on the exhaust valve side is most likely located on the piston, and that the small spot under the upper intake valve is located on the valve face. This latter spot, and the analogous one that first appears on the lower intake valve in cycle #2, we find interesting. These regions grow as near semi-circles, and the LIF intensity indicates that the film is significantly thicker near the leading edge. Recalling that the injector used produces a single spray aimed at the septum dividing the two intake-valve ports, these semi-circular films appear to be the result of fuel that has impacted the back of the valve and moved along the surface around the valve seat and onto the face. Note also that liquid film along the intake valve crevices develops considerably slower than it did for CVI. We believe that the film in the seat crevices is from the flow of liquid films collecting initially on the port side of the valves, and that the semi-circular films on the intake valve faces are from direct impingement of the spray on the back of the valve seat during OVI.

For OVI, the film is slower to develop in the squish region on the intake side than it did with CVI. Note, however, that in cycle #15 the film in this region clearly follows the contour of the valve seat, with this trend continuing into cycle #20, suggesting that the observed LIF is due to fuel that is on the head. However, as was the case with the CVI, post-test visual inspection reveals a residual film on the piston in this region. We suspect that liquid films exist on both the piston and head in this region, but the LIF from the film on the head dominates the image, while the sapphire window in the piston retains a larger post test residue than the head and intake valves.

It is interesting to note that the film under the upper exhaust valve does not significantly change in size or shape from cycle #5 through #20. In fact, the amount of film on the exhaust-valve side of the chamber continually decreases after cycle #5, and is unexpectedly asymmetric, with very little fuel ever visible over the face of the lower exhaust valve. Similarly, the area of the film on the head between the exhaust valves is largest in cycle #2, and is gone by cycle #15, except for very close to the spark plug; this behavior may be indicative of a rapid heating of the head surface between the exhaust valves.

7 CONCLUSIONS

We have demonstrated the LIF imaging of liquid-fuel films on the combustion chamber walls of a production engine head for common gasoline without the need of dopants. Excellent image clarity was achieved using a frequency-tripled Nd:YAG light source expanded into a light field, and an intensified CCD camera with a narrow bandpass filter. Background subtraction was used to enhanced the image quality by removing unfiltered, elastically-scattered light and fluorescence due to contaminants on the combustion chamber surfaces. We found that gasoline, and, to a lesser degree, oil, leave a residue that requires the cleaning of all surfaces between tests. Because commonly-used oils will fluoresce in the same spectral regions as gasoline, this potential interference should be carefully monitored. We also found it useful to employ an overlay feature of the frame-grab interface board to create a diagram of the combustion chamber geometry directly on the LIF images.

For closed-valve injection, liquid fuel is first observed in the intake valve crevices by cycle #2, but the amount is not significant until about cycle #10 (the first measurable IMEP occurred in cycle #9). This film grows to fill the regions of the head surfaces between the intake and exhaust valves, but does not grow in size over the intake valves themselves. Liquid fuel is also observed in the squish region between the intake valves beginning at about cycle #10, from where it continues to grow until it covers more than half of the valve-face area by cycle #20. Liquid fuel is never seen on the exhaust-valve side of head.

For open-valve injection, liquid fuel is seen under one of the exhaust valves in the very first cycle (the first measurable IMEP occurred in cycle #4). This particular film does not change in size or shape from cycles #5-20. Significant liquid fuel is never observed under the other exhaust valve, revealing an unexpected asymmetry. In a manner similar to closed-valve injection, liquid fuel is seen to accumulate in the intake valve crevices. However, additional semi-circular pools of liquid form on the intake valves in the region near the spark plug. This appears to be due to fuel that has impacted the back of the valve, and flows around the valve seat to the valve face. These pools grow fairly large, and appear to be much thicker along the leading edge. Finally, while a liquid film does form in the squish area between the intake valves, it is considerably smaller than the one observed for closed-valve injection.

ACKNOWLEDGMENT

The authors wish to acknowledge the efforts of Duane Sunnarborg in the design and assembly of all mechanical aspects of this work. This work was performed at the Combustion Research Facility of the Sandia National Laboratories, and was funded by the U.S. Department of Energy, Office of Transportation Technologies.

REFERENCES

Almkvist, G., Denbratt, I., Josefsson, G. & Magnusson, I. 1995, Measurements of Fuel Film Thickness in the Inlet Port of an S.I. Engine by Laser Induced Fluorescence, SAE Paper No. 952483.

Felton, P. G., Kyritsis, D. C. & Fulcher, S. K. 1995, LIF Visualization of Liquid Fuel in the Intake Manifold During Cold Start, SAE Paper No. 952464.

Fry, M., Nightingale, C. & Richardson, S. 1995, High-Speed Photography and Image Analysis Techniques Applied to Study Droplet Motion within the Porting and Cylinder of a 4-Valve SI Engine, SAE Paper No. 952525.

Hoult, D. P., Lux, J. P.,Wong, V. W. & Billian, S. A. 1988, Calibration of Laser Fluorescence Measurements of Lubricant Film Thickness in Engines, Trans. SAE, vol. 97, sec. 3, p. 576.

Le Coz, J-F., Catalano, C. & Baritaud, T. 1994, Application of Laser Induced Fluorescence for Measuring the Thickness of Liquid Films on Transparent Walls, Seventh International Symposium on the Application of Laser Techniques to Fluid Mechanics, Lisbon, p. 29.3.1.

Saito, K., Sekiguchi, K., Imatake, N., Takeda, K. & Yaegashi, T. 1995, A New Method to Analyze Fuel Behavior in a Spark Ignition Engine, SAE Paper No. 950044.

Shin, Y., Cheng, W. K. & Heywood, J. B. 1994, Liquid Gasoline Behavior in the Engine Cylinder of a SI Engine, SAE Paper No. 941872.

Smart, A. E. & Ford, R. A. J. 1974, Measurement of Thin Liquid Films by a Fluorescence Technique, Wear 29, p. 41.

V.6   Spray Characteristics in the Port and Cylinder of a Four-Valve Spark-Ignition Engine

B Cousyn[1], F Neveu[1], M Posylkin[2], D S Whitelaw[2] and J H Whitelaw[2]

[1] Peugeot S.A. DRAS/RMP, Route de Gisy, Velizy 78140 France
[2] Thermofluids Section, Department of Mechanical Engineering
  Imperial College of Science, Technology and Medicine, London SW7 2BX England

Abstract.   Measurements of fuel droplet velocity, size and distribution have been obtained inside the cylinder and the inlet port of a four-valve, four-cylinder spark-ignition engine. The local measurements in the port were complemented by photography and were obtained mainly with the engine motored at 1200 rpm. The fuel was added as a double jet from a production injector with a single injection period of constant duration and with the onset of injection as variable. A shaft encoder and computer provided the constant rotational speed, and a motor acted as a driver for the motored experiments and as a dynamometer for normal operation.

Operation with low engine load and injection with the valves closed at crank angles between 150 and 690 degrees led to nearly-constant combustion performance which suggested good homogeneity of the mixture at the time of ignition. The number of droplets detected in the cylinder was small and constituted a small fraction of injected fuel due to the residence time in the port which allowed evaporation of droplets and of liquid films. The evaporation was also assisted by the back flow of hot residual gases from the cylinder to the inlet port which occurred with low load at the early stages of intake.

The photographs and the measurements of the temporal velocity characteristics of fuel droplets in the inlet port showed that the magnitude of the back flow of gases during valves overlap was sufficient to cause reatomisation of the spray and to reverse its trajectory so that no droplets from the injector reached the cylinder in the crank angle interval between 10 to 30 degrees, with only few droplets detected afterwards. The back flow decreased as the throttle was opened and high load conditions led to droplets entering the port from the valve gap a short time before the closure of the inlet valves, that is around 210 CAD after the top-dead-centre of intake.

Keywords. SI engine, inlet valve, fuel, spray, injection, atomization, impingement, liquid films, droplets, combustion

1. Introduction

The present investigation follows from that of Chappuis, Cousyn, Posylkin, Vannobel and Whitelaw (1997) in which the influence of injection timing on the performance of a four-valve per cylinder engine was quantified in terms of mean-effective pressure, drivability and emissions of unburned hydrocarbon. The results were related to measurements of fuel droplet characteristics within the cylinder and to those of previous investigations including those of Vannobel, Robart, Dementhon and Whitelaw (1994), Posylkin, Taylor, Vannobel and Whitelaw (1994) and Cousyn, Posylkin, Vannobel and Whitelaw (1995) in a two-valve engine. Previous experiments of the influence of injection timing include those of Bandel, Fraidl, Mikulich, Carstensen and Quissek (1989), Mikulich, Quissek and Fraidl (1990), Horie, Nishizawa, Ogata, Akazaki and Miura (1992), Winklhofer, Fraidl and Plimon (1992), Alkidas (1994), Posylkin, Taylor and Whitelaw (1995) and Hardalupas, Taylor, Whitelaw, Ishii, Miyano and Urata (1995).

It is known from the investigations of the previous paragraph with four- and two-valve engines that injection with the valves open led to secondary atomisation of droplets as a result of impingement on the valve and port surfaces. Most of the droplets emerged from the valve gap in an expanding cone directed away from the injector and the Sauter mean diameter of droplets inside the cylinder was of the order of 40 μm and at least three times smaller than that for the incident spray produced by a gasoline injector. The velocity of droplets with early injection reached 40 m/s and decreased with valve opening, with the possible advantage of a reduction in the number of droplets impinging on the piston and liner. Under these circumstances, the characteristics and targeting of incident spray upstream of the inlet valve are expected to be important as indicated by results of Brehm et al (1996) who examined the influence of injector types including those with air-assist and, in engines not equipped with mechanisms for swirl enhancement, inhomogeneity of the charge in the cylinder with injection with the valves open led to combustion performance which was generally poor.

In contrast, injection with inlet valves closed led to better combustion performance and the results of Chappuis et al (1997), and also those of Vannobel et al (1994) and Posylkin et al (1994), showed that these injection angles allowed the liquid fuel to remain in the port so that evaporation from hot metallic surfaces occurred to an extent which depended on the residence time, with remaining liquid fuel in the form of a film swept from the surface by the air flow. The resulting fuel droplets in the cylinder were relatively large and slow moving, and constituted a proportion of the fuel which increased with decrease in the residence time of the liquid in the vicinity of the hot surfaces. The performance of the engine, however, remained nearly-constant with the crank angles of injection between 150 and 690 degrees, with a small, but unexpected, improvement when injection occurred very short time before opening of the inlet valves. This improvement is likely to be related to the back flow of hot residual gases from the cylinder into the intake port, provided that $P_{port} < P_{cylinder}$ at the time of inlet valves opening, with the beneficial effects on the homogeneity of mixture and injection with the valves closed suggested by the results of Alkidas (1994) and Shin, Min and Cheng (1995).

The purpose of the present work was to investigate the behaviour of fuel droplets in the inlet port, and in particular with injection angles shortly before the opening of the valves, in order to establish the reason for the improved performance referred to above, and to determine if it provides sufficient justification for suggesting that injection of much of the fuel should be arranged to occur in a range of crank angles close to inlet valve opening.

The engine is described in the following section together with instrumentation and its operation. The results are presented and discussed in the third section and the final section provides summary of the most important conclusions.

2. Experimental Arrangement

The experiments were carried out with a two litre, four-stroke, four-cylinder spark ignition engine (PSA model XU10J4R) described by Chappuis et al (1997) and with the specifications of table 1. The engine was a production model with port injection, four valves per cylinder, and a combustion chamber with a pent roof. Modifications were required to allow optical access to one of the cylinders which was the only one fired during the experiments, with the other three cylinders disabled by disengaging the inlet and exhaust valves from the camshaft, disconnecting their fuel injectors and removing their spark plugs to reduce the pumping work. The crankshaft was coupled to a 25 kW DC motor (Bull Electric) which operated as a dynamometer and the rotational speed was maintained constant to within 5 rpm by a controller (Shackleton System Drivers, model 590).

The engine was lubricated by an internal oil pump and the cooling water was circulated by an external pump through a heater and cooler, with the temperature maintained constant to within 2° C at values from 50 to 95° C by a proportional controller. The fuel pump, throttle, inlet and exhaust manifolds, catalytic converter and silencer were those provided with the production engine. Fuel was injected into the port with production gasoline injector (Bosch, type 0280 155 216) which delivered the fuel in two cones, each directed towards the head of an inlet valves.

An optical shaft encoder was coupled to the crankshaft and provided a sequence of pulses with angular resolution of 0.125 crank angles degrees. A magnetic pick-up was fixed to one of the camshafts to provide a reference reset pulse every 720 crank angle degrees. The signals from the shaft encoder were processed by the custom-built engine controller which replaced the production management system (Bosch Motronic MP 5.1) to control ignition and injection timings and to produce a time-base for the data acquisition systems.

Table 1: PSA XU10J4R engine specifications

Engine type		4 cylinders in-line 4 valves per cylinder
Bore x Stroke		86 mm x 86 mm
Displacement		1998 cm^3
Compression ratio		10.4
Valves diameter	Inlet:	32.8 mm
	Exhaust:	27.5 mm
Valves timing	Inlet:	IVO: 7° B-TDC IVC: 37° A-BDC
	Exhaust	EVO: 37° B-BDC EVC: 7° A-TDC
Valves lift	Inlet:	9.2 mm
	Exhaust:	9.2 mm

The investigation comprised measurements of combustion performance in terms of indicated mean-effective pressure, drivability and exhaust concentrations of unburned hydrocarbon and of measurements of droplet characteristics inside the cylinder and in the inlet port; those in the cylinder were obtained under fired operating conditions of low load and speed of 1200 rpm and those in the inlet port with similar speed and motored engine.

The phase-Doppler velocimeter used for characterisation of the spray inside the cylinder, its arrangement and location of the measurement were similar to these described by Chappuis et al (1997). The in-cylinder pressure traces were measured with a water-cooled pressure transducer (Kistler, type 7061) 'flush' mounted in the cylinder wall and a charge amplifier (Kistler, type 5007) amplified the transducer output. The transducer 'face' was coated with silicon rubber to eliminate thermal strain and it was calibrated according to the procedure described by Lancaster et al (1975). Concentrations of unburned hydrocarbons in the exhaust gases were measured with a fast FID system (Cambustion Ltd. model HFR 400), with the sampling probe located in the exhaust port about 25 mm from the valves. Condensation in the sampling probe was avoided by maintaining the tube temperature at 170° C with ohmic heating.

The output signals of the pressure transducer and fast FID system were interfaced to a microcomputer through a fast multi-channel A/D converter (Amplicon, type Dash-27) to obtain in-cylinder pressure and hydrocarbon concentrations simultaneously, both as functions of crank angle. The maximum and mean-effective pressures were averaged from the pressure traces of 200 engine cycles and the concentrations of unburned hydrocarbon from a similar number of cycles and integrated over the exhaust stroke.

To visualise the spray and allow measurements of droplet distribution and velocity in the inlet port, a second head of the production engine was modified to allow optical access as shown in figure 1. The photographic investigation made use of a copper vapour laser (Oxford Lasers, model Cu-15) with a compatible fiber for illumination, and the images were recorded at a rate of 14,000 frames per second by a synchronised high speed 16 mm cine camera (Hadland Photec IV).

Droplet velocity was measured with a laser-Doppler velocimeter, arranged to detect back-scattering light as shown in figure 1 which also shows the orientation of the measurements and the definition of the positive velocity direction; the laser, optical fiber and diffraction grating transmitting unit were the same as those used with the phase-Doppler velocimeter. A differential piezoresistive pressure transducer (Honeywell, type 26PC) was mounted in the inlet port and recorded the pressure traces relative to the pressure in the plenum chamber of the inlet manifold.

The measurements reported in the following section were obtained at a rotational speed of 1200 rpm and, under fired conditions, with a steady-state water temperature of 85° C. The manifold depressions were 100 and 600 mbar and corresponded to high and low loads, respectively, and with emphasis on the latter. The stoichiometric ratio of gasoline and air was determined by an oxygen sensor and the duration of injection pulse was around 5 ms. The modifications to allow optical access to the inlet port precluded continuous firing so that, under motored conditions, the duration of injection was 5 ms, or 36 CAD at the speed of 1200 rpm, and was maintained constant irrespective of the manifold depression, which was varied from 100 to 600 mbar. The crank angle corresponding to the beginning of injection was varied from zero to 690 CAD after top-dead-centre of intake and the measurements of droplet characteristics were obtained with iso-octane (GPR 2,2,4 Trimethylpentane) rather than commercial gasoline since it reduced fouling of the windows.

The uncertainty of pressure measurements, based on the 12 bit resolution of the A/D converter, was better than ± 1.5 mbar and of concentrations of unburned hydrocarbons, ± 30 ppm. The measured pressure traces were synchronized with the pulse train from the shaft encoder and generated by the engine management system so that uncertainty in the triggering signals, with respect to the real top-dead-centre of the engine, was less than ± 0.25 crank angle degrees and led to an uncertainty of less than 1% in the value of IMEP. Measurements of droplet characteristics were based, wherever possible, on at least 1000 samples, resulting in statistical uncertainties of less than 5% in the mean, 15% in the rms of the velocity signal and 5% in the cumulative size distribution based on the number of droplets.

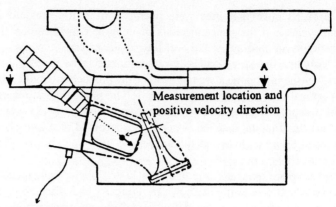

Measurement location and
positive velocity direction

Differential pressure transducer connected
to manifold plenum chamber

Section A-A

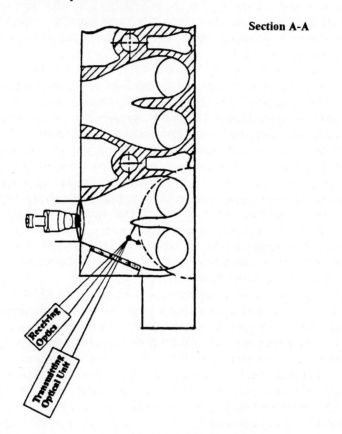

Receiving
Optics

Transmitting
Optical Unit

Figure 1: Engine head, arrangement for optical access to the inlet port and instrumentation

3. Results and Discussion

The effects of injection timing on engine performance, in terms of indicated mean-effective pressure, drivability and exhaust concentrations of unburned hydrocarbons, are shown in figure 2 together with in-cylinder characteristics of droplets for two injection timings corresponding to the closed inlet valves. Most of the large droplets were detected relatively late, some 90 CAD after the top-dead-centre of intake, and the number of droplets in the cylinder was extremely small compared with injection with the valves open; about 10 minutes were required to obtain the sample of the size shown in the figures. The small number of fuel droplets indicates that evaporation in the port led to a mixture in the cylinder which was well premixed and nearly homogeneous and this is in accord with the results of combustion performance. Nevertheless, the data rate of droplets showed that decrease in the residence time for evaporation from 75 to 25 ms with injection at 180 and 540 CAD, respectively, led to an increase by nearly a factor of two in the number of droplets. Although this was expected to have an adverse effect on combustion, it did not materialised; indeed, further reduction in the residence time for evaporation with injection at around 675 CAD led to a slight improvement in performance.

The images of figure 3 show the consequences of interaction between the fuel spray and the air flow in the inlet port with manifold depression of 600 mbar and with injection started at 675 CAD, a short time before opening of the valves and providing improved performance. The leading edge of the spray is visible about 35 CAD after the start of injection and, in these initial stages, the trajectory of the droplets remained nearly constant as expected from the orientation of injector. The images show, however, that about 5 CAD after the top-dead-centre of intake, the trajectory of the spray changed so that most droplets were observed close to the bottom of the port. At the same time, the diameters were considerably reduced and none were visible on the photographs after about 15 CAD, though the duration of injection pulse implied that the spray should be visible until about 30 CAD after the top-dead-centre of intake. The photographs show that the magnitude of the back flow was sufficient to reverse the trajectory of the incoming spray, with some reatomisation caused by the high relative velocities between droplets and the flow from the cylinder to the port at the time of inlet valves opening. As expected, the effects of the back flow decreased with increase in load and, at the same time, high load conditions led to the transport of a small number of droplets into the port from the valve gap a short time before the closure of the inlet valves, that is around 210 CAD after the top-dead-centre of intake. These droplets were probably stripped from the surfaces around the valve gap by the accelerating air flow caused by the upward movement of the piston after 180 CAD and closure of the inlet valve.

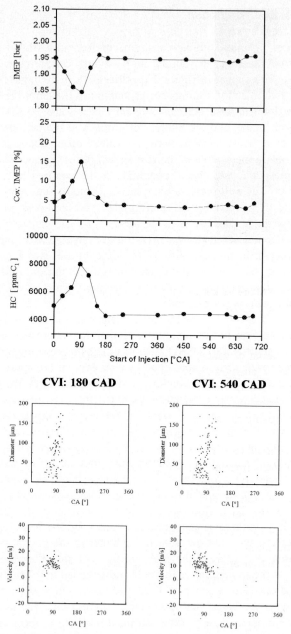

CVI: 180 CAD

CVI: 540 CAD

Figure 2: Effect of injection timing on combustion performance, concentrations of unburned hydrocarbon and droplet size and velocity characteristics. 1200 rpm, 85°C cooling water temperature.

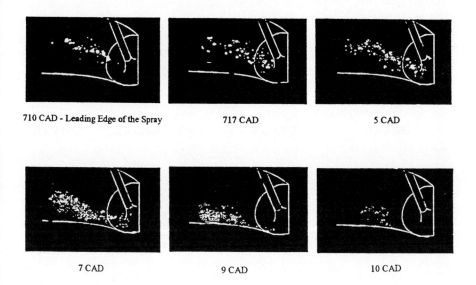

710 CAD - Leading Edge of the Spray   717 CAD   5 CAD

7 CAD   9 CAD   10 CAD

Figure 3: Effects of the back flow on trajectory and size of incoming spray with low load and injection at 675 CAD.

The visualisation of sprays injected at earlier angles, and also with the valves open, showed no changes in trajectory with load. At the same time it was noted that the deposition of fuel on the closed inlet valve with low load, led to the appearance of a white cloud in the port a short time after the top-dead-centre of intake and probably caused by atomisation of the liquid films exposed to the back flow.

The temporal characteristics of droplet velocities in the inlet port, 50 mm downstream of the injector nozzle, are shown in figure 4 for the range of injection angles with the inlet valves closed and low engine load corresponding to the manifold depression of 600 mbar; similar results obtained with the manifold depression of 100 mbar are also shown and, for all cases, no droplets were detected outside the crank angle interval of 180 CAD which commenced with the start of the injection.

With all injection angles, droplets appeared at the measuring location with a similar delay which corresponded to a time of flight of about 5 ms and, with injection at 180 and 540 CAD, the effects of load on velocity, temporal distribution and rate of arrival of droplet were small. With injection angles of 675 and 690 CAD, which implied that the incoming spray was present at the location of the probe volume during the opening of the inlet valves, it is clear that the back flow with the manifold depression of 600 mbar led to detection of a smaller number of droplets and that almost no droplets were present in the crank angle interval between 10 to 30 CAD. The small number of droplets detected in the inlet port after the occurrence of the back flow, suggests their dispersion in the port upstream of the measurement location. With increase in load and a manifold depression of 100 mbar, the effects of the back flow decreased and the number of droplets which passed through the probe volume did not change with injection timing.

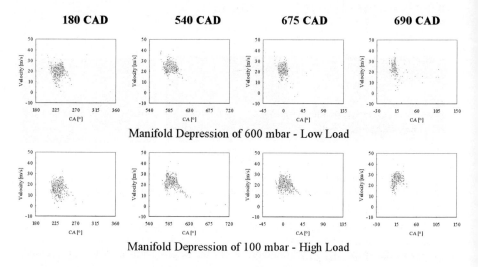

Figure 4: Effects of injection timing with the valves closed on velocities of droplets in the inlet port for two engine loads at 1200 rpm.

The pressure trace of figure 5 corresponds to the manifold depression of 600 mbar and shows an increase just before top-dead-centre of intake when the inlet valves opened, so that the pressure in the inlet port was about 30 mbar higher than in the manifold plenum at about 8 CAD. As a first approximation, the pressure increase of 30 mbar would correspond to the dynamic pressure for a velocity of 80 m/s away from the cylinder, representing a Weber number close to 20 for a droplet diameter of 100 μm and larger than a critical break-up value of 12, Lefebvre (1989). These values are consistent with measurements of air flow velocities in the port and also with the images of figure 3 which showed reatomisation of part of the incoming spray. The break-up of fuel droplets by the back flow and longer residence time in the inlet port led to enhanced evaporation and explains the improvement in combustion performance of figure 2 with low load and injections around 675 CAD.

The effects of the back flow on the process of mixture preparation should not be limited to very late injection angles. With injection with the valves closed, the liquid films on the inlet valve surfaces prior to its opening were exposed to the high velocity burst of the back flow. In addition to altering the thermal

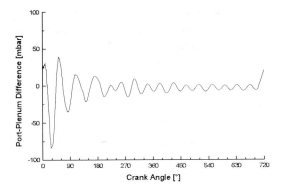

Figure 5: Pressure traces measured between the inlet port and the manifold plenum for depression of 600 mbar with engine motored at 1200 rpm.

environment of the port, the back flow facilitated evaporation by stripping the films, with possible dispersion of the resulting droplets in the inlet port, and further increase in the residence time available for evaporation. This is consistent with the late detection of droplets in the cylinder, figure 2, and also with the results of Posylkin et al (1994), obtained in the cylinder of a two-valve engine where the rate of droplet detection sharply decreased with load. The effects of the back flow on the liquid films explain the nearly-constant performance with injection angles between 150 and 600 CAD, measured with low engine load and the negligible effect of the initial characteristics of spray on combustion performance.

4. Conclusions

The main conclusions from the experiments described above may be summarised as follows:

1.   Injection with the valves closed and crank angles between 150 and 690 degrees led to the nearly-constant combustion performance and suggested good homogeneity of the mixture at the time of ignition. The number of droplets detected in the cylinder was small and constituted a small fraction of injected fuel due to the residence time in the port which allowed evaporation of the liquid films. The evaporation was also assisted by the back flow of hot residual gases from the cylinder to the inlet port which occurred with low load in the early stages of intake. Under those circumstances, the initial characteristics of spray had little effect on combustion performance.

2.   Visualisation and temporal velocity characteristics in the inlet port showed that the spray from a production double-cone injector arrived at the inlet valves in a narrow cone directed at the back of the valve. The trajectory of the spray was unaffected by injection timing with the single exception of low load and injection at 675 CAD for which the incident spray was present in the port at the time of the inlet-valves opening and with back flow sufficient to reverse the trajectory of the incoming spray so that no droplets reached the cylinder in the crank angle interval between 10 to 30 CAD with only few droplets detected afterwards. The break-up of droplets by the back flow and longer residence time for evaporation caused the small improvement in combustion performance.

5.   References

A C Alkidas (1994) The effects of fuel preparation on hydrocarbon emissions of a SI engine operating under steady-state conditions. SAE Paper 941959.

W Bandel, G Fraidl, L A Mikulic, H Carstensen and F Quissek (1989) Investigation of mixture preparation and charge motion effects on the combustion of fast-burn gasoline engines. SAE paper 890160

C Brehm, N Carabateas, B Cousyn, R Mangano, F Neveu, M Posylkin and J H Whitelaw (1996) Evaluation of the influence of injector type in a four-valve engine. SAE Paper 961998.

S Chappuis, B Cousyn, M Posylkin, F Vannobel and J H Whitelaw (1997) Velocity and drop-size distributions in a four-valve production engine. Exp in Fluids, to be published.

B Cousyn, M Posylkin, F Vannobel and J H Whitelaw (1995) Droplet characteristics in two cylinders of a firing spark-ignition engine. SAE Paper 952466

Y Hardalupas, A M K P Taylor, J H Whitelaw, H Miyano, K Ishii and Y Urata (1995) Influence of injection timing on in-cylinder fuel distribution in a Honda VTEC-E engine. SAE Paper 950507

K Horie, K Nishizawa, T Ogawa, S Akazaki and K Miura (1992) Development of a high fuel economy and high performance four-valve lean burn engine. SAE paper 920455

D R Lancaster, R B Krieger and J H Lienesch (1975) Measurement and analysis of engine pressure data. SAE Paper 750026.

A H Lefebvre (1989) <u>Atomization and Sprays</u>. Hemisphere Publishing Corporation, N.Y., USA.

L A Mikulic, F Quissek and G K Fraidl (1990) Development of low emission high performance four valve engines. SAE paper 900227

M Posylkin, A M K P Taylor, F Vannobel and J H Whitelaw (1994) Fuel droplets inside a firing spark-ignition engine. SAE Paper 941989

M Posylkin, A M K P Taylor and J H Whitelaw (1995) Manifold injection and the origin of droplets at the exit of an inlet valve. Applications of Laser Techniques to Fluid Mechanics, 7, 132.

Y Shin, K Min and W K Cheng (1995) Visualisation of mixture preparation in a port-fuel injection engine during engine warm-up. SAE Paper 952481.

F Vannobel, D Robart, J-B Dementhon and J H Whitelaw (1994) Velocity and drop size distributions in a two-valve production firing engine. Proc COMODIA 371-378.

E Winklhofer, G K Fraidl, and A Plimon (1992) Monitoring of gasoline fuel distribution in a research engine. Proc. IMechE, 206, 107.

Author Index

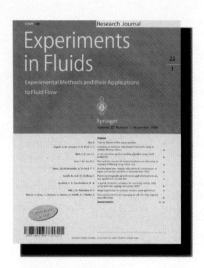

Publishes research papers and technical notes concerned with the development of new measuring techniques and with the extension and improvement of existing techniques, for the measurement of flow properties necessary for the better understanding of fluid mechanics.

Equally it publishes research papers and notes concerned with the application of experimental methods to the solution of problems of turbulence, aerodynamics, hydrodynamics, basic fluid dynamics, convective heat transfer, two-phase flow, combustion, turbomachinery and chemical, biological and geophysical flows. Papers which report on analyses of flow problems are considered, provided they have a substantial experimental content.

Abstracted/Indexed in:
Applied Mechanics Reviews, CAS, Current Contents INIS, SCISEARCH Database, Springer Journals Preview Service, Verfahrenstechnische Berichte BAYER

Please order from
Springer-Verlag Berlin
Fax: + 49 / 30 / 8 27 87- 448
e-mail: subscriptions@springer.de
or through your bookseller

In EU countries the local VAT is effective.

Experiments in Fluids

Experimental Methods and Their Applications to Fluid Flow

ISSN 0723-4864   Title No. 348

Editorial Board:
M. Gharib, Pasadena/CA
W. Merzkirch, Essen
D. Rockwell, Bethlehem/PA
J.H. Whitelaw, London

Springer-Verlag, P. O. Box 31 13 40, D-10643 Berlin, Germany   Gha.